Flavor Chemistry and Technology

Flavor Chemistry and Technology

Henry B. Heath
London, England

and

Gary Reineccius
Department of Food Science
and Nutrition
University of Minnesota
St. Paul, Minnesota

An avi BOOK
Published by Van Nostrand Reinhold Company
New York

An AVI Book
(AVI is an imprint of Van Nostrand Reinhold Company Inc.)

Softcover reprint of the hardcover 1st edition 1986
Library of Congress Catalog Card Number 86-3579

Van Nostrand Reinhold Company Inc.
115 Fifth Avenue
New York, New York 10003

Van Nostrand Reinhold Company Limited
Molly Millars Lane
Wokingham, Berkshire RG11 2PY, England

Van Nostrand Reinhold
480 La Trobe Street
Melbourne, Victoria 3000, Australia

Macmillan of Canada
Division of Canada Publishing Corporation
164 Commander Boulevard
Agincourt, Ontario M1S 3C7, Canada

16, 15, 14, 13, 12, 11, 10, 9, 8, 7, 6, 5, 4, 3, 2

Library of Congress Cataloging-in-Publication Data

Library of Congress Cataloging-in-Publication Data

Heath, Henry B.
Flavor chemistry and technology.

Includes bibliographies and index.
1. Flavor. 2. Flavoring essences. I. Reineccius, Gary. II. Title.
RP372.5.H43 1986 664'.06 86-3579
ISBN 978-94-011-9761-8 ISBN 978-94-011-9759-5 (eBook)
DOI 10.1007/978-94-011-9759-5

ABCDE 5432109876

To my teachers
Frank Fischetti and Marcel Perret

G. A. R.

Contents

Preface

This book combines the essentials of both flavor chemistry and flavor technology. Flavor chemistry is a relatively new area of study which became significant in the 1960s with the availability of gas chromatography and mass spectrometry. Prior to this instrumentation, flavor chemistry focused on only the most abundant chemical constituents. It is a well-documented fact that often the trace constituents of flavors are the most important components.

Flavor chemistry flourished in the late 1960s and early 1970s. Since money was readily available for flavor research great strides were made in understanding the biosynthetic pathways of flavor formation and the chemical constituents that are important to flavor. But the 1970s and early 1980s have not been good years for flavor research, especially in the United States. Since funding agencies have chosen to support research in nutrition and toxicology, many of the research leaders in the flavor area have had to change their research emphasis in order to obtain funding. Today, European researchers turn out the majority of published work in flavor chemistry. While all of the flavor houses conduct some basic flavor research, it is confidential and seldom becomes published. Therefore, the reader will note that a lot of the references are from the late 1960s and early 1970s; and also that European authors dominate the flavor literature in recent years.

Flavor technology is an ancient area of study. Man has searched for a means of making food more pleasurable or palatable since time began. The flavor industry had its beginnings with the trading of spices many centuries ago. There is little published about today's flavor industry because of its highly competitive and secretive nature. It is difficult and sometimes impossible to protect a flavoring via patents, so flavoring materials, synthetic chemicals and manufacturing procedures are kept secret to avoid supporting competitive companies.

The authors compiled the most relevant information on flavor chemistry and technology. While only a sampling of the literature available could be included in many areas, there are references to additional in-depth information sources. This book is intended to serve as a suitable text for both undergraduate and graduate level courses.

G. A. Reineccius

Related AVI Books

Arbuckle: Ice Cream, Fourth Edition
Barker: The Psychobiology of Human Food Selection
Bender, Douglass, and Kramer: Statistical Methods for Food and Agriculture
Farrell: Spices, Condiments and Seasonings
Heath: Source Book of Flavors
Minifie: Chocolate, Cocoa and Confectionery, Second Edition
Mountney: Poultry Products Technology, Second Edition
Pearson and Tauber: Processed Meats, Second Edition
Suderman and Cunningham: Batter and Breading Technology

Part I

Flavor Chemistry

1

Flavor and Its Study

Flavor is a very complex sensation primarily composed of aroma and taste but also complemented by tactile and temperature responses. Taste is limited to the tongue's responses to salty, sweet, sour and bitter sensations. The tongue surface also reacts to the tactile and temperature stimuli, which include the cooling of menthol and the heat of red pepper. There is a tactile response to texture, astringency, etc.—all of which contribute to the overall flavor perception. However, the most important characteristic of flavor is aroma. The importance of odor to flavor perception becomes very obvious when a person catches a cold and can only sense flavor characteristics by the taste, tactile and temperature responses.

Ordinarily the human can discriminate between several thousands of odors. Because of the importance of aroma to flavor perception, the vast majority of analytical flavor studies have focused on the volatile constituents, or the odor, of foods. The task of identifying volatile flavor components is formidable for several reasons:

1. The initial reason is that laboratory instrumentation is not as sensitive to many odors as is the human olfactory system. Stuiver (1958) calculated that as few as 8 molecules of a potent odorant can trigger one olfactory neuron and that only 40 molecules may provide an identifiable sensation. Making a few assumptions about air concentration versus absorption on the olfactory membrane, it is postulated that the nose has a theoretical odor detection limit of about 10^{-19} moles, which rivals or surpasses even the most sensitive analytical abilities.
2. A second complication is that food flavor is distributed throughout a food matrix. Isolation of only the volatile portion of the food is made difficult by the exceedingly low concentrations of flavor compounds and the presence of sugars, carbohydrates, lipids, proteins and water as major constituents. The low concentration of flavors in food products necessitates the isolation of the volatile fraction so it may be concentrated for analytical work, and this isolation may be hindered by flavor binding of food proteins. Proteins are also great emulsifiers, which complicates

a simple flavor extraction process using organic solvents. The presence of lipids lowers flavor compound vapor pressure and the lipids are extracted by organic solvents.

3. Flavor isolation and analysis are made difficult also by the fact that flavors comprise a large number of chemical classes. The flavor chemist cannot focus only on one functional group, and thus greatly simplify the procedure. Rather, the chemist must attempt to effectively extract and concentrate alcohols, aldehydes, acids, ketones, amines, heterocyclics, aromatics, gases, nonvolatiles (or nearly so), etc.

4. The absolute number of flavor compounds in a food further complicates flavor analysis. It is a rather simple, natural flavor that has less than 200 identified constituents. In fact those with less than 200 identified constituents probably have not been adequately researched. It is not uncommon for the browning flavors (e.g., of meats) to be comprised of more than 1000 identified constituents. Research has only started to scratch the surface in identifying volatile flavor constituents. In 1984, 4,300 different flavor compounds had been identified in foods (Maarse 1984). Rijkens and Boelens (1975) estimated that probably 5000–10,000 actually exist.

5. Once the flavor of a food has been extracted, concentrated, separated, and detected, a major question arises concerning each chemical's importance to the flavor. Unfortunately, analytical instrumentation has no sense of taste or smell. Instrument response for the flame ionization detector (used in gas chromatography) is related to the number of carbon–carbon bonds, whereas the human olfactory system varies greatly in response to different odors. For example, 2-methoxy-3-hexyl pyrazine has an odor threshold of 1 part/10^{12} parts water, while pyrazine has an odor threshold of 175,000 parts/10^{12} parts water (Seifert *et al.* 1970). On pyrazines alone, the human threshold varies by nearly 2×10^8. It could be that on the gas chromatograph the smallest peak may be more important to flavor than the largest peak and it must also be recognized that the instrument is providing no appreciation for flavor character of each component. It is not apparent, for example, that peak 3 is buttery while peak 48 contributes oxidized flavor notes.

6. A final problem complicating the study of flavor is instability. The food product being examined is a dynamic system, readily undergoing flavor changes while being stored awaiting analysis to begin. The flavor isolation step may initiate chemical reactions (e.g., thermally induced degradation or oxidations) which alter the flavor profile. The final analysis via gas chromatography may produce isomerizations in the gas chromatograph itself.

There is no question that flavor analysis offers a most challenging analytical problem. After the presentation of the methods for isolation and analysis of flavor in foods, it will be easy to see that each method has its own unique shortcomings. No single method will provide a flavor profile truly representative of the food. This chapter addresses how the analytical chemist uses the most commonly practiced methods for flavor analysis. The interested reader can refer to more detailed discussions of this subject by Reineccius and Anandaraman (1981, 1984), Cronin (1982), Teranishi *et al.* (1981), Schreier (1984a,b) and Maarse and Belz (1981).

A. Sample Preparation

The first step is to select samples of the food most typical of the flavor or off-flavor to be studied. If one is studying an off-flavor problem, the strongest yet characteristically flavored samples need be selected. Recalling the extreme sensitivity of the human olfactory system demands that the most intense samples be selected in order to improve the probability that relatively insensitive machines can pick out the volatiles of interest.

One cannot simply put an apple or a pie into an instrumentation and expect a response (at least a desirable response!) but one must somehow extract the flavor from the food and concentrate the flavor isolate. This generally requires that the food be crushed, homogenized, blended or extracted in some manner (Sugisawa 1984). Most fresh plant and animal tissues contain active enzyme systems which may quickly alter the flavor profile once cellular disruption has occurred (Drawert *et al.* 1965; Fleming *et al.* 1968; Kazeniak and Hall 1980; Dirinck *et al.* 1981). Singleton *et al.* (1975) demonstrated how sample handling during flavor isolation may influence the flavor profile of peanuts. Peanuts ground in water following immersion in liquid N_2 contained 62% less pentane and 8% less carbonyls than the control which was ground in water at room temperature. Peanuts ground in liquid N_2 showed an 81% decrease in pentane and an increase in total carbonyls compared to dry grinding at room temperature. Blending time, temperature and pH all were shown to have a pronounced effect upon the flavor profile of peanuts.

The inactivation of enzymes of fresh plant and animal tissue when the isolation procedure exceeds only a few minutes is essential. A common method is to homogenize the food in methanol (Drawert *et al.* 1969; Schreier *et al.* 1976). This does, however, dilute the sample, decrease the polarity of an aqueous food slurry and may interfere with later isolation methods. Thermal processes may also be employed if the product is a juice and may be rapidly passed through a high temperature short-time heat exchanger. One must be aware of the artifacts or interferences contributed by the means of enzyme inactivation.

Long isolation procedures may even permit fermentation to occur. Ribereau-Gayon *et al.* (1975) added sodium fluoride to crushed grapes to inhibit microbial growth.

In addition to enzyme or microbially induced changes in flavor profile during the isolation procedure, one must also be aware of chemical changes. Long isolation times may permit oxidative changes to occur. Thought must be given to doing a procedure under CO_2 or N_2. Some researchers have chosen to add antioxidants such as BHA, BHT or ascorbic acid (Ribereau-Gayon *et al.* 1975). High temperatures (greater than 60°C) for extended periods can promote nonenzymatic browning reactions. Reduced temperatures (e.g., with vacuum distillations) should be used whenever possible.

B. Isolation of Food Flavors

1. Headspace Methods

a. Direct Injection

Direct analysis of the headspace vapors above a food product seems to be the ideal method of flavor isolation. This method is simple (i.e., draw 10 ml of headspace into a syringe and inject it into a gas chromatograph) and rapid, and samples only what the nose receives. The problem with direct headspace analysis is that too little sample is available for instrumental analysis. Note in Fig. 1.1 that only the most volatile esters are detected by the conventional split headspace analysis. Since direct headspace injections are generally limited to 10 ml or less, only those volatiles present in the headspace at concentrations exceeding 10^{-7} g/L (headspace) will be detected by gas chromatography (GC) and only those exceeding 10^{-5} g/L (headspace) will be adequate for mass spectrometry (MS) (Schaefer 1981). Since the concentration of volatiles above a food product ranges from about 10^{-11} to 10^{-4} g/L (Weurman 1974), only the most abundant volatiles above a food may be analyzed using direct headspace sampling. While the use of splitless or on-column headspace techniques substantially improves the sensitivity of the headspace methods, the technique is still not adequate for the analysis of trace volatiles in foods (Leahy and Reineccius 1984). Additional problems with direct sampling include condensation of volatiles inside the sampling syringe, absorption of volatiles in the septum of the sampling jar, irreproducibility of injection sizes (e.g., syringe leaks and vacuum in the sample bottle) and, finally, relating headspace volatile concentration to actual concentration in the food itself (Gregoire 1985).

Despite the many problems and limitations of direct headspace sampling, it has found substantial application in flavor studies. Some applications of headspace sampling as a means of flavor isolation may be found in the work of Crain and Tang (1975) on macadamia nuts; Williams *et al.* (1972) on dimethyl sulfide in processed foods; Seo and Jeol (1980) on a study of lipid oxidation; and Sullivan *et al.* (1974) on the flavor quality of dehydrated potatoes. If trace analysis is not necessary, direct headspace injection may be the method of choice.

b. Headspace Concentration

Trace analysis of food volatiles may be accomplished via headspace concentration techniques. The equilibrium headspace vapors above a food or the food itself may be purged with an inert gas, in order to obtain a large volume of headspace gas for analysis. The most common methods for isolating and concentrating volatiles from this headspace gas include the use of cryogenic, porous polymer or on-column vapor traps.

i. Cryogenic trapping The simplest means of concentrating headspace vapors is by passing the headspace or purge gas through a series of cold traps. Assuming proper design and operation of the traps, organic volatiles will be condensed from the purge gas. A major problem with cryogenic trapping is that water is the most abundant volatile in most foods and, therefore, the trap condensate is primarily water. Also an additional step is generally then necessary to extract the flavor constituents from the water.

A typical headspace cryogenic trap system is shown in Fig. 1.2 (Chang *et al.* 1977). The sample chamber of this system has a volume of 18.5 L and will hold 27.5 lb of french fries. Inert gas flows into the bottom of the sample chamber, through the food sample and then out into the series of condensers. The initial condenser is generally cooled with ice water while the other condensers may be cooled with dry ice–acetone or liquid N_2. The condensates are typically pooled and then extracted with diethyl ether, dichloromethane or a some other organic solvent. The solvent extract is dried with anhydrous magnesium sulfate or sodium sulfate and then concentrated for GC analysis. The concentration step is often accomplished by gently warming the extract under a stream of N_2 and permitting the solvent to evaporate. However, more sophisticated concentration techniques (e.g., Kuderna-Danish or spinning band distillation) may also be used.

While headspace analysis using cryogenic trapping meets the objective of sampling a large headspace volume, it does require 2–4 h of

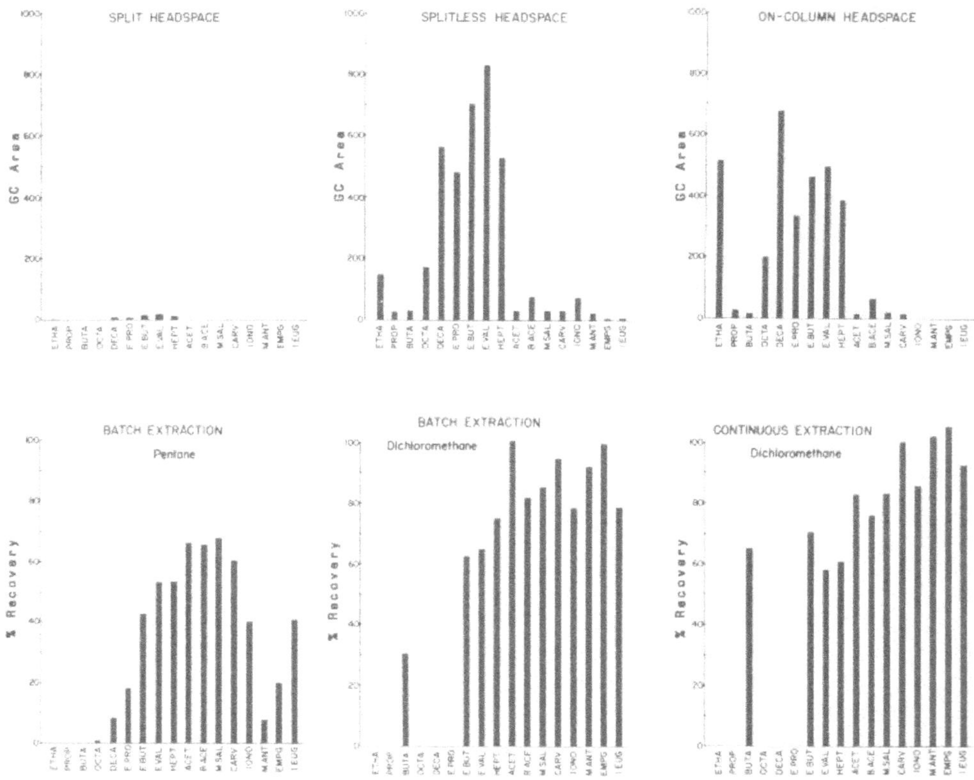

analysis time and subsequent solvent extraction of the aqueous distillate. The long sampling time increases the opportunity for sample changes during flavor isolation. The need for solvent extraction adds concerns for solvent impurities (artifacts), efficiency of extraction (uniformity) and flavor losses during solvent concentration.

Badings and Jong (1984) proposed a simpler method for cryotrapping headspace volatiles. They designed an inlet system for the GC which included a pretrap at −10°C to eliminate the majority of water from the purge gas followed by a second trap which was 10 cm long (1 mm, i.d.) and was cooled to −130°C. This second trap effectively collected the flavor components during purging and then could be heated quickly to desorb the volatiles into the GC for analysis. This system permitted short purge times (20 min) and yet provided adequate material for analysis. Badings *et al.* (1985) have further refined this technique for concentrating headspace volatiles and it is now available commercially (Chrompack, Middleburg, Netherlands).

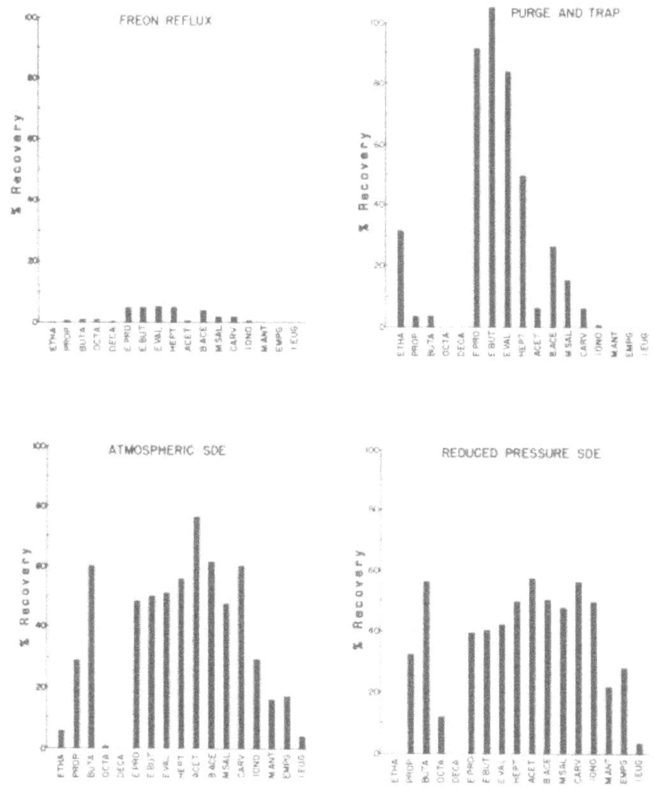

Fig. 1.1. Comparison of different methods for the recovery of volatile flavor compounds from dilute aqueous solutions (ethanol, propanol, butanol, octane, decane, ethylproprionate, ethyl butyrate, ethyl valerate, 2-heptanone, acetophenone, benzyl acetate, methyl salicylate, carvone, β-ionone, methyl anthranilate, ethyl methyl phenyl glycidate and isoengenol.
Leahy and Reineccius 1984

ii. Adsorbent trapping Volatile flavors are often trapped on adsorbent materials to avoid co-condensation of water with the flavor volatiles. The trapping of volatile flavors on an adsorbent which has minimal affinity for water eliminates the need for solvent extraction and associated problems.

Activated carbon is one of the earlier adsorbents used for the trapping of flavors from headspace vapors (Tang and Jennings 1967; Jennings and Nursten 1967). Activated carbon will strongly adsorb nonpolar substances yet only weakly adsorb water vapor. Therefore, as headspace

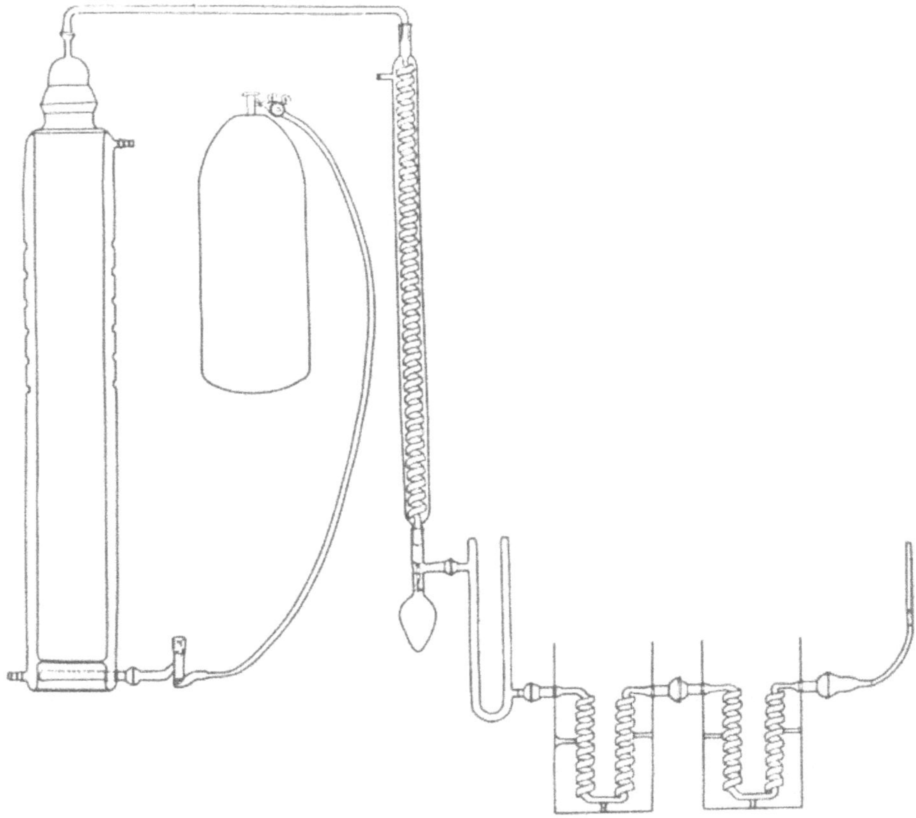

Fig. 1.2. Apparatus for the isolation of trace volatile constituents in headspace gas of foods.
Reprinted with permission from Chang et al. (1977). Copyright 1977, American Chemical Society

vapors are passed over the active carbon, the organic flavor volatiles are preferentially adsorbed while water vapor passes through. Carbon will adsorb an amount of organics approximately equal to its weight (Schaefer 1981). However, once the carbon trap (1–10 mg carbon) becomes saturated (10–100 L of headspace), compounds of greater affinity will displace the more weakly adsorbed compounds. Therefore, overloading of the adsorbent will result in atypical GC profiles. Adsorbed flavor compounds may be desorbed from the charcoal by either thermal desorption (Paillard 1965) or solvent extraction (Grob 1973; Jennings and Nursten 1967).

A major advantage of carbon over the adsorbents to be discussed shortly is carbon's very large adsorption capacity (see Table 1.1) which far exceeds that of the synthetic polymers. However, concerns have appeared in the literature for artifact formation on carbon traps (Baigrie *et al.* 1985). Apparently, the use of impure carbon may cause artifacts to form during thermal desorption of the trapped volatiles (Palamand *et al.* 1968). Despite Grob's (1973) work showing that carbon can be an excellent trapping material if the carbon is desorbed by solvent extraction (CS_2), a high quality carbon is used (coconut charcoal) and a suitable adsorbate-to-adsorbent ratio maintained <1:10,000), there have been very few applications of charcoal-based traps (Roeraade and Enzell 1972) until the recent development of microwave desorption systems. The use of microwave heating permits very rapid heating of the charcoal trap and therefore desorption of trapped volatiles. The desorption is sufficiently rapid that excellent capillary chromatography can be accomplished without cryofocusing the desorbed volatiles. This simplifies the GC inlet system and provides exceptional reproducibility (Liardon and Spadone 1985). Recent applications have included the work of Toulemonde and Beauverd (1985) and Reineccius and Liardon (1986).

The use of synthetic porous polymers for the trapping of volatiles from headspace vapors has become the most commonly used adsorbent for this purpose (Charlambous 1978). While many different polymers have been evaluated, Tenax GC has become the adsorbent of choice. While this choice is not obvious from the data in Table 1.1, Tenax GC does permit superior recovery of adsorbed volatiles and exhibits relatively good thermal stability in comparison to the other synthetic polymers. These two advantages outweigh the poor adsorption capacity displayed by Tenax GC. Withycombe *et al.* (1978) compared Chromosorb 105, Porapak Q and Tenax GC for the recovery of headspace volatiles

Table 1.1. Absorbents Used for Trapping of Organic Volatiles from Vapor Phase[a]

Absorbent	Structure	Surface Area (m^2g)	Absorbent Capacity[b] Ethanol	Benzene
Charcoal	Coconut carbon	1150–1250	7.9	24.7
Porapak Q	Ethyl vinyl benzene–divinyl benzene	550–650	0.18	NA[c]
Tenax GC	Diphenyl-phenylene oxide	18.6	NR[d]	0.53
XAD-4	Styrene divinyl	849	0.40	2.9
XAD-7	Acrylic ester	445	0.90	1.8
XAD-9	Sulfoxide	70	0.70	0.82

[a] From Sydor and Pietryzk (1978).
[b] Percentage of weight of adsorbent.
[c] NA, not available.
[d] NR, not retained.

over hydrolyzed vegetable protein. They found Tenax GC to give the most characteristic flavor isolate by aroma judgment.

The limited adsorption capacity of Tenax is quite evident from the work of Buckholz *et al.* (1980) on peanut aroma. They used three traps in series to evaluate "breakthrough" losses of peanut volatiles. The highly volatile components in peanut headspace vapors broke through the first two traps after only 15 min of purging at 40 ml/min. Jennings and Filsoof (1977) have also done work on the effectiveness of porous polymers for headspace trapping. They found poor retention of ethanol, pentanol, hexanol and heptane on both Tenax and Porapak traps. The data presented in Fig. 1.1 (purge and trap data) also show poor recovery of the alcohols and higher boilers. The higher boilers show poor recovery due to low volatility rather than "breakthrough" as observed for the alcohols.

While one can see that there are problems with applying headspace adsorption techniques to flavor analysis, the method has become quite popular in the literature. The work of Simon *et al.* (1980); Ruen *et al.* (1982); Liardon *et al.* (1984) and Noble *et al.* (1979, 1980) has shown adsorption headspace methods to be exceptionally reproducible and accurate. This method is commonly the method of choice for quantitative flavor studies.

2. Isolation of Flavors by Distillation Methods

Distillation methods take advantage of the volatility of flavor components and nonvolatility of the major food constituents. Aroma components, by definition, must be volatile to make a contribution to odor. It is not unexpected then that distillation is one of the oldest methods for flavor isolation from foods.

Distillation is a rather broad term which includes techniques such as steam distillation and high vacuum stripping of oils. Steam distillation typically involves a source of steam, the sample vessel and a series of cold traps. Steam enters the sample so that the food sample is continuously agitated. Volatile components steam distill from the food and are condensed in the cold traps. The initial trap is generally cooled with ice water, secondary traps cooled with dry ice/acetone and the final traps with liquid N_2. The distillate is a very dilute solution of volatile flavors and water. The flavor must then be isolated from aqueous solution. This is generally done by extracting the distillate with an organic solvent (e.g., pentane, diethyl ether or dichloromethane). The organic solvent containing the flavor isolate is dried via the addition of anhydrous $MgSO_4$ (or other drying agents), filtered and concentrated for instrumental analysis.

A simpler method for isolating flavors via distillation is to use a simultaneous distillation/extraction (SDE) method. While there have been several modifications to the initial distillation head designed by Likens and Nickerson (1964), the basic idea remains unchanged. The apparatus used by MacLeod and Cave (1975) in Fig. 1.3 is an evolution of several other workers (Schultz *et al.* 1977; Picardi and Issenberg 1973). Note that this design permits very intimate mixing of the product steam (no outside steam) and the distilled extracting solvent. This provides for very efficient flavor extraction (Table 1.2). The apparatus shown in Fig. 1.3 could readily be modified to permit vacuum operation by replacing the water bubbler by a vacuum line. This would permit the use of lower product temperatures and thereby minimize thermally induced artifact formation.

Since the co-distillation of water is necessary to effectively transfer volatiles into the distillation head,the design in Fig. 1.3 is not suitable for lipid-based samples. However, Au-Yeung and MacLeod (1981) modified the sample flask to permit the addition of steam. Lipid samples could then be steam distilled and effectively extracted using this system (Table 1.3).

A major disadvantage of steam distillation–solvent extraction is that an additional step (i.e., solvent extraction of the distillate) must be in-

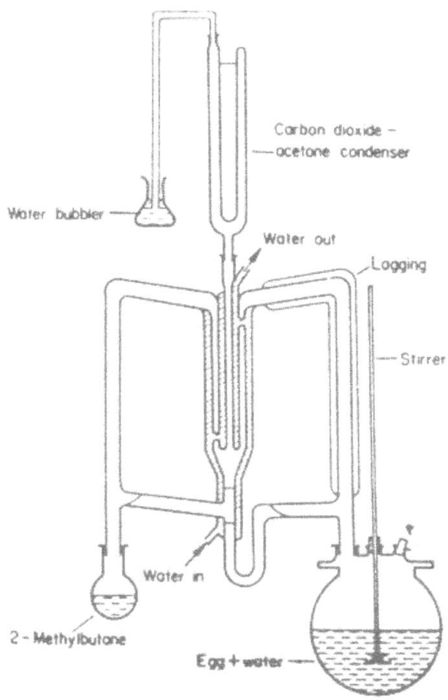

Fig. 1.3. Apparatus for the simultaneous distillation/solvent extraction from eggs.
MacLeod and Cave 1975.

Table 1.2. ***Recovery of Components by SDE from the Second Model Mixture***[a,b]

Time of SDE:	1 h									4 h
Pressure:	Atmospheric pressure								100 mm	Atm
Volume of Solvent:	125 ml							10 ml[c]	125 ml	125 ml
Solvent:	Hexane					Pentane	Ether	Hexane	Hexane	Hexane
pH:	3.4	5.0	6.5	7.8	5.0[d]	5.0	5.0	5.0	5.0	5.0
Ethyl acetate	0	0	0	0	0	59	89	19	0	0
Ethyl butyrate	98	99	99	91	99	101	97	84	100	98
Ethyl hexanoate	100	101	101	95	101	102	99	97	103	99
Ethyl octanoate	99	99	100	95	100	102	100	99	100	99
Ethyl 3-hydroxy hexanoate	41	41	41	19	42	44	49	30	6	90
Ethanol	0	0	0	0	0	0	58	0	0	0
l-Hexanol	101	101	103	98	100	102	100	96	98	100
Linalool	73	99	100	96	99	99	97	97	99	98
Octanal	102	102	103	98	102	103	101	99	103	101
Citronellal	59	78	98	94	81	81	79	77	95	80
Carvone	98	97	98	95	98	99	97	97	92	99

[a] Source: Schultz *et al.* 1977.

[b] At a concentration of 165 ppm (w/v) for each compound recovery as percentage of initial amount.

[c] For this run, additional hexane (13 ml) was added through the vent to fill the solvent overflow arm before the distillation was started, and the extract was not concentrated after SDE.

[d] 1.0 ml of acetic acid (glacial), titrated in solution to pH 5 with sodium hydroxide, also was present in this run in addition to the usual citrate buffer at 0.05 *M*.

Table 1.3. Percentage Recoveries from Lipid Medium Using Modified Likens and Nickerson Apparatus for Steam Distillation[a]

Time (h)	Ethyl acetate[b] isop.[c]	Ethyl acetate[b] CH_2Cl_2	Ethanol[b] isop.	Ethanol[b] CH_2Cl_2	Methyl pyrazine[b] isop.	Methyl pyrazine[b] CH_2Cl_2	Allyl isothiocyanate[b] isop.	Allyl isothiocyanate[b] CH_2Cl_2
0.5	69	85	0	[f]	27	69	46	77
	(6[d])	(17)	(tr)[e]		(tr)	(6)	(tr)	(tr)
	[7]	[27]	[tr]		[tr]	[7]	[tr]	[tr]
1.0	73	81	0	[f]	29	63	64	81
	(10)	(18)	(7)		(tr)	(9)	(tr)	(tr)
	[38]	[56]	[11]		[11]	[22]	[8]	[17]
3.0	83	87	0	[f]	44	74	72	79
	(45)	(80)	(8)		(14)	(12)	(10)	(16)
	[45]	[83]	[13]		[17]	[24]	[13]	[18]

[a] From Au-Yeung and MacLeod (1981).
[b] Concentration of 10^{-2} *M*.
[c] isop. = isopentane.
[d] Recoveries using the apparatus without applied steam at 100°C are in parentheses and those at 150°C are in brackets.
[e] tr = trace.
[f] Ethanol unresolved from solvent on GC.

corporated into the method. Now a method whose isolation efficiency depends on volatility has been combined with a method whose isolation efficiency depends on extraction efficiency. This second step adds additional selectivity and error to the method.

Sugisawa *et al.* 1984 have combined steam distillation with adsorption to recover volatiles. Basically, product steam is passed through a Tenax or charcoal trap, the steam is condensed and then returned to the product. The trap is typically desorbed by solvent extraction. This technique eliminates some of the disadvantages of solvent extraction but then adds the shortcomings of adsorption techniques.

Concern must also be given to the potential for artifact formation during flavor isolation. Several workers have commented on artifacts produced during flavor isolation via steam distillation. Sources of artifacts include antifoam agents, vacuum greases on glass joints, contaminated water supplies, solvent impurities (Jeon *et al.* 1976); rubber tubing (Ferretti and Flanagan 1972); and thermally induced chemical changes (Azzouz *et al.* 1976; Schultz *et al.* 1980; Koedam *et al.* 1980; Sugisawa 1984).

Despite these disadvantages, steam distillation is a very popular method for flavor isolation. The method is very efficient (Fig. 1.1, atmospheric SDE and reduced pressure SDE) and artifacts may be minimized with adequate care.

High vacuum stripping (molecular distillation) of oils is often used for the isolation of flavors from lipid-based foods. Molecular distillation involves the direct transfer of a volatile from a food sample to a cold condenser. The distance a molecule must travel to the condenser surface must be small in comparison to the mean free path of the molecule. This requires very short distances between the condenser and food sample and the use of high vacuum systems ($<10^{-3}$ Torr). The vacuum requirement limits this method to the isolation of volatiles from pure fats and oils (i.e., using essentially no water). A very efficient molecular still is shown in Fig. 1.4. The heated oil sample is placed in the sample flask (A), the cold finger (F) chilled with liquid N_2 and then oil is slowly allowed to enter the still. The thermally jacketed swept surface still provides a large surface area for the loss of volatiles. After the oil has passed through the still, the cold finger is removed and the volatiles are recovered using a small volume of solvent in the apparatus labeled B.

Recovery of volatiles via molecular distillation can be very good. Nawar *et al.* (1969) reported 85–99% recovery of hydrocarbons (C_9–C_{19}) from corn oil using a similar apparatus. At a still temperature of 55°C, Badings (1970) found 72–89% recovery of C_5–C_{19} alkanals from butterfat. While elevated still temperatures may result in artifact formation, a

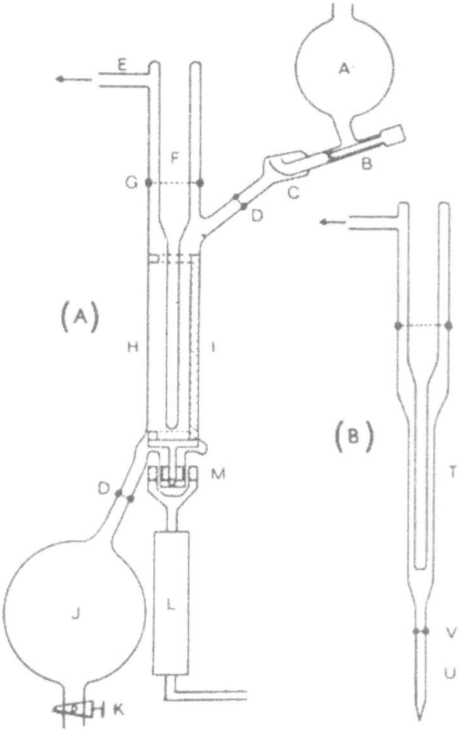

Fig. 1.4. Apparatus for combined high vacuum degassing and falling film molecular distillation. Key parts: A—sample reservoir. C—drip catcher. D—O-ring seal. E—vacuum connection. F—cold finger. H—glass jacket. I—teflon wipers. T—glass tube. U—sample tube.
Brandt and Jensen 1975

significant advantage of molecular distillation is the decreased opportunity for artifact contamination relative to alternate methods. The most common artifacts found in flavor isolates obtained via molecular distillation are silicone compounds from the high vacuum grease used on glass joints. The apparatus used by Brandt and Jensen (1975) avoids this source of contamination via the use of silicone O-rings to make seals.

3. Isolation of Flavors by Solvent Extraction

Most flavor compounds exhibit substantial solubility in organic solvents such as diethyl ether, dichloromethane and pentane. Therefore, nonfat-containing foods and aqueous distillates (steam distillation methods) may be solvent extracted to obtain a flavor isolate. Unfortunately, when one extracts a fat-containing food one obtains a flavor/fat extract. This solvent extract (fats and flavors) may be further treated (e.g., by steam distillation, molecular distillation or dialysis) in order to obtain a flavor isolate. This additional handling does enhance the probability of artifact production and eliminates a primary advantage of direct solvent extraction: the quantitative advantages of solvent extraction. Therefore, solvent extraction is generally practiced only on fat-free foods. Foods such as wines, breads (if no-lipid), fruit and berry juices, some vegetables and alcoholic beverages are often solvent extracted to obtain flavor isolates.

a. Equipment and Procedures

The method used for solvent extraction is determined by the physical state of the food (i.e., solid vs. liquid), quantity of food (small batch process or large continuous operation) and extracting solvent (heavier or lighter than water).

The simplest method of solvent extraction is batch extraction using separatory funnels. The need to use multiple extractions using limited solvent quantities rather than a single extraction using a large solvent volume is obvious from the concern for artifacts introduced by the extracting solvent and extraction efficiency as evident from the following equation:

$$W_r = W_0 \left(\frac{V_w}{DV_0 + V_w} \right)^N$$

where W_r is weight of solute remaining following extraction; W_0, weight of solute in original solution; V_w, volume of aqueous phase; V_0, volume of extracting solvent; D, partition coefficient; and N, number of extractions.

Since the term in parentheses is always a fraction, raising this term to a power decreases its value quickly. As an example, if 100 ml aqueous solution containing a component with $D = 100$ were extracted five times with 20 ml (5×20) of organic solvent, only 2.4×10^{-7} the original amount would remain in the aqueous phase following extraction. A single extraction of the same efficiency would require 4000 *liters* of solvent. It is quite obvious that multiple extractions must be used for flavor isolation. If proper care is given to provide adequate solvent-sample shaking and multiple extractions are used, excellent recovery of flavor compounds may be achieved [see Fig. 1.1., batch extraction and Reineccius *et al.* (1972)].

The major problem encountered with batch solvent extraction is emulsion formation in the separatory funnel. Emulsions may be broken by the use of centrifugation, salt addition (only if solvents lighter than water are used), slow stirring and/or pH adjustment. Care must be exercised if one is centrifuging a flammable organic solvent. A second problem with batch extraction is that it is tedious and labor intensive.

Continuous liquid/liquid (L/L) extraction may also be used to obtain flavor isolates from foods and steam distillates of foods. Continuous batch L/L generally provides more efficient extraction (see Fig. 1.1., continuous extraction) and reduced labor requirements compared to batch separatory funnel but at greater equipment costs and extraction time. A typical continuous L/L extraction would take 2–4 h. Continuous batch L/L extractors are commercially available for solvents lighter than water (Fig. 1.5) and heavier than water. In Fig. 1.5, solvent (e.g., diethyl ether or pentane) is placed in the sidearm flask and an aqueous food in the main body of the apparatus. The solvent is heated and vapors travel up the sidearm into the condenser. The solvent condenses, drops into the center distribution tube and is forced by gravity to displace the aqueous food in this tube. The solvent then exits from the bottom of the distribution tube and percolates up through the food to return to the solvent pot. This design permits continuous extraction with a small volume of solvent which is always being distilled for extraction.

Jennings (1981) has designed a quick solvent extraction procedure suitable for quality control work (Fig. 1.6). He has modified a Babcock bottle to permit rapid sample addition through the large side tube and a small center tube for solvent to collect for GC sampling. Basically, the aqueous food sample is added to the bottle, 100 μl of solvent is added and the bottle is shaken well. Water is then added to bring the solvent into the bottle neck, the bottle centrifuged to promote solvent recovery and the solvent extract sampled by syringe for direct injection into a GC.

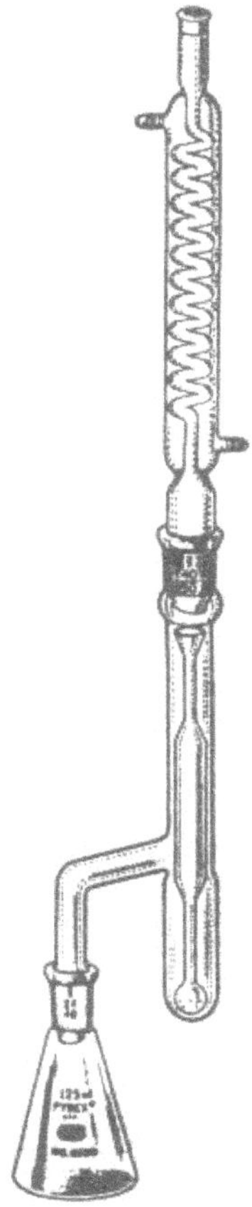

Fig. 1.5. Continuous liquid/liquid extractor for the extraction of flavors from aqueous systems.
Courtesy of Ace Glass Vineland, NJ

Solvent extraction with liquid CO_2 has found both industrial and research application. Specialized equipment is necessary, since 320–800 psig pressure is needed to effect a phase change and produce desirable solvent properties. Jennings (1979) has developed a rather simple system for the extraction of solids with liquid CO_2. The solids may be a food material or a "loaded" porous polymer headspace trap. He essentially put a Soxhlet extractor inside a stainless steel bomb. This method is particularly useful when a solvent-free flavor isolate is needed for sensory evaluation or extremely low boilers are being studied.

Solids may be solvent extracted using a Soxhlet extractor at atmospheric pressures, if a suitable low boiling solvent is used. Care must be taken to use an efficient solvent condenser for flavor isolation. A problem with Soxhlet extractors is their unfavorable solvent-to-sample ratio. There generally needs to be nearly a 2:1 volume ratio. This results in a

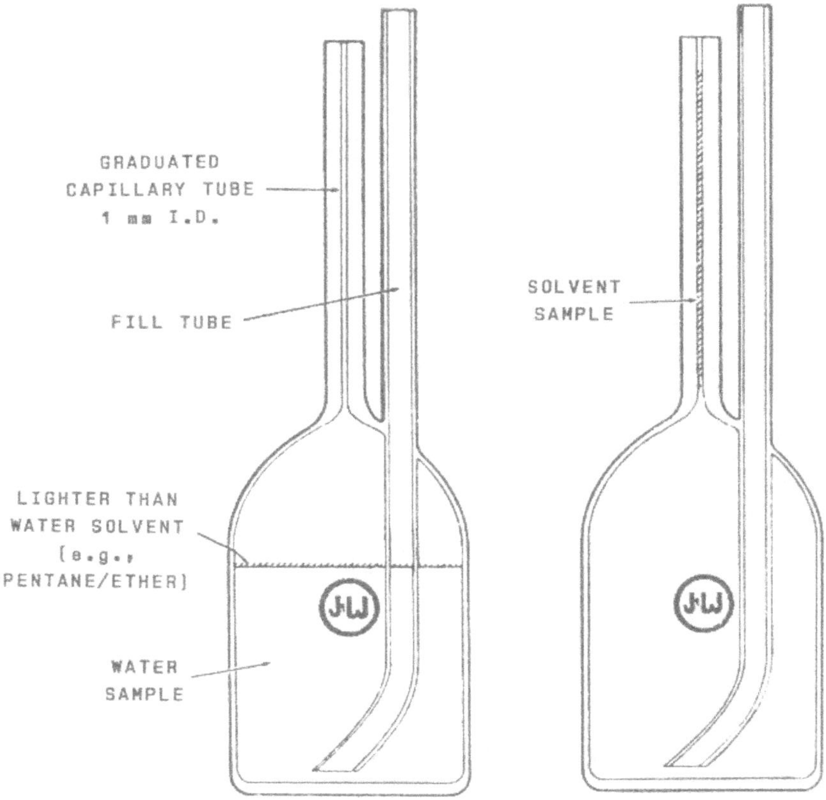

Fig. 1.6. Modified Babcock bottle for the extraction of flavors from aqueous essences. *Courtesy of J & W Scientific Rancho Cordorva, CA*

large volume of solvent if a large quantity of sample is to be extracted. Large solvent volumes necessitate extreme solvent purity to minimize artifacts or solvent contaminants.

An alternate means of extracting a solid would be to grind the dry food with an inert bulking agent such as Celite, packing this blend into a glass chromatography column and then eluting this column with organic solvent (Reineccius *et al.* 1972; Ferretti and Flanagan 1973).

b. Solvent Selection

Solvent selection is determined by several factors: the most important are the components one desires to extract and the food system being extracted. Nonpolar solvents such as pentane, hexane, isopentane or trichlorofluoromethane should be used for alcoholic beverages since alcohol has limited solubility in these solvents (Schreier 1980; Schreier *et al.* 1979; Williams and Tucknott 1973). The same solvents are recommended when methanol is used for enzyme inactivation of fresh plant tissue. Nonalcoholic food systems are often extracted with a more polar solvent such as diethyl ether. Diethyl ether is a very common solvent for flavor extractions since it has a low boiling point and good solvent properties for most flavor compounds.

The components being extracted also influence solvent choice. Very low boiling flavors require the use of a low boiling solvent, e.g., Freon 11 (bpt 23.7°C), CO_2 (−78°C) or Freon 12 (bpt −29°C). The isolation of high boiling flavors permits the use of hexanes and perhaps even chloroform as extracting solvent (Kazeniak and Hall 1980). Since the flavor isolate must generally be concentrated via evaporation of the solvent, the solvent should have a lower boiling point than the components of interest.

c. Solvent Impurities

A major problem with solvent extraction is sample concentration of solvent impurities. Solvent impurities may be unintentional such as by-products of solvent synthesis, unclean glassware or packaging migrants. The impurities may be intentional, for example, antioxidants added to peroxidizable solvents such as diethyl ether. Solvents are generally purified by distillation using a high reflux ratio and efficient fractionating columns (Badings 1970). Hydrocarbon solvents may be "cleaned up" using additional techniques such as adsorption (on activated silica gel) and oxidation (strong acid). Despite the best solvent clean up and storage, a solvent blank should always be run to determine impurities.

d. Solvent Extraction of Fatty Foods

As was mentioned in the introduction to this section, fatty foods are generally not solvent extracted since the solvent co-extracts food lipids and flavors. When a fatty food is solvent extracted, the flavor must be isolated from the fat/flavor extract via molecular distillation, steam distillation or dialysis. Steam and molecular distillation have been discussed earlier in this chapter. Dialysis is a relatively new alternative to distillation (Benkler and Reineccius 1979, 1980; Chang and Reineccius 1985). The dialysis method uses a tubular countercurrent continuous dialysis system (Fig. 1.7). The inner tube is made of perfluorosulfonic acid and is the dialysis tubing. The outer tube is Teflon. The fat/flavor extract is dissolved in 1% water in diethyl ether and is passed through the center tubing (30 m long) at a rate of about 1 ml/min. Pure 1% water in diethyl ether is passed through the outer tubing at a rate of approximately 2 ml/min countercurrent to the sample flow. Recoveries of short chain

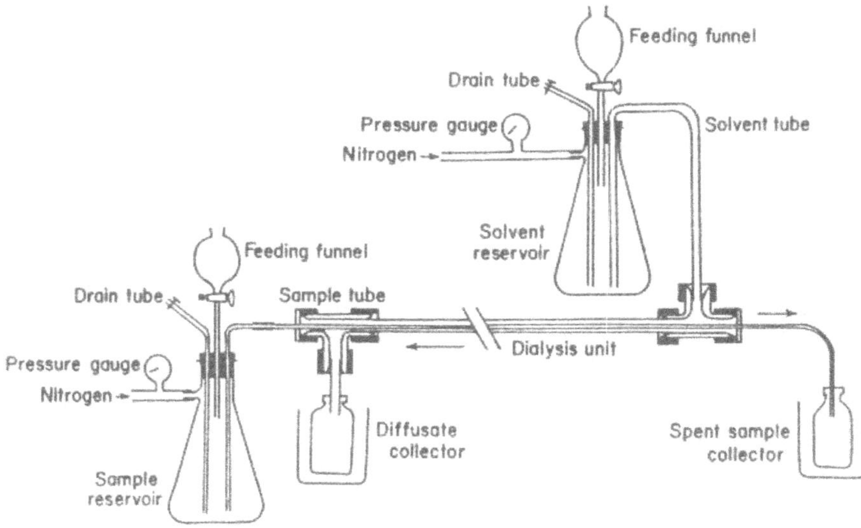

Fig. 1.7. Countercurrent continuous dialysis system for the recovery of flavors from solvent extracts of fatty foods.
Chang and Reineccius 1985

volatiles (up to C_6) is excellent in a single pass through the system. Recovery drops off quickly as molecular size of the volatiles increases to C_{10}–C_{14} (ca 20% recovery). Therefore, the method is well suited to the isolation of short chain aliphatics and most aromatics from fat but does not perform well for larger molecular weight volatiles.

4. Isolation of Individual Classes of Volatile Flavors

Isolation of individual classes of flavor components from foods generally takes advantage of the chemical reactivity of the class's functional group. This may involve derivative formation (e.g., 2.4 dinitrophenyl hydrazones of carbonyls) or complex formation (e.g., lead precipitates of thiols).

Class isolation as a chemical derivative offers several advantages. The derivate often imparts UV absorbing properties which then may permit high pressure liquid chromatography (HPLC) separation and UV detection. Many flavor compounds are highly volatile and may be unstable. The derivate is generally stable and of very low volatility. The specificity of the derivatizing reagent generally greatly simplifies the problems of separation and detection. Some derivates may be regenerated and then sensory evaluations can be made. This permits the study of individual classes of compounds.

A disadvantage of derivative formation or individual class study is that only a small part of the overall flavor is being considered (e.g., carbonyls or alcohols). At one time analytical techniques (i.e., separation and detection) were not adequately advanced to handle the complex mixture of volatiles found in a flavor isolate. Researchers had to do individual class separations in order to simplify the analytical task. Today gas chromatographic techniques are sufficiently advanced to permit resolution of very complex mixtures. The use of thick film, narrow bore fused silica glass capillary columns has provided tremendous separation capabilities. Detection levels have also been reduced due to advances in electronics, the computer age and the use of glass capillaries. The need to simplify flavor analysis is no longer as acute as it once was. Therefore, we are seeing much less class separation work in flavor research today. Due to this declining need for class separations, the following discussion will be rather brief. More detailed discussions of this topic may be found in the reviews of Reineccius and Anandaraman (1984), Maarse and Belz (1981) and Bemelmans (1979).

a. Sulfur Compounds

Most schemes for the isolation of sulfur compounds depend upon the slightly acidic nature of sulfur compounds or their binding to heavy metals (e.g., lead or mercury). Wilson *et al.* (1973) isolated sulfur compounds from beef by first making an ether solvent extract of the beef and then extracting the ether extract with aqueous potassium hydroxide. The acidic compounds (e.g., sulfur compounds and acids) went into the aqueous phase. The aqueous phase was then acidified and re-extracted with ether. Washing this ether phase with water removed the water soluble acids and left primarily sulfur compounds in the ether phase.

The classical isolation method for sulfur compounds depends upon heavy metal precipitation. Lead acetate is used to trap out hydrogen sulfide, mercuric chloride for sulfides and disulfides and mercuric cyanide for mercaptans (Grill *et al.* 1966). Gold has also been shown to be effective in trapping volatile sulfur compounds (Shaw *et al.* 1980). Farbood and MacNeil (1979) warn that class separation based on heavy metal precipitation may not be complete unless proper technique is observed.

b. Acids

The most common isolation procedure for volatile acids starts with steam distillation, followed by an acid/base separation (Fig. 1.8). The isolated short chain volatile acids may be analyzed directy by GC while

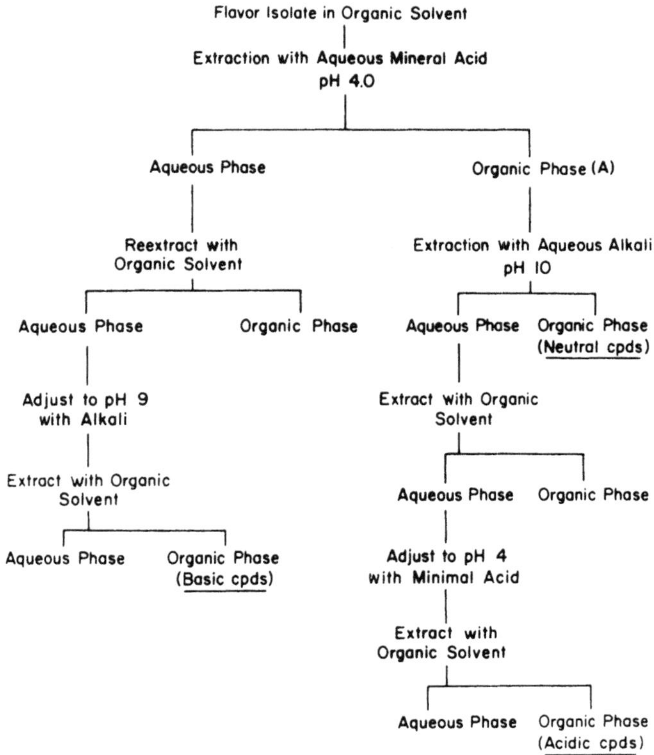

Fig. 1.8. Fractionation scheme for the separation of volatiles into acidic, neutral and basic components.
Reineccius and Anandaraman 1984

longer chain fatty acids are typically converted to the corresponding methyl ester.

Acids may be recovered from fat-free foods by direct solvent extraction. The food would first be acidified (pH 2.5) and then extracted with an organic solvent. The organic solvent would extract both neutral components and acids, so further fractionation would be necessary. The fractionation scheme shown in Fig. 1.8 would be followed starting at point A.

Silicic acid chromatography has been used for the isolation of acids from cheese (Nieuwenhof and Hup 1971). Cheese was ground with silicic acid and then packed into a chromatographic column. The column was then eluted with ether to remove fat, neutral and basic flavor components. Acids were eluted from the column using diethyl ether/phosphoric acid eluant.

A final alternative approach for the isolation of acidic flavor compounds involves the use of ion exchange chromatography. An aqueous homogenate of a food may be passed over an anion exchange column. The column would be rinsed with water to remove all materials except for acids and then the acid would be eluted with dilute formic acid.

c. Alcohols

Class isolation of alcohols is generally accomplished via derivative formation. Since alcohols are neutral compounds of low chemical reactivity, quite reactive reagents are needed for derivative formation. Some common derivatives used for this purpose are *p*-nitro benzoate esters (Connell and Strauss 1972), pyruvyl chloride-2,6-dinitrophenyl hydrazones (Schwartz and Brewington 1967) and benzoates (Gasco and Barrera 1972).

d. Carbonyls

Carbonyls are highly reactive. Therefore, chemical derivatives can be formed and the free carbonyl regenerated with ease. While 2,4-dinitrophenyl hydrazine, hydroxylamine, bisulfite, *O*-aminodiphenyl, 2,4,6-

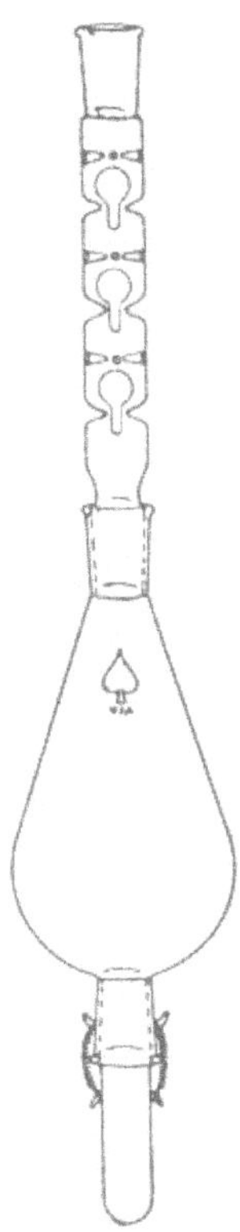

Fig. 1.9. Kuderna-Danish apparatus for the concentration of dilute solutions of flavors in organic solvents.
Courtesy of Ace Glass

trichlorophenyl hydrazine and carboxymethyl trimethyl ammonium chloride hydrazide (Girard's Reagent T) readily form stable derivatives with carbonyls, 2,4-dintirophenyl hydrazine (DNPH) has been the reagent of choice. The DNPH derivatives have been used in carbonyl studies for over 55 years. There is a vast amount of literature reporting on the use of DNPH derivatives in flavor studies. A sampling of some of the literature includes MacDonald *et al.* 1960; Moerck and Ball 1979; Reineccius *et al.* 1978 and Smagula and Bramloge 1977.

e. Amines

Amines are organic bases which are readily isolated by acid/base separations discussed earlier in this chapter. Amines are recovered in the basic fraction (Fig. 1.9) along with pyrazines, pyridines and thiazoles.

Amines may also be derivatized to aid in class separation and analysis. Moffat *et al.* (1972) have described the use of perfluorobenzene derivatives for the analysis of primary and secondary amines. The perfluorobenzene group permits analysis by electron capture detectors. This enhances detection threshold and minimizes interferences.

C. Concentration of Dilute Organic and Aqueous Flavor Isolates

Virtually all of the methods used for the isolation of volatile flavors from foods discussed produce a dilute solution of volatiles for GC analysis. These solutions typically must be concentrated to permit GC detection.

Differences in boiling point, freezing point or polarity between the solvent and flavor compounds typically form the bases for concentration. The method of choice depends upon the type of flavor involved, solvent used, as well as purpose of the study. For example, a decision to study high boilers only would permit a different type of concentration as opposed to a study of low boilers.

The major problems encountered in the concentration process include the loss of volatile flavor compounds, and the introduction of artifacts.

1. Evaporation

Evaporative techniques take advantage of the difference in boiling point between the flavor compounds and solvent. Evaporative techniques are typically used for flavor isolates in organic solvents (e.g., pentane, dicholoromethane, diethyl ether and isopentane). The solvent must be low in boiling point, inert, thermally stable and of extreme purity.

A disadvantage of evaporative techniques is that volatiles may be lost by co-distillation. Unfortunately, the loss of different components may not be the same nor predictable. Therefore, quantitative results may be in error even when multiple internal standards are employed. Since the flavor isolate typically contains a small amount of water (from the food product or distillate), care must be taken to remove water prior to concentration. While this is typically done via the addition of dessicants (e.g., anhydrous magnesium sulfate or sodium sulfate), an alternative would be to freeze out the water (Buttery and Kamm 1980). Failure to remove the water prior to concentration will result in steam distillation of the flavor components and substantial flavor losses.

Concern also has to be given to oxidation of the flavor components. Times and temperatures used in evaporative methods may be adequate to oxidize the more labile flavor compounds. This problem is often minimized by evaporating under vacuum or in an inert atmosphere.

Equipment used for evaporative concentration may be very simple or quite sophisticated. The simplest approach is to evaporate the solvent from an Erlenmeyer flask using gentle heating under a stream of nitrogen. Assuming one is using a low boiling solvent and the most volatile flavor constituents are not of interest, this is a very suitable method for concentration. However, if one is interested in the low boiling flavors, a reflux system should be employed. A very efficient reflux setup is the Kuderna-Danish (Fig. 1.9). A high reflux ratio provides effective solvent removal with minimal loss of volatile flavors. A more efficient system is to use a spinning band fractionating column (Coleman and Ho 1980a,b). Proper care in operation will result in nearly complete recovery of volatiles that differ in boiling point from the solvent by less than 1°C.

Vacuum systems are occasionally used for sample concentration (MacLeod and Cave 1975; Peacock *et al.* 1980; MacLeod and Pieris 1981). While vacuum systems may be used routinely, they are particularly suited to flavor isolates containing relatively higher boiling and heat sensitive components.

2. Freeze Concentration

Freeze concentration may be used to concentrate flavor compounds in an aqueous system (e.g., steam distillate). Partial concentration of the aqueous system prior to solvent extraction has the advantage of reducing solvent volumes for extraction. Freeze concentration is based on the property that water will freeze out of solution as a pure crystal. Therefore, as the water freezes, solutes become concentrated in the unfrozen matrix.

A simple method for freeze concentration is to place the sample in a 10°F cold room overnight and then centrifuge out the ice chunks (Hale and Cole 1963). More elaborate systems include the use of a cold finger or stainless steel beaker to partially freeze the sample (Kepner *et al.* 1969). While freeze concentration seems to be an ideal process and should be used whenever the flavor chemist has an aqueous flavor isolate, it seldom is employed in flavor studies. The primary reason freeze concentration is not used is that only minimal concentration is achievable (5–20-fold). Beyond a 5–20-fold concentration effect, volatile losses due to surface evaporation, occlusion of volatiles in the freezing ice mass and entrapment of volatiles as vapors in ice air pockets become limiting. This limited concentration does not appear to justify the extra handling and time expenditure.

3. Adsorption

Adsorption techniques for the concentration of flavors in aqueous solutions may utilize charcoal, silica gel, alumina, porous polymers or reverse phase packings as adsorbents. While most of the early work used charcoal (Heinz and Jennings 1966; Steinke and Paulson 1964), concern for artifact formation and irreversible binding of flavor compounds to the charcoal has made the porous polymers [Porapak Q (Vitzthum and Werkhoff 1975)] and reverse phase packings [C_{18} (Parliment 1981)] more desirable. Typically, an aqueous distillate is passed through a column of adsorbent. The adsorbent is rinsed and then eluted with an organic solvent. This process accomplishes both concentration and transfer of the flavor components to a nonpolar solvent for analysis.

D. Flavor Analysis by Direct Injection

It is theoretically possible to determine flavor compounds in food (>10 ppb) by directly injecting the food into a GC. One microliter of a food containing a flavor component at 10 ppb would be adequate (0.01 ng) for analysis. The practical problem is that water and nonvolatiles such as lipids, carbohydrates, proteins and minerals are also injected into the GC. These components may either move onto the column and cause chromatographic problems (e.g., column deterioration, poor resolution and changing column characteristics) or thermally degrade to produce artifacts. Previous efforts to circumvent these problems have involved the use of specialized inlet systems to isolate lipids (Dupuy *et al.* 1971) and other nonvolatiles from the GC column and absorb water (Fischer *et al.* 1979). The system used by Legendre *et al.* (1979) to analyze volatiles from aqueous and nonaqueous systems is shown in Fig. 1.10. Ground sample is placed in the inlet assembly and then heated while purging with carrier gas. Desorbed volatiles pass through the Na_2SO_4 trap while moisture is absorbed. The volatiles travel through the valving system into the column for analysis.

The availability of bonded phase capillary columns and on-column injection provides the opportunity to do direct injection on a smaller scale. One could directly inject 1 μl of oil, slurried food sample or distillate into a bonded phase column and chromatograph normally. A buildup of food materials at the column head would require column removal and backflushing with solvent at regular intervals.

Direct injection offers an ideal method for quantitation of volatiles in foods. Also, there is little sample handling or opportunity for artifacts

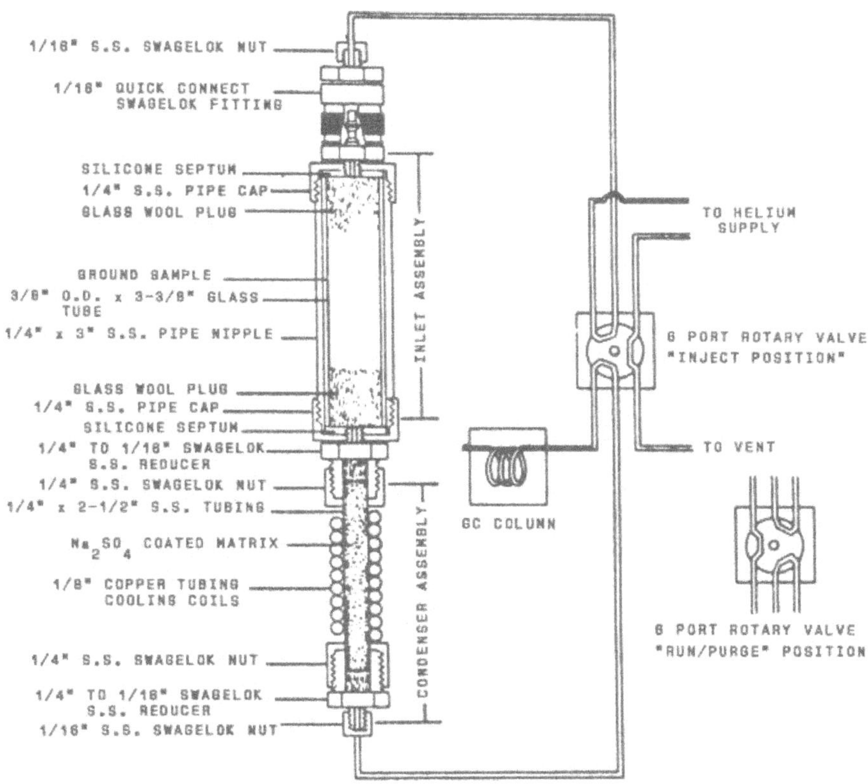

Fig. 1.10. Gas chromatographic inlet system for the direct analysis of volatiles in foods. *Reprinted with permission of Legendre et al. (1979) Copyright 1979. American Chemical Society*

to enter the analysis. A limitation is that trace analysis could not be accomplished; sample size is too small. Concern would also have to be given for absolute sample uniformity. Any phase separation or non-uniformity in flavor distribution would lead to errors.

E. Gas Chromatography

Due to the large number of flavor compounds and their diversity, gas chromatographic separation and measurement is a most challenging problem. While GC capillary columns offer tremendous resolving powers, it is often highly desirable to prefractionate a flavor isolate prior to GC analysis.

1. Fractionation of Flavor Isolates

Some of the more common methods to prefractionate flavor isolates prior to GC analysis include acid/base separations, high pressure liquid chromatography (HPLC), silicic acid and preparative GC.

Acid/base separations have been discussed (see Fig. 1.8), and there is no need to go into this method any further.

The HPLC method is attractive for flavor fractionation since it uses a different set of physical properties for separation than GC does. A flavor isolate may be separated by adsorption and reverse/normal phase chromatography. Adsorption chromatography is a good initial choice since it has the greatest column capacity and can handle the widest range of types of compounds (Teitelbaum 1977). Fractions based on adsorption affinity could then be further fractionated on a normal or reverse phase column (Parliment 1981). The GC task is greatly simplified by a prefractionation scheme such as this.

A simple inexpensive method for flavor fractionation is via silicic acid (Moshonas and Shaw 1979; Schreier *et al.* 1984; Seifert and Buttery 1978). Basically, the flavor concentrate would be passed through a column of silicic acid and then eluted with a solvent gradient. This effectively fractionates the flavor isolate by compound polarity. The major problem with this method is potential artifact formation. However, careful control of the activity of the silicic acid and limiting contact time minimizes artifact formation.

Preparative GC offers the greatest resolving power of the fractionation methods. Typically a flavor concentrate is chromatographed on a $\frac{1}{8}$-in. or $\frac{1}{4}$-in. o.d. packed column and the effluent collected via cold trap in numerous fractions. The solvent-free fractions may be subjected to sensory evaluation at this point to focus on a specific sensory property (e.g., an off-flavor or desirable note). Proper selection of chromatographic cuts can make minor components now major components of the fraction. The primary criticisms of preparative GC as a fractionation method are that the sample may be subjected to high temperatures causing chemical changes and some compounds may be irreversibly bound to the GC column. Despite these concerns, preparative GC has commonly been used for flavor fractionation (Ho and Coleman 1981; Ho *et al.* 1982; Hsieh *et al.* 1982; Thompson *et al.* 1978).

2. Gas Chromatography of Flavor Concentrates

Flavor research has benefited tremendously from the development of GC. In 1963, only 500 flavor compounds had been identified in foods. The development of GC in the mid-1960s and subsequent application to flavor research has resulted in over 4300 compounds being identified to date.

Gas chromatography is ideally suited to flavor studies since it has excellent separation powers (up to 200,000 plates/column) and extreme sensitivity (picogram detection levels). The resolution and sensitivity are essential for analysis of the complex flavor isolates encountered routinely in flavor work.

In this chapter, we shall not go into a general treatment of GC since there are many excellent books on this subject (e.g., Jennings 1980a,b; Freeman 1981; van Straten 1981). However, we shall look into some aspects of GC which are particularly relevant to flavor research.

a. Capillary Column GC

Capillary column GC has become the standard for work in the flavor area. A fused silica column will typically provide about 3000 theoretical plates per meter compared to about 1500 plates per meter for a packed column. This difference may not seem particularly significant until one considers that a packed column is typically limited to about a 7-m length while a capillary column may be more than 50 m long. Therefore, resolving power of a capillary column may be 10–20-fold better than the best-packed column. This vastly improved resolution comes at little or no cost in analysis times. The very thin uniform film coatings in capillary columns permit very rapid mass transfer and thereby allow very high linear gas velocities. Analysis times using gas capillaries are typically much shorter than the corresponding analysis on packed columns. Time becomes quite important in quality control situations where large numbers of samples are being run.

Capillary columns provide very reproducible retention times. This is essential for computerized pattern recognition of GC profiles and identifications based on Kovats indices. The fused silica columns offer exceptional durability and inertness. Their flexibility and small diameter permit insertion into the injection port and detector of the GC. This minimizes flavor compound contact with metal surfaces and dead volume.

Capillary columns enhance detection limits. This is accomplished by reducing column bleed (background) and increasing the concentration per unit time of a compound in the detector. The new bonded phase columns exhibit very little bleed which permits greater sensitivity. The very minimal band spreading (i.e., sharp GC peaks) results in less sample being needed to produce a detector response.

The only disadvantage to capillary columns has been their limited capacity. Column capacity is being improved upon by the development of wide bore thick film bonded phase columns. While a typical fused silica column is limited to less than 100 ng of each component, the new thick film columns will handle about 500 ng of each component without overloading (Grob and Grob 1983). While this capacity is still insufficient for some needs (e.g., trapping for IR and NMR), it is sufficient for splitting and smelling or multiple detectors.

Jennings (1980b) has reviewed the use of two-dimensional, SECAT and multiple pass GC for improving GC resolving power. Two-dimensional GC involves passing a section of one GC run over a second column of different polarity. This is done in order to improve separation of one part of a GC run and is accomplished by valving a part of a run onto a second column. One would then generate two separate chromatograms simultaneously. While this "heart cutting" was initially accomplished by cold trapping a portion of the GC run and then reinjecting this cut into a second analytical system, automated systems utilizing a cold trap between a dual-oven GC have been developed (Oreans *et al.* 1984).

The SECAT GC involves sequentially connecting two GC columns of different polarity and operating the columns at different temperatures. Apparently quite dramatic changes in elution time and order can be accomplished by controlling the temperature differential between the columns.

Multiple pass GC is recycling the run on the same column. Shorter capillary columns have greater resolution than the longer capillaries so resolution can be gained by recycling rather than extending column length. Over 2,000,000 theoretical plates have been developed in less than 16 min on a multiple pass GC system.

b. GC Detectors

The detectors commonly used for flavor work include flame ionization (FID), thermal conductivity (TCD), electron capture (ECD), nitrogen–

phosphorus (NPD), flame photometric (FPD) and the human nose. The FID, TCD and nose are considered universal detectors while the NPD, FPD and ECD are specific detectors. The FID is most commonly employed in flavor work since it exhibits excellent sensitivity and responds to nearly all flavor compounds. The TCD was the first GC detector to see common usage. However, it has lacked enough sensitivity to be useful in trace analysis of flavors. Recent developments in TCD have permitted the use of TCD with capillary columns (Dirinck *et al.* 1977).

Specific detectors are exceptionally valuable in flavor studies. Specific detectors provide information about an individual peak (e.g., whether it contains N, P, S or capture electrons), aid identifications, and simplify otherwise complex chromatograms.

The nose is by far the most valuable detector for the flavor chemist. Either the GC column effluent is split such that about 10 parts go to a sniffing port and 1 part goes to a FID or a TCD (nondestructive) is used. The excellent sensitivity of the FID has generally dictated the use of an effluent splitter/FID system to provide a sample for smelling. However,

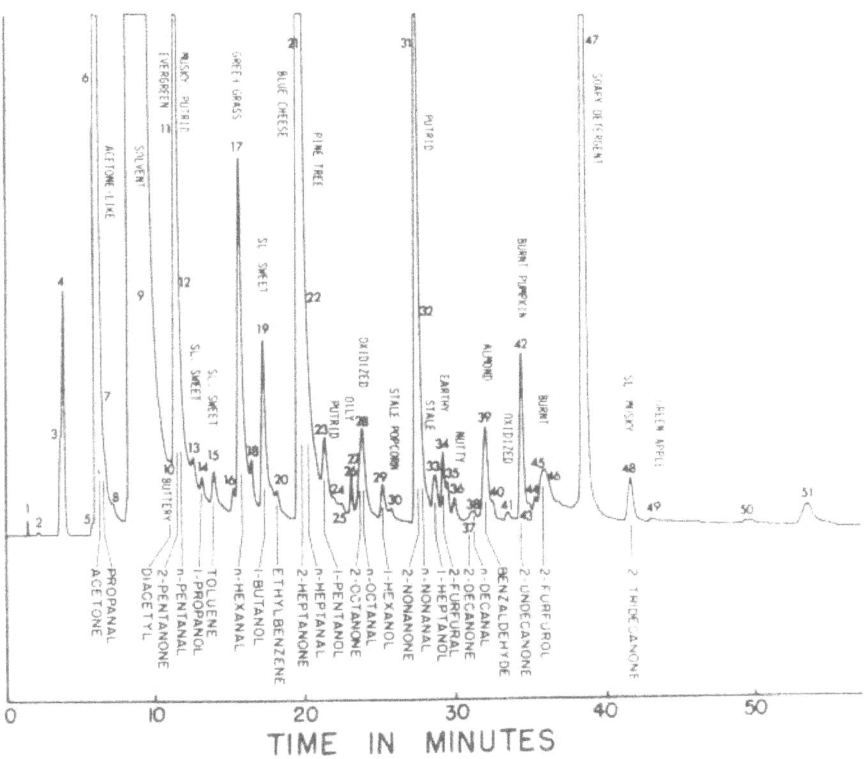

Fig. 1.11. Aromagram of the volatiles isolated from oxidized UHT milk.
Jeon et al. 1978

improved sensitivity and low volume TCDs will make the TCD more attractive in the future since there is no need for splitting with this system. The entire sample passes through the detector and is available for odor evaluation. An experienced chemist can provide a great deal of information about a chemical eluting from a GC column. The chemist can typically give a molecular weight estimate and functional group or compound class. It is relatively easy to distinguish between a pyrazine and a methoxy pyrazine by odor. Sniffing produces what is called an "aromagram," i.e., a documentation of sensory response to each peak or area of a chromatogram (Fig. 1.11). An aromagram is extremely valuable since it indicates where the chemist should focus efforts. If one is studying a particular off-flavor, once it is located in the aromagram, efforts can then be directed towards the identification of the compound or compounds of interest. The aromagram may show that the most important area of the chromatogram is where there are no GC peaks. This would suggest going "back to the drawing-board" and trying to do a better job in isolation. Acree *et al.* (1984) have developed a computerized technique for preparing "aromagrams."

F. High Pressure Liquid Chromatography

High pressure liquid chromatography is not particularly well suited for general flavor studies. The columns lack resolving power, and detectors lack the sensitivity to detect nonultraviolet absorbing compounds.

However, HPLC has been applied to several specific areas of flavor study. HPLC is particularly well suited to the study of compounds which have low vapor pressures. Substantial work exists on the study of pungent principles of pepper (Lego 1984; Kawada *et al.* 1985; Johnson *et al.* 1979) and ginger (Smith 1982; Baronoski 1985). The pungent principles have traditionally been determined using sensory techniques (Scoville heat value). HPLC has also found application in the determination of hop constituents (Schulze *et al.* 1979; Forester *et al.* 1984), licorice (Hiraga *et al.* 1984; Hurst *et al.* 1983) and vanilla (Dalang *et al.* 1983).

G. Identification of Volatile Flavors

In most flavor studies it becomes necessary to identify some or all of the flavor components. While there are a large number of methods available for identification of unknowns, only a few are useful in flavor studies. In most cases, the researcher is attempting to identify one compound in a complex mixture. Therefore, the compound must be separated from the mixture for analysis. This may mean high resolution GC and a maximum of 500 ng unknown per GC run. Minute sample amounts limit the researcher to GC, mass spectrometry (MS), infrared (IR), nuclear magnetic resonance (NMR) and Fourier transform infrared (FTIR) for providing the necessary clues for identification. A discussion of these methods is outside the scope of this book so it is suggested that the reader go to the review by Reineccius and Anandaraman (1984) and Schreier (1984b) for a general discussion of how these methods are applied in flavor research and to individual texts for detail on each of the methods.

H. Summary

This chapter should provide an appreciation for the complexity of food flavors and the task faced by the analytical flavor chemist. While the challenge is indeed formidable, quite significant progress has been made in flavor chemistry in recent years. This chapter closes with some reminders for the chemist.

The chemist must pick representative samples for analysis and be certain that the flavor profile is not altered by factors such as microbes or chemical or enzymatic changes during storage or analysis. The next step is to choose a method for flavor isolation. The sample type (containing fat, aqueous system, etc.), time available (quality control procedure or research technique), research goal (identify a phenolic off-flavor or low boiling fresh notes, etc.) and available equipment (CO_2 extractor, automated purge and trap, etc.) all contribute to choosing a particular method for flavor isolation. In any event, the chemist *must evaluate the flavor isolate by sensory methods* to be certain that the flavor character desired is present in the flavor isolate. If not, a lot of time may be wasted looking for some flavor notes that were not even isolated. The next step is concentration of the flavor isolate. Moderate and high boilers may be concentrated by simple evaporation under a stream of nitrogen. Studies on very low boilers (e.g., hydrogen sulfide or acetaldehyde) require special care in this step. Analysis then proceeds with GC of the flavor concentrate. Capillary column GC is generally a must. The sample should be sniffed by an experienced person to give direction to the study. Identification of all the flavor components in the isolate should not be attempted unless that is the research goal. The task of unnecessarily identifying every flavor component in a flavor isolate should be left only to the young and foolish. One must focus efforts on the flavor components of interest. Identifications will typically be made using a combination of GC and MS. Identification will be confirmed by MS, odor and co-chromatography on two different polarity columns.

References

ACREE, T. E., BARNARD, J. and CUNNINGHAM, D. G. 1984. The analysis of odor-active volatiles in gas chromatographic effluents. *In* Analysis of Volatiles, Methods and Applications, P.Schreier, ed. de Gruyter, New York.

AU-YEUNG, C. Y. and MACLEOD, A. J. 1981. A comparison of the efficiency of the Likens and Nickerson extractor for aqueous, lipid/aqueous, and lipid samples. J. Agric. Food Chem. *29*, 502.

AZZOUZ, M. A., REINECCIUS, G. A. and MOSHANAS, M. G. 1976. Comparison between cold pressed and distilled lime oils through the application of gas chromatography and mass spectrometry. J. Food Sci. *41*, 324.

BADINGS, H. T. 1970. Cold storage defects in butter and their relation to the autoxidation of unsaturated fatty acids. Thesis Agric., Univ. of Wageningen, Netherlands.

BADINGS, H. T. and DE JONG, C. 1984. Headspace analysis for the study of aroma compounds in milk and dairy products. *In* Analysis of Volatiles, Methods and Applications, P. Schreier, ed. de Gruyter, New York.

BADINGS, H. T., DE JONG, C., DOOPER, R. P. M., and DE NIJS, R. C. M. 1985. Rapid analysis of volatile compounds in food products by purge and cold trapping/capillary gas chromatography. *In* Progress in Flavour Research 1984, J. Adda, ed. Elsevier, Amsterdam.

BAIGRIE, B. D., LAURIE, W. A. and McHALE, D. 1985. Artifact formation during headspace analysis. *In* Progress in Flavour Research 1984, J. Adda. ed. Elsevier, Amsterdam.

BARONOWSKI, J. D. 1985. High performance liquid chromatographic separation of pungency components of giner. J. Chromatogr. *319*(3), 471.

BEMELMANS, J. M. H. 1979. Review of Isolation and Concentration Techniques. *In* Progress in Flavour Research, D. G. Land and H. E. Nursten, eds. Appl. Sci. Publ., London.

BENKLER, K. F. and REINECCIUS, G. A. 1979. Separation of flavor compounds from lipids in a model system by means of membrane dialysis. J. Food Sci. *44*, 1525.

BENKLER, K. F. and REINECCIUS, G. A. 1980. Flavor isolation from fatty foods via solvent extraction and membrane dialysis. J. Food Sci. *45*, 1084.

BRANDT, P. E. and JENSEN, B. H. 1975. Recovery of volatiles from oils. J. Amer. Oil Chem. Soc. *52*, 278.

BUCKHOLZ, L. I., WITHYCOMBE, D. A., and DAUN, H. 1980. Application and characteristics of polymer adsorption method used to analyze flavor volatiles from peanuts. J. Agric. Food Chem. *28*, 760.

BUTTERY, R. G. and KAMM, J. A. 1980. Volatile components of alfalfa: Possible insect host plant attractants. J. Agric. Food Chem. *28*, 978.

CHANG, S. S., VALLESE, F. M., HUANG, L. S., HSIEH, O. A. L., and MIN, D. B. S. 1977. Apparatus for the isolation of trace volatile constituents from foods. J. Agric. Food Chem. *25*, 450.

CHANG, Y. I. and REINECCIUS, G. A. 1986. Isolation of flavor compounds in model systems by countercurrent continuous dialysis. J. Agric. Food Chem. *33*(6), 1168.

CHARLAMBOUS, G. 1978. Analysis of Foods and Beverages, Headspace Techniques. Academic Press, New York.

COLEMAN, E. C. and HO, C. T. 1980a. Chemistry of baked potato flavor. 1. Pyrazines and thiazoles identified in the volatile flavor of baked potato. J. Agric. Food Chem. *28*, 66.

COLEMAN, E. C. and HO, C. T. 1980b. Chemistry of baked potato flavor: Further identification of heterocyclic compounds in the volatile flavor of heterocyclic compounds in the volatile flavor of baked potato. J. Food Sci. *45*, 1094.

CONNELL, D. W. and STRAUSS, C. R. 1972. Chromatographic examination of the *p*-nitrobenzoates and alpha-naphtylurethans of lower alphatic alcohols. J. Chromatogr. *72*, 391.

CRAIN, W. O. and TANG, C. S. 1975. Volatile components of roasted macadamia nuts. J. Food Sci. *40*, 207.

CRONIN, D. A. 1982. Techniques of analysis of flavours: chemical methods including sample preparation. *In* Food Flavours, Part A: Introduction, p. 15, I. D. Morton and A. J. MacLeod, eds. Developments in Food Science, Elsevier, London.

DALANG, F., MARTIN, E. and VOGEL, J. 1982. High-performance liquid chromatographic study of vanilla flavors and vanilla-flavored food products, Mitt. Geb. Lebensmittelunters. Hyg. *73*(4), 371.

DIRINCK, P., SCHREYEN, L., DESCHOEMAKER, L., WYCHUYSE, F. and SCHAMP, N. 1977. Volatile components of crude palm oil. J. Food Sci. *42*(3), 645.

DIRINCK, P. J., DEPOOTER, H. L., WILLAERT, G. A. and SCHAMP, N. M. 1981. Flavor quality of cultivated strawberries: The role of sulfur compounds. J. Agric. Food Chem. *29*, 316.

DRAWERT, F., HEIMANN, W., EMBERGER, R., and TRESSL, R. 1965. Enzymatische Veranderung des Naturlichen Apfelaromas bei der Aufarbeitung. Naturwissenschafter. *52*, 304.

DRAWERT, F., HEIMANN, W., EMBERGER, R., and TRESSL, R. 1969. Gas-chromatographische Untersuchung Pflanzlicher Aromen. II. Anreicherung, Trennung und Identifizierung von Apfelaromastoffen. Chromatographia *2*, 57.

DUPUY, H. P., FORE, S. P., and GOLDBLATT, L. A. 1971. Elution and analysis of volatiles in vegetable oils by gas chromatography. J. Amer. Oil Chem. Soc. *48*, 876.

FARBOOD, M. I. and MACNEIL, J. H. 1979. Retention and stability of sulfur compounds in chicken meat. J. Food Sci. *44*, 652.

FERRETTI, A. and FLANAGAN, V. P. 1972. Steam volatile constituents of stale nonfat dry milk. The role of the Maillard reaction in staling. J. Agric. Food Chem. *20*, 695.

FERRETTI, A. and FLANAGAN, V. P. 1973. Characterization of volatile constituents of an *N*-alpha-formyl-L-lysine-D-lactose browning system. J. Agric. Food Chem. *21*, 35.

FISCHER, G. S., LEGENDRE, M. G., LOVGREN, N. V., SCHULLER, W. H., and WELLS, J. A. 1979. Volatile constituents of southern pea seed. [*Vigna unguiculata* (L.) Ualp]. J. Agric. Food Chem. *27*, 7.

FORESTER, A., KOEBERLEIN, A. and SCHALLER, G. 1984. HPLC studies on hop constituents in yeast heads and new beer. Monatsschr. Brauwiss. *37*(4), 162.

FREEMAN, R. R. 1981. High Resolution Gas Chromatography. 2nd edition. Hewlett Packard Co., Avondale, Pa.

GASCO, L. and BARRERA, R. 1972. The use of gas-chromatographic identification of alcohols, primary and secondary amines and thiols in food aromas. Anal. Chim. Acta. *61*, 253.

GREGOIRE, J. 1985. DCI: The way to demonstrate and to eliminate quantitative errors of static headspace based methods. *In* Progress in Flavour Research 1984, J. Adda, ed. Elsevier, Amsterdam.

GRILL, H., PATTON, S. and CONE, J. F. 1966. Aroma significance of sulfur compounds in surface-ripened cheese. J. Dairy Sci. *49,* 409.

GROB, K. 1973. Organic substances in potable water and in its precursor. Part I. Methods for their determination by gas–liquid chromatography. J. Chromatogr. *84,* 255.

GROB, K. and GROB, G. 1983. Capillary columns with very thick coatings. J. High Resol. Chrom. and Chrom. Comm. *3,* 133.

HALE, W. S. and COLE, E. W. 1963. A freezing technique for concentrating preferments. Cereal Chem. *40,* 287.

HEINZ, D. E. and JENNINGS, W. G. 1966. Volatile components of Bartlett pear. J. Food Sci. *31,* 69.

HIRAGA, Y., ENDO, H., TAKAHASHI, K. and SHIBATA, S. 1984. High-performance liquid chromatographic analysis of licorice extracts. J. Chromatogr. *292*(2), 451.

HO, C. T. and COLEMAN, E. C. 1981. Halogen compounds identified in volatile constituents of baked potatoes. J. Agric. Food Chem. *29,* 201.

HO, C. T., LEE, M. H. and CHANG, S. S. 1982. Isolation and identification of volatile compounds from roasted peanuts. J. Food Sci. *47,* 127.

HSIEH, O. A. L., HUANG, A. S. and CHANG, S. S. 1982. Isolation and identification of objectionable volatile compounds in defatted soybean flour. J. Food Sci. *47,* 16.

HURST, W. J., McKIM, J. M. and MARTIN, R. A., Jr. 1983. High-performance liquid chromatographic determination of glycyrrhizin in licorice products. J. Agric. Food Chem. *31*(2), 387.

JENNINGS, W. G. 1979. Vapor-phase sampling. J. High Resol. Chrom. & Chrom. Comm. *2,* 221.

JENNINGS, W. G. 1980a. Gas Chromatography with Glass Capillary Columns, 2nd edition. Academic Press, New York.

JENNINGS, W. G. 1980b. Recent advances in the separation of volatile components. *In* The Analysis and Control of Less Desirable Flavor in Foods and Beverages. G. Charalambous, ed. Academic Press, New York.

JENNINGS, W. G. 1981. Recent developments in high resolution gas chromatography. *In* Flavour '81, P. Schreier, ed. de Gruyter, New York.

JENNINGS, W. G. and FILSOOF, M. 1977. Comparison of sample preparation techniques for gas chromatographic analysis. J. Agric. Food Chem. *25,* 440.

JENNINGS, W. G. and NURSTEN, H. E. 1967. Gas chromatographic analysis of dilute aqueous systems. Anal. Chem. *39,* 521.

JEON, I. J., REINECCIUS, G. A., and THOMAS, E. L. 1976. Artifacts in flavor isolates produced by steam vacuum distillation and solvent extraction of the distillate. J. Agric. Food Chem. *24,* 433.

JOEN, I. J., THOMAS, E. L. and REINECCIUS, G. A. 1978. Production of volatile flavor compounds in ultrahigh-temperature processed milk during aseptic storage. J. Agric. Food Chem. *26*(5), 1183.

JOHNSON, E. L., MAJORS, R. E., WEURUM, L. and REICHE, P. 1979. The determination of naturally occurring capsaicins by HPLC. *In* Liquid Chromatographic Analysis of Foods and Beverages, Vol. 1. G. Charalambous, ed. Academic Press, New York.

KAWADA, T., WATANABE, T., KATSURA, K., TAKAMI, H. and IWAI, K. 1985. Formation and metabolism of pungent principle of capsicum fruits. 15: Microdetermination of capsaicin by HPLC with electrochemical detection. J. Chromatogr. *329*(1), 99.

KAZENIAK, S. J. and HALL, R. M. 1980. Flavor chemistry of tomato volatiles. J. Food Sci. *35,* 519.

KEPNER, R. E., STRATEN, S. V., and WEURMAN, C. 1969. Freeze concentration of volatile components in dilute aqueous solutions. J. Agric. Food Chem. *17,* 1123.

KOEDAM, A., SCHEFFER, J. J. C. and SVENDSEN, A. B. 1980. Monoterpenes in the volatile leaf oil of *abies x arnoldiana* Nitz. J. Agric. Food Chem. *28*, 662.

LEAHY, M. M. and REINECCIUS, G. A. 1984. Comparison of methods for the analysis of volatile compounds from aqueous model systems. *In* Analysis of Volatiles: Methods and Application, P. Schreier, ed., de Gruyter, Berlin.

LEGENDRE, M. G., FISHER, G. S., FULLER, W. H., DUPUY, H. P. and RAYNER, E. T. 1979. Novel technique for the anlysis of volatiles in aqueous and nonaqueous systems. J. Am. Oil Chem. Soc. *56*, 552.

LEGO, M. C. 1984. HPLC in the flavour/spice industry. Food Technol. *38*(4), 84.

LIARDON, R., OTT, U. and DAGET, N. 1984. Analysis of coffee headspace profiles by multivariate statistics. *In* Analysis of Volatiles, Methods and Applications, P. Schreier, ed. de Gruyter, New York.

LIARDON, R. and SPADONE, J. C. 1985. Coffee aroma investigations by combined capillary GC headspace analysis and multivariate statistics. Proc. of the 11th Int. Scientific Conference on Coffee (Lome, Togo) ASIC.

LIKENS, S. T. and NICKERSON, G. B. 1964. Detection of certain hop constituents in brewing products. Amer. Soc. Brew. Chem. Proc., p. 5.

MAARSE, H. 1984. Volatile Compounds in Food, Quantitative Data. Vols. 1–3, Division for Nutrition and Food Research, TNO, Zeist, Netherlands.

MAARSE, H. and BELZ, R. 1981. Hanbuch der Aroma Forschung. Akademie-Verlag, Berlin, West Germany.

MacDONALD, B., GRAY, J. I., KAKUDA, Y. and LEE, M. L. 1960. Role of nitrite in cured meat flavor. Chemical Analysis. J. Food Sci. *45*, 889.

MacLEOD, A. J. and CAVE, S. J. 1975. Volatile flavour components of eggs. J. Sci. Food Agric. *26*, 351.

MacLEOD, A. J. and PIERIS, N. M. 1981. Volatile flavor components of wood apple (*Feronia limonia*) and a processed product. J. Agric. Food Chem. *29*, 49.

MOERCK, K. E. and BALL, H. R. 1979. Influence of microorganisms on the carbonyl compounds of chicken tissue. J. Agric. Food Chem. *27*, 854.

MOFFAT, A. C., HORNING, E. C., MATIN, S. B. and ROWLAND, M. 1972. Perfluorobenzene derivatives as derivatizing agent for the gas chromatography of primary and secondary amines using electron capture detection. J. Chromatogr. *66*, 255.

MOSHONAS, M. G. and SHAW, P. E. 1979. Composition of essence oil from overripe oranges. J. Agric. Food Chem. *27*, 1337.

NAWAR, W. W., CHAMPAGNE, J. R., DUBRAVCIC, M. F. and LeTELLIER, P. R. 1969. Recovery and measurements of volatiles from lipids. J. Agric. Food Chem. *17*, 645.

NIEUWENHOF, F. F. J. and HUP, G. 1971. Gas chromatographic determination of free fatty acids in cheese. J. Dairy Sci. *41*, 397.

NOBLE, A. C., MURAKAMI, A. A. and COOPE, G. F. III. 1979. Reproducibility of headspace analysis of wines. J. Agric. Food Chem. *27*, 450.

NOBLE, A. C., FLATH, R. A. and FORREY, R. R. 1980. Wine headspace analysis. Reproducibility and application to varietal classification. J. Agric. Food Chem. *38*, 346.

OREANS, M., MÜLLER, F., LEONHARD, D. and HEIM, A. 1984. The analysis of trace components using the total transfer technique in coupled column systems. *In* Analysis of Volatiles, Methods and Applications, P. Schreier, ed. de Gruyter, New York.

PAILLARD, N. 1965. Analyse des produits volatils emis par les pommes. Fruits *20*, 189.

PALAMAND, S. R., MARKL, K. S. and HARDWICK, W. A. 1968. Some techniques used for the concentration of volatile compounds in flavor research. Amer. Soc. Brew. Chem. Proc. (Annu. Meet.), p. 75.

PARLIMENT, T. H. 1981. Concentration and fractionation of aromas on reverse phase adsorbents. J. Agric. Food Chem. *29*, 841.

PEACOCK, V. E., DEINZER, M. L., McGILL, L. A. and WROLSTAD, R. E. 1980. Hop aroma in American beer. J. Agric. Food Chem. *28*, 774.

PICARDI, S. M. and ISSENBERG, P. 1973. Investigation of some volatile constituents of mushrooms (*Agaricus bisporus*). Changes which occur during heating. J. Agric. Food Chem. *21*, 959.

REINECCIUS, G. A. and ANANDARAMAN, S. 1984. Analysis of volatile flavors. *In* Recent Advances in the Chromatographic Analysis of Organic Compounds in Foods, J. Lawrence, ed. Dekker, New York.

REINECCIUS, G. A. and LIARDON, R. 1985. The use of charcoal traps and microwave desorption for the analysis of headspace volatiles above heated thiamine solutions. *In* Topics in Flavour Research, R. Berger, S. Nitz, and P. Schreier, eds. H. Eichorn Publishing, Marzling-Hangenham.

REINECCIUS, G. A., KEENEY, P. G., and WEISBERGER, W. 1972. Factors affecting the concentration of pyrazines in cocoa beans. J. Agric. Food Chem. *20*, 202.

REINECCIUS, G. A., ANDERSON, H. C., and FELSKA, B. J. 1978. Influence of butterfat content on methyl ketone retention during spray drying as determined by high pressure liquid chromatography of 2,4-dinitrophenyl hydrazones. J. Food Sci. *43*, 1494.

RIBEREAU-GAYON, P., BOIDRON, J. N., and TERRIER, A. 1975. Aroma of Muscat Grape Varieties. J. Agric. Food Chem. *23*, 1042.

RIJKENS, F., AND BOELENS, H. 1975. The future of aroma research. *In* Proc. Int. Symp. Aroma Research, p. 203, H. Maarse and P. J. Groenen, eds. Pudoc, Wageningen, Netherlands.

ROERAADE, J. and ENZELL, C. R. 1972. Tobacco chemistry: XIV. Sampling, concentration and examination of tobacco headspace vapors. J. Agric. Food Chem. *20*, 1035.

RUEN, W. W., REINECCIUS, G. A. and THOMAS, E. L. 1982. Analysis of the fruity off-flavor in milk using headspace concentration capillary column gas chromatography. J. Agric. Food Chem. *30*(3), 512.

SCHAEFER, J. 1981. Isolation and concentration from the vapour phase. *In* Handbuch der Aroma Forschung, p. 44, H. Maarse and R. Belz, eds. Akademie-Verlag, Berlin, West Germany.

SCHREIER, P. 1980. Wine aroma composition. Identification of additional volatile constituents of red wine. J. Agric. Food Chem. *28*, 926.

SCHREIER, P. 1984a. Chromatographic Studies of Biogenesis of Plant Volatiles. Huethig Verlag, Heidelberg, Basel, New York.

SCHREIER, P. 1984b. Analysis of Volatiles, Methods and Applications. de Gruyter, New York.

SCHREIER, P., DRAWERT, F. and JUNKER, A. 1976. Identification of volatile constituents from grapes. J. Agric. Food Chem. *24*, 331.

SCHREIER, P., DRAWERT, F. and WINKLER, F. 1979. Composition of neutral volatiles constituents in grape brandies. J. Agric. Food Chem. *27*, 365.

SCHREIER, P., IDSTEIN, H. and HERRES, W. 1984. The use of HRGC-FTIR in tropical fruit flavour analysis. *In* Analysis of Volatiles, Methods and Applications, P. Schreier, ed. de Gruyter, New York.

SCHULTZ, T. H., FLATH, R. A., MON, T. R., EGGLING, S. B., and TERANISHI, R. 1977. Isolation of volatile components from a model system. J. Agric. Food Chem. *25*, 446.

SCHULTZ, T. H., BLACK, D. R., MON, T. R. and TERANISHI, R. 1980. Turkey mullein volatile constituents. J. Agric. Food Chem. *28*, 1255.

SCHULZE, W. G., TING, P. L., and GOLDSTEIN, H. 1979. Separation of hop compounds by reverse-phase HPLC. *In* Liquid Chromatographic Analysis of Food and Beverages, Vol. 2, G. Charalambous, ed. Academic Press, New York.

SCHWARTZ, D. P. and BREWINGTON, C. R. 1967. Methods for the isolation and characterization of constituents of natural products. II: Separation of homologous series of esters of pyruvic acid 2,6-dinitrophenyl hydrazone by thin layer chromatography. Microchem. J. *12*, 1.

SEIFERT, R. M. and BUTTERY, R. G. 1978. Characterization of some previously unidentified sesquiterpenes in carrot roots. J. Agric. Food Chem. *26*(1), 181.

SEIFERT, R. M., BUTTERY, R. G., GUADAGNI, D. G., BLACK, D. R. and HARRIS, J. G. 1970. Synthesis of some 2-methoxy-3-alkylpyrazines with strong bell pepper-like odors. J. Agric. Food Chem. *18*, 246.

SEO, C. W. and JOEL, D. L. 1980. Pentane production as an index of rancidity in freeze-dried pork. J. Food Sci. *45*, 26.

SHAW, P. E., AMMONS, J. M. and BRAMAN, R. S. 1980. Volatile sulfur compounds in fresh orange and grapefruit juices: Identification, quantitation, and possible importance to juice flavor. J. Agric. Food Chem. *28*(4), 778.

SIMON, P. W., LINDSAY, R. C., and PETERSON, C. E. 1980. Analysis of carrot volatiles collected on porous polymer traps. J. Agric. Food Chem. *38*, 549.

SINGLETON, J. A., PATTEE, H. E. and SANDERS, T. H. 1975. Some parameters affecting volatile production in peanut homogenates. J.Food Sci. *40*, 386.

SMAGULA, J. M. and BRAMLOGE, W. J. 1977. Determination of acetaldehyde in apple tissue by 2,4-dinitrophenyl hydrazone formation. J. Food Sci. *42*, 48.

SMITH, R. M. 1982. Analysis of pungent principles of ginger and grains of paradise by HPLC using electrochemical detection. Chromatographia. *16*, 155.

STEINKE, R. D. and PAULSON, M. C. 1964. The production of steam-volatile phenols during the cooking and alcoholic fermentation of grain. J. Agric. Food Chem. *12*, 381.

STUIVER, M. 1958. Doctoral Thesis, Rijks University, Groningen, Netherlands. Cited in Teranishi *et al.* (1981).

SUGISAWA, H. 1984. Prerequisites for sample preparation in flavour analysis *In* Analysis of Volatiles, Methods and Applications. P. Schreier, ed. de Gruyter, New York.

SUGISAWA, H., CHEN, C. and NABETA, K. 1984. Simultaneous distillation adsorption and its application. *In* Analysis of Volatiles, Methods and Applications, P. Schreier, ed. de Gruyter, New York.

SULLIVAN, J. F., KONSTANCE, R. P., CALHOUN, M. J., TALLEY, F. B., CORDING, J. JR. and PANASIUK, O. 1974. Flavor and storage stability of explosion-puffed potatoes. Nonenzymatic browning. J. Food Sci. *39*, 58.

SYDOR R. and PIETRYZK, D. J. 1978. Comparison of porous copolymers and related adsorbents for the stripping of low molecular weight compounds from a flowing air stream. Anal. Chem. *50*, 1842.

TANG, C. S. and JENNINGS, W. G. 1967. Volatile components of apricot. J. Agric. Food Chem. *15*, 24.

TEITELBAUM, C.L. 1977. A new strategy for the analysis of complex flavors. J. Agric. Food Chem. *25*, 466.

TERANISHI, R., FLATH, R. A. and SUGISAWA, H. 1981. Flavor Research: Recent Advances. Dekker, New York.

THOMPSON, J. A., MAY, W. A., PAULOSE, M. M., PETERSON, R. J. and CHANG, S. S. 1978. Chemical reactions involved in deep-fat frying of foods: VII. Identification of volatile decomposition products of trilinolein. J. Amer. Oil Chem. Soc. *55*, 897.

TOULEMONDE, B. and BEAUVERD, D. 1985. Headspace analysis: Trap desorption by microwave energy application to the volatile components of some tropical fruits. *In* Progress in Flavour Research 1984, J. Adda, ed. Elsevier, Amsterdam.

VAN STRATEN, S. 1981. Principles of chromatography. *In* Handbuch der Aroma Forschung. H. Maarse and R. Belz, eds. Akademie-Verlag, Berlin, West Germany.

VITZTHUM, O. G. and WERKHOFF, P. 1975. Cycloalkapyrazines in coffee aroma. J. Agric. Food Chem. *23*, 510.

WEURMAN, C. 1974. Sampling in airborne odorant analysis. *In* Human Responses to Environmental Odors, A. J. Turk, W. Johnston, and D. G. Moulton, eds. Academic Press, New York.

WILLIAMS, A. R. and TUCKNOTT, O. G. 1973. The selective extraction of aroma components from alcoholic distillates. J. Sci. Food Agric. *24*, 863.

WILLIAMS, M. P., HOFF, J. E. and NELSON, P. E. 1972. A precise method for the determination of dimethyl sulfide in processed foods. J. Food Sci. *37*, 408.

WILSON, R. A., MUSSINAN, C. J., KATZ, I. and SANDERSON, A. 1973. Isolation and identification of some sulfur compounds in pressure cooked beef. J. Agric. Food Chem. *21*, 873.

WITHYCOMBE, D. A., MOOKERJEE, B. D. and HURZA, A. 1978. Isolation of trace volatile constituents of hydrolyzed vegetable protein via porous polymer entrainment. *In* Analysis of Foods and Beverages, Headspace Techniques. G. Charalambous, ed. Academic Press, New York.

2

Biogenesis of Flavor in Fruits and Vegetables

Knowledge of precursors and pathways leading to the formation of flavor in fruits and vegetables is lacking in many areas. Emphasis has been placed on determining what constitutes flavor rather than on the mechanisms of flavor formation, and the task of elucidating metabolic pathways in very complex biological systems is difficult at best. Other than academic interest, there is little incentive to determine metabolic pathways leading to flavor formation. However, there is a strong interest in determining what chemicals make up a flavor since this knowledge could be used to study the appearance of an off-flavor, produce an artificial substitute or determine species flavor differences. The result is rather long lists of flavor chemicals found in each plant but little understanding of how or why they are there (Maarse 1984).

In this chapter we examine some of the known mechanisms of flavor formation in fruit and vegetables (see Schreier 1984). This discussion is generally limited to the formation of volatile flavors or aroma and excludes the taste sensations.

A. Biogenesis of Fruit Aroma

The typical flavor of fruit such as bananas, peaches, pears, and cherries is not present during early fruit formation but develops entirely during a rather brief ripening period. This flavor development period, or ripening, occurs during the climacteric rise in respiration (Fig. 2.1). During this period, metabolism of the fruit changes to catabolism and flavor formation begins. Minute quantities of lipids, carbohydrate, protein and amino acids are enzymatically converted to volatile flavors.

The rate of flavor formation reaches a maximum during the postclimacteric ripening phase. Flavor formation may not proceed uniformly

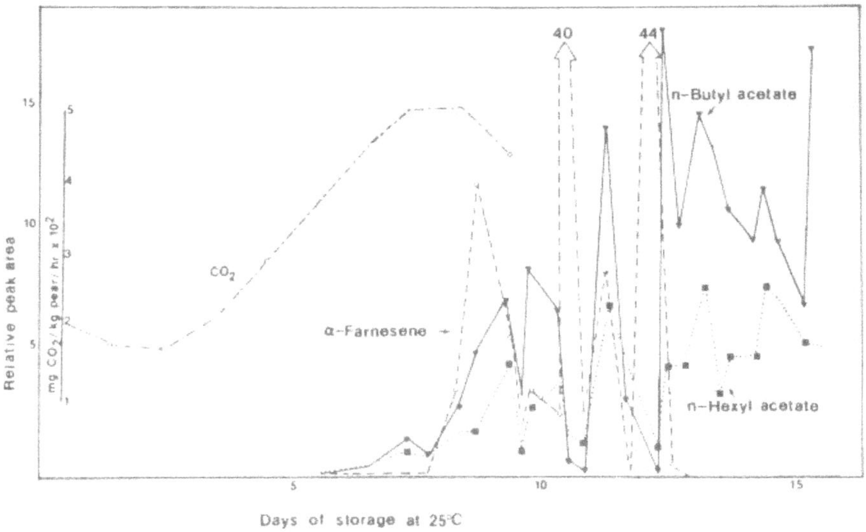

Fig. 2.1. Production of individual volatiles during the ripening of Bartlett pear.
Tressl et al. 1975

but in a cyclic manner. One pathway may be activated while another pathway is inhibited and then the cycle may reverse itself.

An example of the dynamic nature of flavor formation in ripening pears is shown in Fig. 2.1. Alpha-farnesene and ester formation (butyl acetate and hexyl acetate) were found to be produced out of phase with each other during ripening.

An overall view of flavor formation during ripening of fruit is shown in Fig. 2.2. Flavors are formed from major plant constituents (e.g., carbohydrates, lipids and proteins) under genetic control. Each metabolic pathway is connected to other metabolic pathways. As direct products of a metabolic pathway or as a result of interactions between pathways or end products, a host of volatile flavors are produced which contribute to the flavor of a ripe fruit. Table 2.1 presents typical flavor compounds which are formed from each major food constituent (Salunkhe and Do 1976).

1. Flavors from Fatty Acid Metabolism

Volatile flavor compounds may be formed from lipids via several different pathways. These include β-oxidation, hydroxyacid cleavage (lactones) and oxidation via lipoxygenase enzymes. While the primary products of these pathways are aldehydes and ketones, various oxidations,

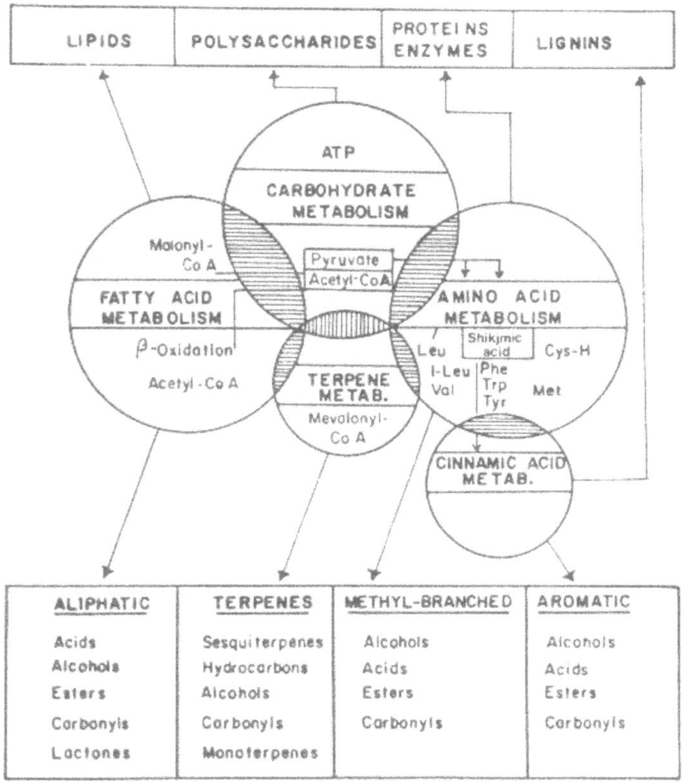

Fig. 2.2. Biosynthesis of fruit volatiles.
Tressl et al. 1975

reductions and esterifications also yield substantial quantities of acids, alcohols, lactones and esters.

In growing fruit, metabolic pathways exist to produce a variety of fatty acids. However, once ripening commences, this metabolic pathway functions to produce numerous fatty acids, ketones, aldehydes and alcohol (Fig. 2.3). These primary volatiles can then be converted to various esters by indigenous enzyme systems. Tressl and Drawert (1973) have demonstrated the conversion of labeled acetate to acetate esters and labeled butanoate to butanoate esters by postclimacteric banana slices. They have further shown the conversion of hexanoic acid to hexanol by these tissues.

The formation of flavors via β-oxidation is exemplified by considering flavor formation in pears. The decadienoate esters are generally considered carriers of the flavor of pear (Jennings 1967). These esters are be-

Table 2.1. Predominant Human Nutrients in Production of Aroma Components in Foods[a]

Nutrient	Aroma Component
Carbohydrates	
Glucose	Organic acids:
Fructose	Pyruvic acid, acetic acid, propionic acid, acetoacetic acid,
Sucrose	butyric acid, hexanoic acid, octanoic acid
	Esters:
	Pyruvates, acetates, propionates, butyrates, acetonacetates, hexanoates, octanoates
	Alcohols:
	Ethanol, propanol, butanol, hexanol, octanol
	Aldehydes:
	Acetaldehyde, propanal, butanal, hexanal, octanal
	Terpenes:
	Monoterpene, linalool, limonene, α-pinene, citronellal, citral, geranial
Amino acids	
Alanine	Pyruvic acid, acetaldehyde, ethanol
Valine	Isopropanal, isopropanol, α-keto-isobutyric acid
Leucine	3-Methylbutanal, 3-methylbutanol, α-keto-isocaproic acid
Isoleucine	2-Methylbutanal, 2-methylbutanol
Phenylalanine	Benzaldehyde, phenylacetaldehyde, cinnamaldehyde
	Hydrocinnamaldehyde, *p*-hydroxybenzaldehyde
	p-Hydroxy phenylacetaldehyde, *p*-hydroxy cinnamaldehyde, *p*-hydroxy cinnamaldehyde
Serine, Threonine, Glycine, Cystine/cysteine, Serine	Pyruvic acid Thiazoles Glyoxal
Fatty acids	
Linoleic acid	*trans*-2-*trans*-4-Decadienal, hexanal, *trans*-2-octenal
	trans-2-Pentanal, *trans* 2-hexenol, hexanal
	cis-3-Hexenal, *cis*-3-hexenol
	trans-2-*trans*-4-Heptadienal, propanal
Organic acids	
Citric acid, Oxaloacetic acid, Malic acid, Lactic acid	Glyoxylic acid, glyoxal, pyruvic acid, acetaldehyde Ethanol
Vitamins	
Carotenoids	
β-Carotene	β-Ionone

[a] From Salunkhe and Do (1976).

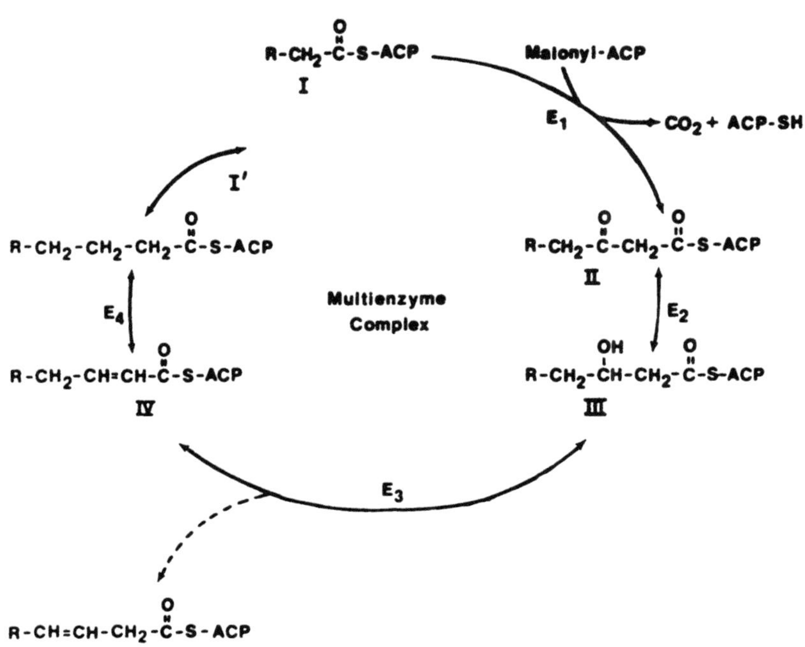

Fig. 2.3. Fatty acid pathway in plants yielding aroma.
Drawert 1975

lieved to be formed via β-oxidation of linoleic acid (see Fig. 2.4). Linoleic acid is metabolized, two carbons at a time, to shorter chain CoA derivatives which may react with alcohols to yield esters. During this process, isomerizations may occur to yield the trans–cis isomers. It is interesting that the β-oxidation pathway can account for all of the volatiles which have been identified in pear.

The production of lactones presents a third means of obtaining flavor from lipid sources. Lactones are very important to the flavor of peaches, apricots and coconuts. Numerous investigators have proposed oxidative pathways which would result in the lactone formation (Maga 1976b). A major pathway involves lipoxygenase action on linolenic acid. There is little information on the formation of lactones by mechanisms other than oxidative degradation. Tang and Jennings (1968) did propose that lactones could be formed during fatty acid synthesis. Their proposal involved hydration of a γ–δ double bond to yield the corresponding lactone. They conceded that there is no known mechanism to place a double bond in the γ–δ position but suggested a double bond shift (from β–γ position) followed by hydration.

The widest variety of flavor compounds formed from lipids arises via lipoxygenase activity. Many of the aliphatic esters, alcohols, acids and

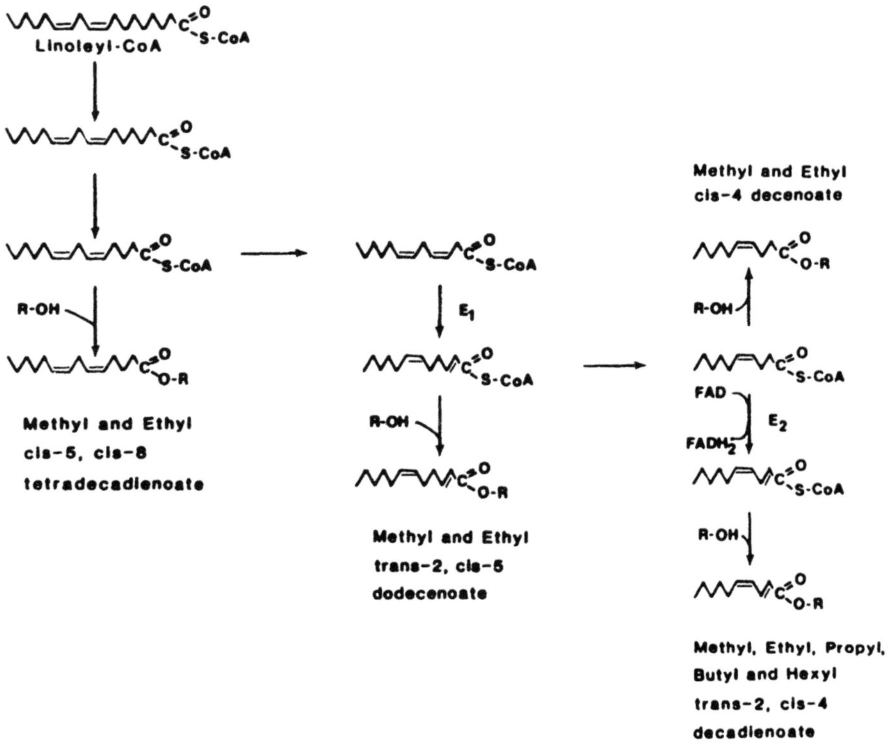

Fig. 2.4. Proposed pathway for the production of unsaturated esters in pears, E_1 = Δ3*cis*-Δ2*trans*-enoyl-CoA isomerase; E_2 = Acyl–CoA dehydrogenase.
Tressl et al. 1975

carbonyls found in fruit are derived from the oxidative degradation of linoleic and linolenic acids. Studies on the generation of flavor in banana (Tressl *et al.* 1970; Tressl and Jennings 1972; Tressl and Drawert 1973) and tomato (Kazeniac and Hall 1970; Schormuller and Grosch 1965; Stone *et al.* 1975; Jadhav *et al.* 1972; Galliard and Matthew 1977; Schreier and Lorenz 1981) elucidate and point out the importance of lipoxygenase activity to flavor development in fruit.

The actual mechanism of lipid oxidation via lipoxygenase in tomato has been studied by several workers. Jadhav *et al.* (1972) proposed that 9-, 12-, 13- and 16-hydroperoxides were formed via enzymatic action on linolenic acid. The mechanism proposed by Stone *et al.* (1975) suggested that the 13-hydroperoxide was the major end product. Later work by Galliard and Matthew (1977) indicated that the 9- and 13-hydroperoxides are the major end products. These two hydroperoxides were found in a ratio of 95:5, respectively. A hydroperoxide cleavage system was found by Galliard *et al.* (1977) only for the 13-hydroperoxide (Fig. 2.5).

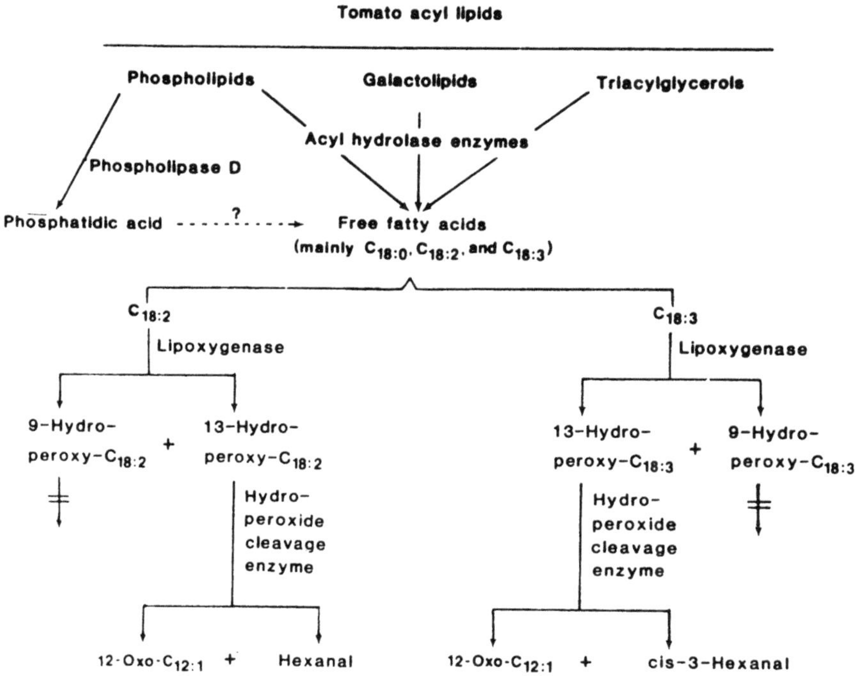

Fig. 2.5. Proposed pathway for the formation of short chain carbonyls by enzymatic degradation of acyl lipids in disrupted tomato fruits.
Galliard et al. 1977

A detailed study on lipoxygenase and the hydroperoxide cleavage-enzyme in tomatoes is presented by Schreier and Lorenz (1981). It appears that the 9-hydroperoxide is not further degraded enzymatically but only slowly decomposes to volatile compounds. The pathway presented in Fig. 2.5 outlines only the formation of hexanal and *cis*-3-hexenal, two of the more abundant volatiles in tomato. In fact, the oxidation of linoleic and linolenic produces a large number of volatiles. Acids and ketones as well as other intermediates in the oxidation process are readily converted to alcohols, aldehydes and esters by other enzyme systems in the plant (Fig. 2.6). Both α- and β-oxidation pathways have been demonstrated to exist thereby providing a wide range of volatiles for further conversion to flavor compounds (Tressl *et al.* 1975).

2. Flavors from Amino Acid Metabolism

Amino acid metabolism generates aliphatic and branched chain alcohols, acids, carbonyls and esters which are important to the flavor of

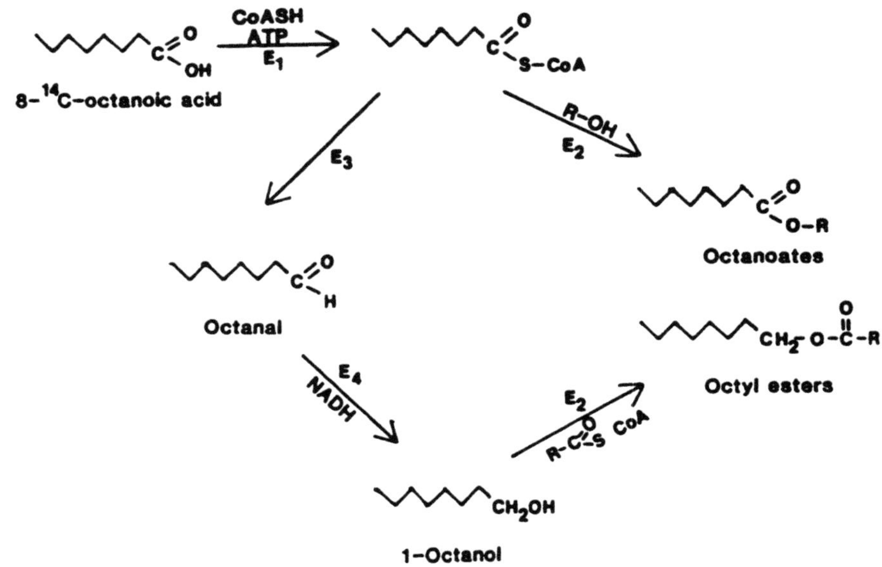

Fig. 2.6. Reaction scheme for conversion of octanoic acid into esters, E_1, acyl-thiokinase; E_2, acyl-CoA-alcohol-transacylase; E_3, acyl-CoA reductase; E_4, alcohol-NAD oxydoreductase.
Reprinted with permission of Tressl and Drawert (1973). Copyright 1973. American Chemical Society

fruit. Yu *et al.* (1968a,b,c) demonstrated that valine, leucine, alanine and aspartic acid could be converted to short chain carbonyls by tomato extracts. The enzymes involved in these transformations were found to be located in different sites in the tomato. The soluble tomato fraction isolated via centrifugation acted on leucine while the mitochondrial fraction metabolized both alanine and aspartic acid. The activities of these enzymes were substantially higher in field-grown vs. hothouse tomatoes (Dalal *et al.* 1968). This undoubtedly accounts for some of the flavor differences between hothouse and field-grown tomatoes.

Studies on banana tissue slices have shown that leucine and valine concentrations increase about threefold following the climacteric rise in respiration (Tressl and Drawert 1973). Radioactive labeling studies have shown that these two amino acids are transformed into branched chain flavor compounds which are essential to banana flavor. As can be seen in Fig. 2.7, the initial step is deamination of the amino acid followed by decarboxylation. Various reductions and esterifications then lead to a number of volatiles which are significant to fruit flavor.

An unusual flavor compound, 2-isobutylthiazole, has been found to be important to the flavor of tomato. It is hypothesized that this com-

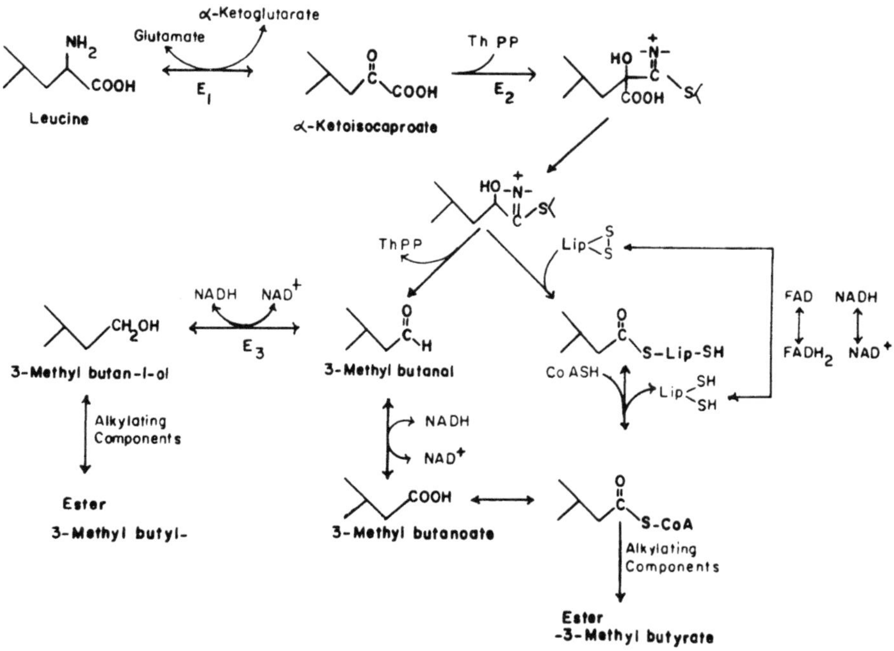

Fig. 2.7. Conversion of amino acids into aroma components of banana as illustrated by leucine. E_1, L-leucine aminotransferase; E_2, pyruvate decarboxylase; E_3, aldehyde dehydrogenase; ThPP, thiamin pyrophosphate; Lip$<^{S}_{S}|$, oxidized lipoic acid; Lip$<^{SH}_{SH}$, reduced lipoic acid; FAD flavin–adenine dinucleotide; NAD^+, oxidized nicotinamide–adenine dinucleotide; CoA–SH, coenzyme A.
Drawert 1975

pound is formed from the reaction of 3-methyl-1-butanal (from leucine) with cysteamine (Schutte 1973).

Aromatic amino acids also may serve as important precursors to fruit flavor. Some of the aromatic flavor compounds have been shown to come from tryosine and phenylalanine (Fig. 2.8). Odors characterized as "phenolic" or "spicy" could arise via this pathway.

It is suggested that the cinnamic acids serve as intermediates for many of the aromatic flavor compounds existing in foods (Schreier 1984). Simple esterification to methyl and ethyl cinnamate has been reported in strawberry (Drawert *et al.* 1973) and guava (Torline and Ballschmieter 1973). Ethyl cinnamate has been found in cranberry (Anjou and von Sydow 1967) and passion fruit (Winter and Kloti 1972). Removal of an acetate group from the cinnamic acids yields benzoic acids. The benzoic

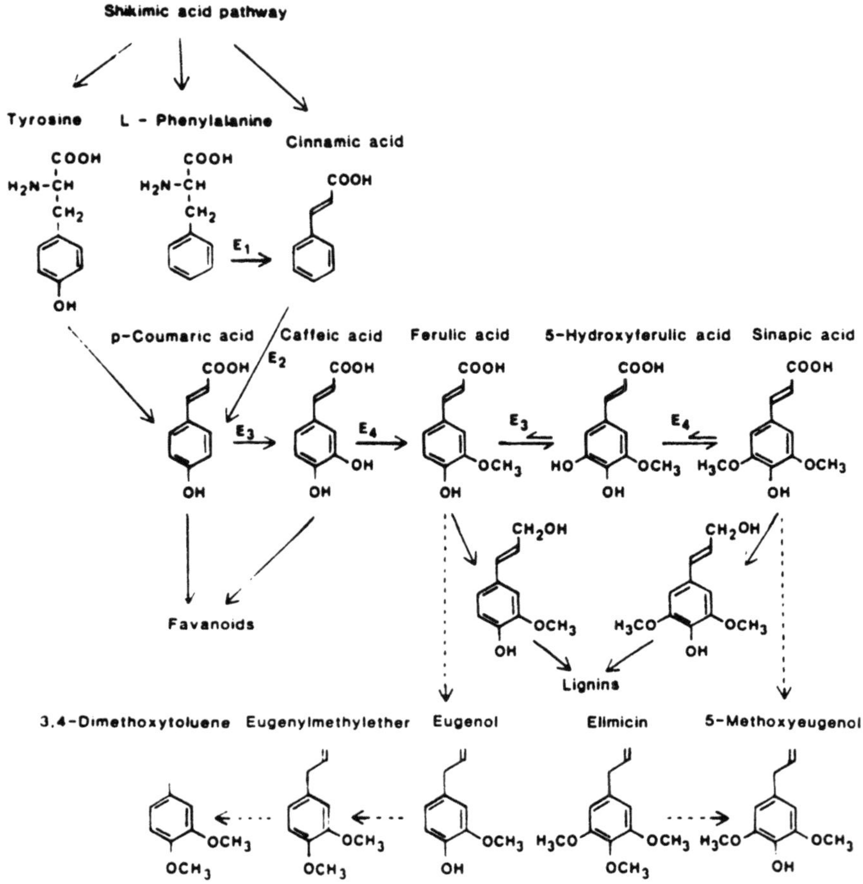

Fig. 2.8. Pathways for the formation of phenolic acids and phenol esters in banana. E_1, L-phenylalanine–ammonia–lyase; E_2, cinnamic-acid-hydroxylase; E_3, phenolase; E_4, methyltransferase.

Reprinted with permission of Tressl and Drawert (1973). Copyright 1973. American Chemical Society

acids may be further transformed by esterification to yield benzyl esters and by reduction to yield various benzaldehydes and benzyl alcohols (Zenk 1979). Decarboxylation would yield phenols.

3. Flavors Formed from Carbohydrate Metabolism

Returning briefly to Table 2.1, one can see that a large variety of volatile flavors can be traced to carbohydrate metabolism. If one considers plant metabolism, it is clearly established that plants obtain all of their energy directly from photosynthesis. The photosynthetic pathways involve turning CO_2 into sugars which are metabolized into other plant needs, e.g., lipids and amino acids. Therefore one may state that nearly all plant flavors come indirectly from carbohydrate metabolism since all of the other flavor precursors come from carbohydrate metabolism. However, there are few flavor constituents that come directly from carbohydrate metabolism.

An exception are the terpenes which may arise from either carbohydrate or lipid metabolism. Citrus products obtain their characteristic flavor from terpenes. Limonene, a monoterpene hydrocarbon possessing little odor, is the major terpene in most citrus oils often accounting for nearly 90% of the oil. The oxygenated terpenes, often comprising less than 5% of the oil, generally provide the characteristic flavor of different citrus species. For example, citral is considered the flavor impact component of lemon oil yet it is unusual to find more than 2% citral in a single fold oil.

The metabolic pathways leading to terpene formation are only partially known. While it is postulated that terpene formation involves mevalonic acid, the mechanism of mevalonic acid synthesis is unknown. Most plants synthesize mevalonic acid from acetate, however key enzymes leading from acetate to mevalonic acid have not been found in citrus fruit (Potty 1969). The incubation of mevalonic acid with orange extracts (and ATP) resulted in the formation of mevalonate 5-phosphate (MA5P) and mevalonate 5-pyrophosphate [MA5PP (Potty and Bruemmer 1970a)]. This activation step precedes decarboxylation and dehydration of the mevalonate to form the intermediates isopentyl pyrophosphate (IPP) and dimethylallyl pyrophosphate [DMAPP (Potty and Bruemmer 1970a)]. The next step in terpene synthesis involves a head-to-tail condensation of DMAPP and IPP to form linaloyl pyrophosphate and other acyclic and cyclic monterpenes (Fig. 2.9). Linalool was the major monoterpene formed from incubation of ^{14}C-labeled mevalonate with orange juice vesicles (Potty and Bruemmer 1970c). While neither nerol nor geraniol were detected by Potty and Bruemmer (1970c), George-Nascimento and Cori (1971) did detect significant quantities of these terpenoids when mevalonic acid was incubated with cell-free extracts or orange flavedo.

The interconversion of terpenes can occur via several pathways. The conversion of neryl pyrophosphate and linaloyl pyrophosphate to terpinolene, α-terpineol and limonene is shown in Fig. 2.9. Note that the product is determined by the means of stabilization of the carbonium ion intermediate. Additional terpene conversions have been traced to oxidoreductase systems in citrus fruits (Potty and Bruemmer 1970b). Figure 2.9 shows the interconversions of geraniol, geranial, citronellol

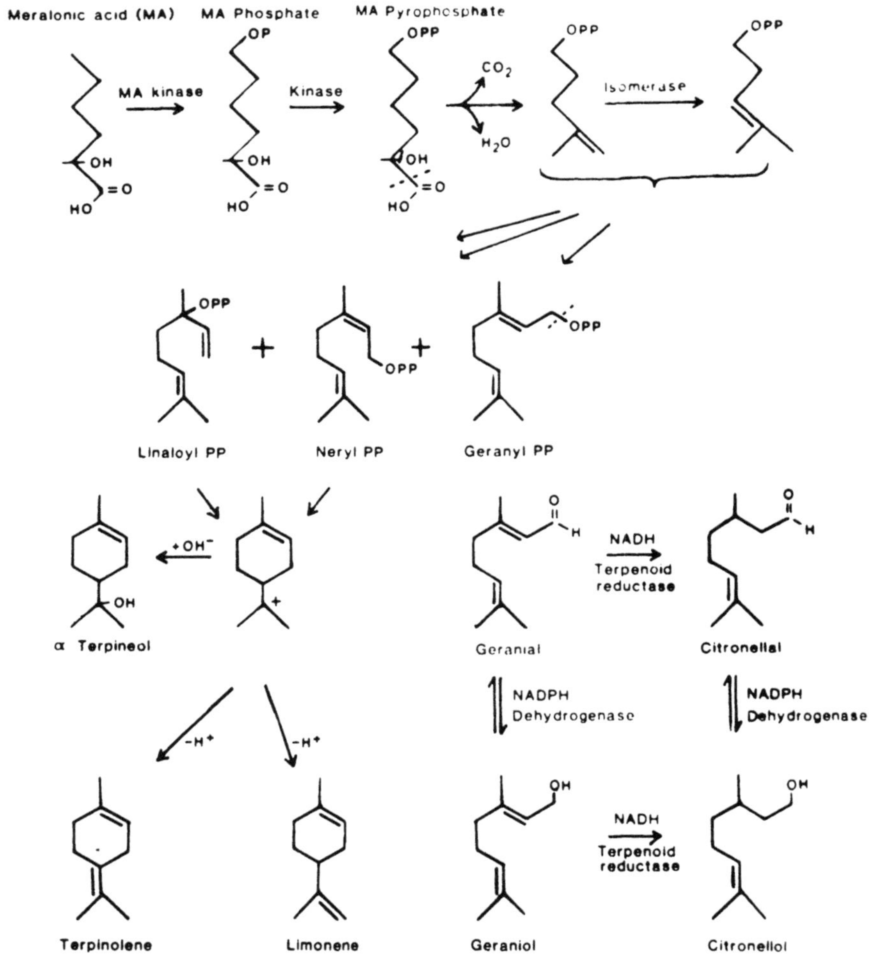

Fig. 2.9. Metabolic pathways leading to the formation of terpenes in citrus fruits. *Eskin (1979) and Potty and Bruemmer (1970b)*

and citronellal. Enzyme systems capable of oxidation/reduction reactions could account for a host of terpenes in citrus products.

The peel of citrus fruits contains a large number of flavone glycosides, some of which are bitter. The most common flavone in oranges and lemons is hesperidin, while naringin predominates in grapefruit. Hesperidin is bland, while naringin is very bitter. The bitterness of naringin is acceptable at low concentrations but can become quite objectionable at increased levels.

The A ring of naringin is formed from a head-to-tail condensation of two malonyl–CoAs and one acetyl–CoA. The B ring and C_3 middle ring

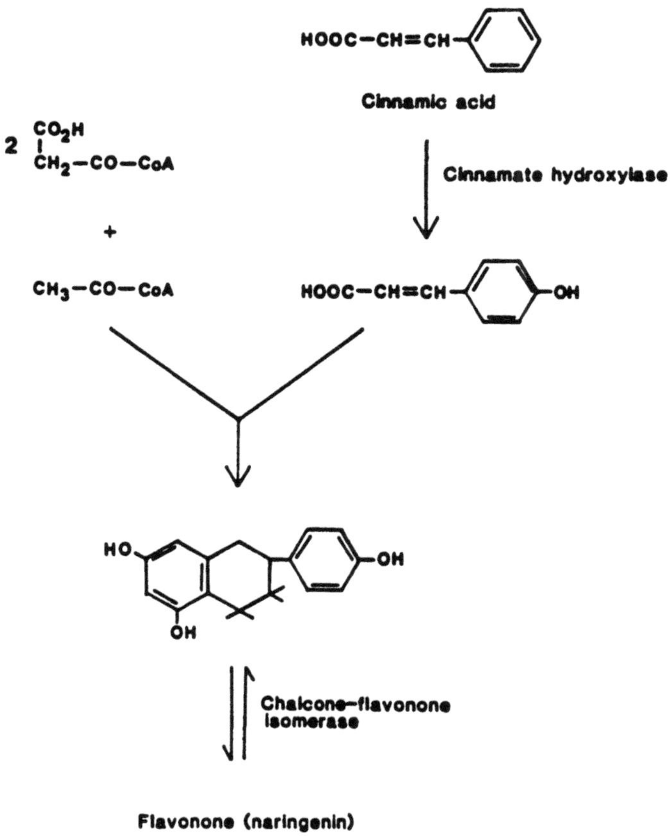

Fig. 2.10. Formation of naringenin from cinnamic acid.
Eskin 1979

are believed to come from a C_6–C_3 precursor such as cinnamic acid. The proposed pathway for naringenin biogenesis is shown in Fig. 2.10. Naringin is formed from naringenin via the addition of rhamnose and glucose to the A ring.

B. Biogenesis of Vegetable Aroma

Flavor formation in vegetables is quite different than in fruits. Vegetables do not have a ripening period as fruits do. While some vegetables will develop partial flavor during growth and the remaining flavor during cellular disruption, most vegetables develop all flavor constituents during cellular disruption. Cellular disruption may be due to cutting a vegetable (e.g., onion) prior to use or to the chewing process. Cellular disruption permits the mixing of enzymes and substrates which had been separated within the cell thereby resulting in the generation of volatile flavor substances. A few examples of vegetables which contain flavor prior to cellular disruption are celery (phthalides and selines), asparagus (1,2-dithiolane-4-carboxylic acid) and bell pepper (2-methyl-3-isobutyl pyrazine).

An overview of vegetable flavor formation is presented in Fig. 2.11. Similar to flavor development in fruit, fatty acid, carbohydrate and amino acid metabolism serve to provide the precursors of vegetable flavor. Also, the action of lipoxygenase on linoleic and linolenic acids to produce volatile carbonyls and alcohols is similar in vegetable flavor formation to that in fruit flavor formation. The major difference in flavor development lies in the importance of sulfur-containing flavor precursors to vegetable flavor. The thioglucosinolates and cysteine sulfoxides serve as precursors to numerous volatile sulfur compounds. An excellent review on the role of sulfur compounds in food flavor has been written by Shankaranarayana *et al.* (1982).

1. Flavor Formation from Cysteine Sulfoxide Derivatives

Undoubtedly the best illustration of flavor formation from cysteine sulfoxide precursors comes from studies on onion flavor. Intact onion cells have virtually no characteristic flavor until cellular disruption (cutting, blending, chewing, etc.). Only seconds after cell damage, the onion develops full flavor. This rapid formation of flavor is characteristic of the genus *Allium* as a whole. This genus includes onions, garlic and chives. *S*-Alk(en)yl-L-cysteine sulfoxides serve as precursors of flavor in the genus. An investigation of 27 *Allium* species found that they could be placed into three groups based on the predominance of certain amino acid precursors (Freeman and Whenham 1975). The predominant precursors were *S*-1-propyl (onion), *S*-2-propenyl (garlic) and *S*-methyl (*A. aflaturense* B. Fedtschenko) cysteine sulfoxides. A pathway for the formation of *S*-1-propenyl cysteine sulfoxide has been proposed by Granroth (1970). This pathway starts with the deamination of valine followed by decarboxylation to yield methacrylate. Reaction of methacrylate with L-cysteine followed by decarboxylation yields 1-propenyl-cysteine sulfoxide (Fig. 2.12).

The mechanism of flavor formation from the alkyl cysteine sulfoxide (alliin) is outlined in Fig. 2.12. While the initial steps of this pathway are enzymatic, reactions beyond sulfenic acid are purely chemical in nature. Sulfenic acid is extremely reactive, readily forming the unstable thiosulfinate intermediate by reaction with a second sulfenic acid mol-

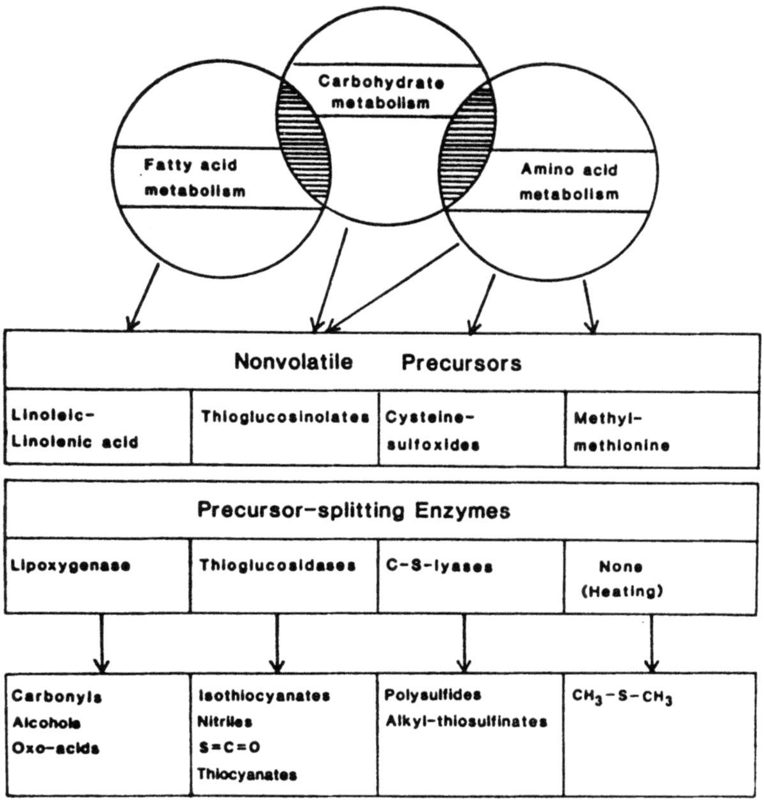

Fig. 2.11. Formation of flavor in vegetables.
Tressl et al. 1975

ecule. This thiosulfinate decomposes to form a relatively stable thiosulfonate and mono, di- and trisulfides. Considering that several different alkyl precursors are available in each *Allium* species, a host of different mono-, di-, and trisulfides can be formed via different sulfenic acid combinations. It is these mono-, di-, and trisulfides which are most important in determining typical *Allium* flavor.

The lachrymatory factor in onion (thiopropanal-*S*-oxide) is considered to come from the *S*-1-propenyl-cysteine sulfoxide precursor (Brodnitz and Pascale 1971). Thiopropanal-*S*-oxide ($CH_3{-}CH_2{-}CH{=}S{=}O$) is somewhat unstable reacting with pyruvate to form propanol, 2-methyl pentanal and 2-methyl pent-2-enal. (Virtanen 1962). It is of interest that no lachrymatory properties were found in 25 species of wild onions (Saghir and Mann 1965). This would suggest that the 1-propenyl precursor may be absent from the wild onion.

Matikkala and Virtanen (1967) found that nearly half of the cysteine

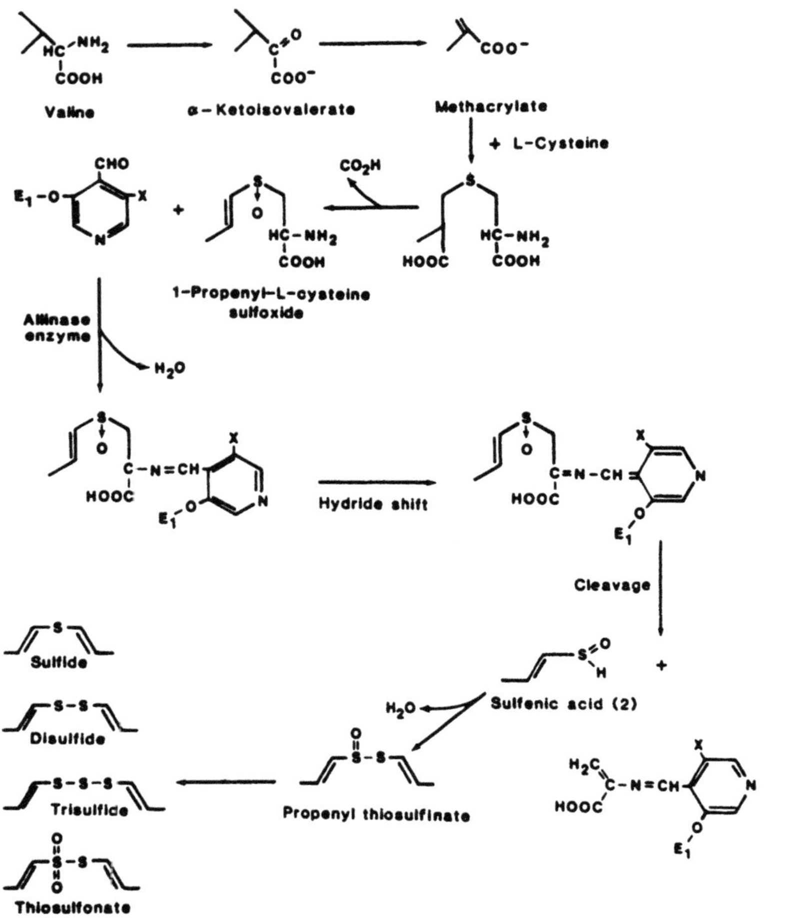

Fig. 2.12. Biosynthesis of S-1-propenyl-L-cysteine sulfoxide, and subsequent flavor formation in onions.
Adapted from Granroth (1970)

sulfoxide precursors of onion are bound up as a γ-glutamyl peptides. The alliinase enzyme cannot act upon these cysteine sulfoxide flavor precursors until they are freed from the glutamyl peptide. This can be accomplished only via the action of a γ-L-glutamyl transpeptidase enzyme. This enzyme transfers the glutamyl moiety to another amino acid. The glutamyl transpeptidases are present only in sprouted onions, not mature onions (Schwimmer and Austin 1971). Therefore, nearly one-half of the flavor potential of the onion cannot be utilized for flavor formation.

The Brassica family also obtains a flavor contribution from alkyl cysteine sulfoxide precursors. The Brassica family includes broccoli, brussel sprouts, cabbage, cauliflower and rutabagas. *S*-Methyl-L-cysteine sulfoxide is the primary cysteine sulfoxide derivative found in this family. An investigation of fresh cabbage flavor found that 6% of the volatiles was dimethyl disulfide, 6% dimethyl trisulfide, 3% dimethyl tetrasulfide and 1.5% methyl ethyl trisulfide (Tressl *et al.* 1975).

Rather unusual sulfur-containing volatiles are formed from cysteine sulfoxide precursors by the Shiitake mushroom (Fig. 2.13). Similar to the onion, the cysteine sulfoxide flavor precursor is bound to a γ-glu-

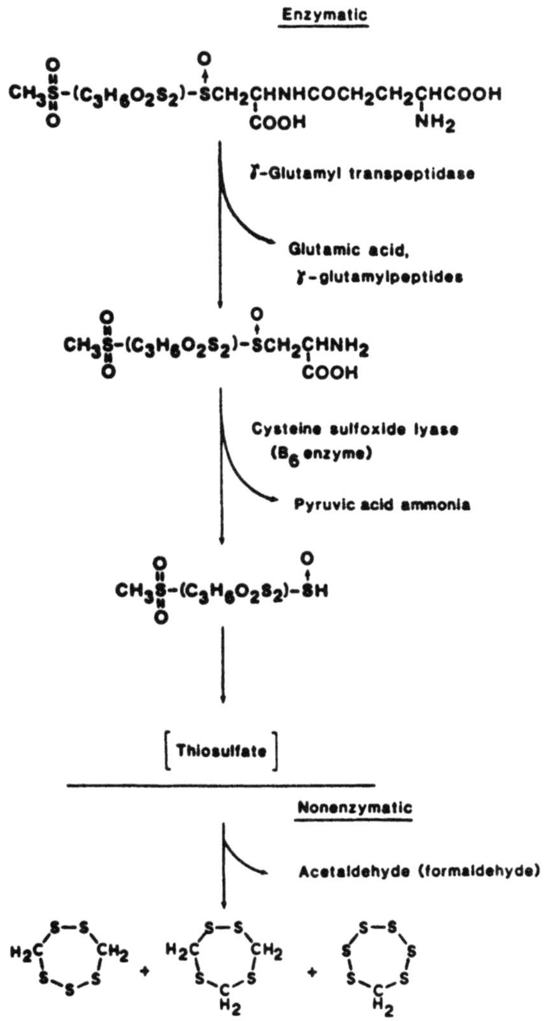

Fig. 2.13. Pathway proposed for the enzymatic and nonenzymatic formation of lenthionine from lentinic acid.
Yasumoto et al. 1971

tamyl peptide. Therefore, the first step in flavor formation is the removal of glutamic acid by γ-glutamyl transpeptidase. The remaining steps are also similar to onion but yield lenthionine, the major flavor component in this mushroom.

2. Flavor Formation from Glucosinolates

Most of the vegetables containing glucosinolates belong to the Cruciferae or Brassica family. More than 50 different glucosinolates have been identified to date (Tressl *et al.* 1975). Glucosinolates are nonvolatile flavor precursors which are enzymatically hydrolyzed to volatile flavors when cellular structure is disrupted. The volatile products initially are isothiocyanates and nitriles. Secondary reactions lead to the formation of several other classes of flavor compounds. Figure 2.14 outlines the pathways proposed to explain the formation of flavor compounds in the radish. The glucose moiety is initially hydrolyzed from the glucosinolate. This results in an unstable molecule which readily splits out HSO_4^-. Depending upon molecule rearrangements, either a isothiocyanate or a

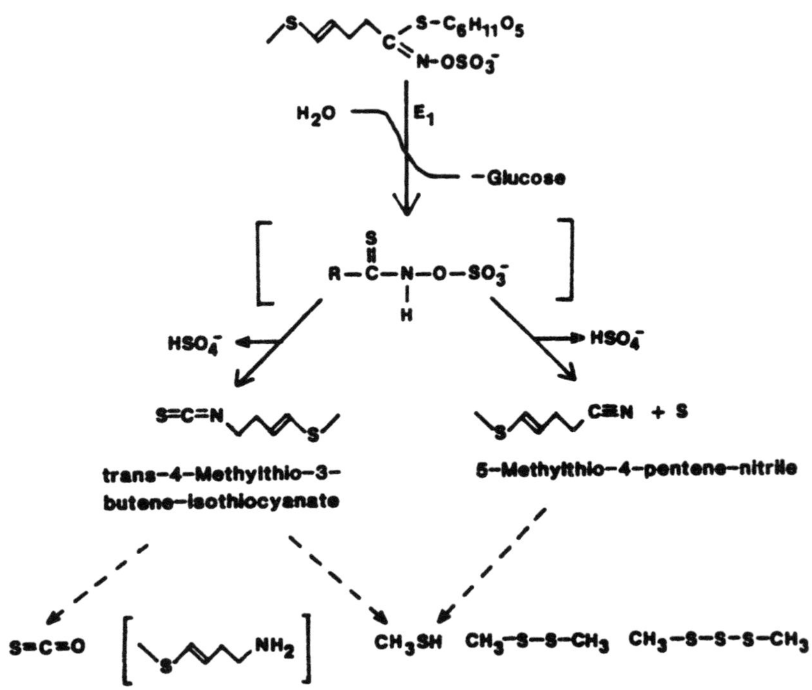

Fig. 2.14. The formation of volatile flavors in radish root.
Tressl et al. 1975

nitrile result. Further reactions result in the formation of thiols, sulfides, disulfides and trisulfides. The contribution of isothiocyanates to the flavor of cabbage has been known for many years. Jensen *et al.* (1953) were the first to identify allyl isothiocyanate in cabbage. Bailey *et al.* (1961) added methyl, *n*-butyl, butenyl, and methyl thiopropyl isothiocyanates to the list of isothiocyanates of cabbage. Allyl isothiocyanate is the most important isothiocyanate to the flavor of cabbage.

3. Additional Pathways for Vegetable Flavor Formation

Carbonyls and alcohols may be present in significant concentrations in vegetables. These carbonyls and alcohols typically arise via fatty acid degradation by lipoxygenase. The mechanism of formation is similar to that discussed previously for carbonyl formation in fruit. However, each plant has a unique fatty acid composition and ancillary enzyme system. Therefore, although the overall mechanism of flavor formation from fatty acids is similar between plants, each plant does develop its own unique carbonyl profile.

A similar analogy could be made for the formation of terpenes in vegetables. Terpenes are rather common to vegetables (Maarse 1984) and arise from mevalonic acid similar to the mechanism of formation in fruit. Yet each vegetable containing terpenes has a unique terpene profile.

C. Location of Flavor in Plant

As one might expect, flavor is typically nonuniformly distributed in the plant. MacLeod and MacLeod (1970) have shown that allyl isothiocyanate is more abundant in the outer leaves than in the inner tissues of cauliflower. Freeman and Whenham (1975) dissected onion bulbs separating the bulb into individual leaves and the stem. They examined each section for lachrymator and pyruvate. They found that the dry outer scales had virtually no lachrymator nor pyruvate but levels increased progressively in the leaves moving towards the plant center and stem. Total terpenes are more concentrated in the crown of carrots than in the midsection or tip (Simon *et al.* 1980a). Not only is there a difference in total flavor depending upon the plant part being examined, but there may also be a difference in flavor profile (Simon *et al.* 1980b). Therefore, if one dissected a plant and ate different individual parts it would not be unusual to find a slightly different flavor character for each part.

D. Influence of Genetics, Nutrition, Environment, Maturity and Storage on Development of Flavor

There is little question that not all apples, oranges or coffee taste the same. We often select a particular variety of apple (e.g., Delicious) because we prefer the taste. From advertising and experience we know that geographical location influences flavor, e.g., the "robust" Columbia coffees and the "fruity" character of German wines. We are aware that climate and soil conditions influence flavor and that often the best fruit comes via handpicking fresh and fully ripened right from the tree. We know that each piece of fruit or vegetable is an individual in appearance and flavor.

There is a great deal of literature reporting on differences in flavor due to genetics, environment, harvesting time and postharvest treatment (Paillard 1981; Eskin 1979; Salunkhe and Do 1976; Freeman 1979). This chapter is devoted to sampling some of this literature to illustrate and explain the flavor differences we observe between foods of the same type. This chapter primarily deals with differences in the flavor of plant foods. While animal products are often affected by similar factors, variations in plant flavor are accepted as natural while variations in animal flavor are generally considered as off-flavors. One does not accept the varietal differences in beef, for example, as one does in apples.

1. Plant Foods

a. Genetics

There is ample evidence both in our daily lives and in the literature to show that plant genetics influence flavor (Simon *et al.* 1980a,b; Simon and Peterson 1979; Maga 1975, 1976a; Salunkhe and Do 1976; Paillard 1981). Genetic composition determines enzyme systems and their activity in flavor formation.

It is most common that varietal differences in flavor are due to quantitative differences in flavor composition rather than qualitative differences. Quantitative differences in 2-isobutylthiazole between tomato varieties are shown in Table 2.2. Two to three times as much 2-isobutylthiazole is in the 146 variety as in the 1327 variety. Since the threshold for 2-isobutylthiazole is 1–3 ppb, the concentration differences as observed would be obvious by sensory analysis.

An additional example of genetic effects on flavor comes from studies on apple flavor. Paillard (1967) found that the same volatile constituents were present in all apple varieties studied but differences in proportions

Table 2.2. Tomato Variety and 2-Isobutylthiazole Concentration

Variety	Concentration (ppb)
Campbell 146[a]	140
Campbell 146[b]	190
Campbell 1327[a]	59
Campbell 1327[b]	60
Campbell accession number 1306[b]	90

[a] From Stevens (1970).
[b] From Kazeniac and Hall (1970).

of these volatiles between varieties was observed. The yellow varieties examined tended to produce more acetate esters while the red varieties produced more butyrate esters. He (Paillard 1979) found Golden Delicious parenchyma to rapidly and nearly completely convert butyric acid to acetic acid which resulted in the generation of acetate esters rather than butyrate esters. It was of interest that peels of apples with waxy cuticles (Calville blanc and Starkling) converted fatty acid substrates primarily to short chain alcohols and esters while apples with a corky cuticle (Canada gris) converted the fatty acids primarily to ketones and a small amount of alcohols.

These differences in metabolism are often responsible for small but yet significant changes in flavor profile. Varietal differences in flavor have come to be accepted as natural and, in fact, one enjoys the many options provided.

b. Environmental Effects on Flavor Development

Simon *et al.* (1980a,b) demonstrated that the same variety of carrots differs in flavor when grown in Florida, Texas and California. The California-grown carrots were found to have the most overall carrot flavor with less harsh notes and more sweetness than carrots grown in Texas or Florida. The Florida-grown carrots were characterized as being low in sweetness and overall flavor. Simon *et al.* (1980a,b) made no effort to determine whether these flavor differences were due to growing temperature, soil nutrition, sunlight or water availability. Each of these factors probably differed between locations and may have contributed to the observed flavor differences.

Somogyi *et al.* (1964) demonstrated that McIntosh apples grown with nitrogen fertilizers produced increased quantities of volatiles. Flavor was actually intensified by the use of a commercial fertilizer. Very striking changes in the flavor of onion, garlic, *A. vineale*, cabbage, mustard and watercress have been noted to be dependent on sulphate content of the growth medium (Freeman and Mossadeghi 1970, 1971, 1972a,b). This is not unexpected since these plants are very dependent on sulfur-containing precursors for their characteristic flavor. Onions grown under extreme sulphate deficiency lacked typical flavor and lachrymatory potency. Both flavor and lachrymatory activity were found to increase with sulphate availability up to the genetically controlled limits of the plant. Freeman (1979) has suggested that the flavor precursors [*S*-alk(en)yl-L-cysteine sulfoxides in the *Allium* species and glucosinolates in the cruciferae] are not essential to plant growth but that a certain level of sulphate nutrition is required for plant synthesis of essential amino acids and proteins. The ability to control flavor via plant nutrition could provide the opportunity to produce vegetables of selected flavor strength. It may be desirable to produce a more bland vegetable, e.g., cabbage, to reduce cooking odors or radishes to reduce harshness.

Another factor known to influence flavor is the water content of the soil during plant growth. Plentiful rainfall often results in large lush vegetables which are unfortunately lacking in flavor. Years with inadequate rainfall often result in fruit and vegetables of smaller size and less attractive appearance but yet with more intense flavor. Freeman (1979) suggested that this more intense flavor may be due to increased amounts of flavor precursors caused by stress put on the plant by restricted water availability. There are numerous examples in the literature of plants accumulating low molecular weight metabolites (e.g., sugars, amino acids and organic acids) when under stress. These metabolites serve as flavor precursors.

While the stress may be due to limited water availability, stress may also be due to temperature. Plants subjected to low temperatures are known to accumulate similar low molecular weight metabolites. This presumably is an evolutionary response directed towards lowering freezing point of the plant tissue to enhance survival of frost damage.

Additional temperature effects on flavor development may be due to temperature dependency of enzymatic reactions. Generally speaking, enzyme reaction rate will increase with temperature until enzyme denaturation begins. However, enzyme reactions differ in extent of temperature enhancement: one reaction may increase only slightly with a 10°C increase in growing temperature while a different reaction may double with this temperature increase. This may result in differential generation of flavor components.

c. Influence of Maturity on Flavor Development

The trend in recent years has been to pick fruit prior to ripening and then control ripening during distribution. This often results in the fruit being picked at a very early stage. The literature generally indicates that flavor development proceeds differently once the fruit has been picked and ripened away from the plant. An exception would, of course, be those foods which only ripen once removed from the parent plant, e.g., cantaloupe and some varieties of avocado (Bliss and Pratt 1979). They would ripen abnormally if left on the parent plant.

Do *et al.* (1969) investigated flavor development in peaches as influenced by harvest maturity and artificial ripening. Peaches obtain most of their characteristic flavor from lactones and, of course, these lactones increase as stage of maturity advances. Artificial ripening of the peaches does not result in similar levels of lactones in the ripe fruit. Artificially ripened peaches contain only about 20% the total lactone content of tree ripened peaches. Benzaldehyde and total esters in the artificially ripened peaches reach only 20% and 50% the respective concentration found in tree-ripened fruit.

Studies on artificially ripened tomatoes show that flavor may not only be reduced in comparison to the vine-ripened product, but also may be abnormal in character. Shah *et al.* (1969) found more butanol, isopentanal, 2,3-butanedione, propyl acetate, isopentyl butyrate and additional unidentified carbonyls in artificially ripened tomatoes than in the vine-ripened fruits. The vine-ripened fruits contained more nonanal, decanal, dodecanal, neral, benzaldehyde, citronellyl propionate, citronellyl butyrate, geranyl acetate and geranyl butyrate than the artificially ripened tomatoes. Studies by Yu *et al.* (1967, 1968a,b,c) and Yu and Spencer (1969, 1970) have shown that immature tomatoes lack equivalent enzyme systems as the vine-ripened tomatoes. Apparently as ripening progresses, more enzyme systems develop which are significant to flavor generation. Therefore, the artificially ripened tomatoes lack the enzyme systems necessary to produce the characteristic tomato flavor.

There is little in the literature on the effect of plant age on the development of volatile flavor components, however Freeman (1979) has reported on flavor formation during onion seed germination and growth. It appears that after approximately 20 days, onion flavor potential is completely developed. The seed itself contains no flavor precursors (cysteine sulfoxide derivatives) and only about 3% the alliinase activity (enzyme) of mature onion bulbs. However, the plant quickly develops alliinase activity and a maximum activity is reached after 15–20 days. Therefore, flavor is developed well before the plant is of suitable size for consumption.

While not the subject of this chapter, the effect of plant age on *overall* vegetable flavor is well documented. Young vegetables generally are softer and sweeter than mature or overly mature vegetables. Overall vegetable quality decreases quickly once maturity is achieved.

d. Effects of Postharvest Storage Conditions on Flavor Development

Since flavor of a fruit develops rapidly during the climacteric period and this period commonly occurs away from the plant, storage conditions during this time will influence flavor development (Paillard 1981). An extreme example of the importance of storage conditions on flavor development during ripening comes from studies on banana flavor. Banana volatile production is an exponential function of storage temperature from 5° to 25°C. Prolonged storage at 10°–12°C results in substantially reduced (60%) flavor production. Temperatures less than 5°C result in no characteristic flavor development. Chill injury occurs which alters lipids in subcellular membranes and irreversibly alters normal fruit respiration. Storage temperatures greater than 27°C also result in abnormal flavor development. Characteristic flavor production decreases while products of fermentation, e.g., ethanol and ethyl acetate, predominate. Therefore, storage temperature must be controlled to permit proper flavor development.

Paillard (1981) has summarized additional factors which may influence flavor development during storage. Controlled atmosphere storage of apples generally decreases volatile production during ripening. A 5% CO_2 + 2% O_2 atmosphere had the greatest effect on flavor generation, nearly completely inhibiting flavor production after $5\frac{1}{2}$ months of storage. Low humidity storage of apples increased the emission of hexyl, isopentyl and butyl acetate and decreased the corresponding alcohols. The low humidity favored ester formation. Illumination of the apples increased volatile production during storage. This increase in volatiles is postulated to be due to the light catalyzing lipid oxidation in the peel. The lipid oxidation products then are metabolized to short chain esters.

2. Animal Products

As was mentioned in the chapter introduction, most variations in flavor of meats are not well accepted by the consumer. A T-bone steak or a hamburger should have a certain flavor without variation. There is ample literature to show that the flavor of meat is particularly dependent upon the animal's diet. However, these variations (e.g., the "rotten egg" or mercaptan odor of lambs pastured on oats) are generally considered as off-flavors and, therefore, will be discussed in the off-flavor chapter of this text.

One of the few exceptions where flavor differences are accepted between animal species is in fish. The consumer has come to appreciate the wide variety of flavors available from sardines, salmon, sole, catfish, bluefish, etc.

E. Conclusions

While flavor development in fruits and vegetables proceeds along many common paths, there are also several contrasts. Fruit flavor develops during a short ripening period, while vegetable flavor develops primarily during cellular disruption. Fruits depend heavily upon esters for their characteristic flavor while sulfur compounds are more important to vegetable flavor.

Similarities exist in the formation of alcohols, aldehydes, ketones and acids by enzymatic pathways. The enzymatically catalyzed oxidation of lipids is a particularly important pathway for flavor formation in both fruits and vegetables.

While the pathways leading to flavor formation are known for many compounds, in other cases many pieces of the synthesis puzzle are missing. It is unfortunate that most of the research on biosynthesis pathways of flavor formation was done in the early 1970s and that this area of study has given way to the more easily fundable areas of toxicology and nutrition. There currently is only a very limited amount of flavor research directed towards elucidating pathways of flavor formation.

It is obvious from this chapter that many factors influence the flavor of a fresh fruit or vegetable. These factors include plant genetics, soil nutrition, growing environment, stage of maturity and conditions of storage from harvesting to consumption. Genetics determine the enzyme systems and precursors involved in flavor formation. Soil nutrition provides some of the essentials for flavor development. A radish, which depends upon sulfur compounds for flavor, will not develop characteristic flavor if there is little sulfur in the soil. Growing conditions influence activity of different enzyme systems and can significantly alter flavor development. Certainly the stage of maturity and storage conditions can further influence flavor.

While a number of variables are outside of the control of the plant grower, some are within his control. The grower should consider the influence of plant genetics, soil nutrition and water regime in determining the flavor of his crop. It seems that appearance and yield per acre are the two major criteria of success in growing produce. Flavor should become a more significant consideration.

References

ANJOU, K. and von SYDOW, E. 1967. The aroma of cranberries. II. *Vaccinium macrocarpon*. Acta Chem. Scand. *21*, 2076.

BAILEY, S. D., BAZINET, M. L., DRISCOLL, J. L., and McCARTHY, A. I. 1961. The volatile sulfur components of cabbage. J. Food Sci. *26*, 163.

BLISS, M. L. and PRATT, H. K. 1979. Effect of ethylene, maturity and attachment to the parent plant on production of volatile compounds by muskmelons. J. Amer. Soc. Hort. Sci. *104*(2), 273.

BRODNITZ, M. H. and PASCALE, J. H. R. 1971. Thiopropanol-S-oxide—a lachrymatory factor in onion. J. Agric. Food Chem. *19*, 269.

DALAL, K. B., SALUNKHE, D. K., OLSON, L. E. and DO, J. Y. 1968. Volatile components of developing tomato fruit grown under field and greenhouse conditions. Plant Cell Physiol. *9*, 389.

DO, J. Y., SALUNKHE, D. K., and OLSON, L. E. 1969. Isolation, identification and comparison of the volatiles of peach fruit as related to harvest maturity and artificial ripening. J. Food Sci. *34*, 618.

DRAWERT, F. 1975. Biochemical formation of aroma components. *In* Aroma Research, p. 13. H. Maarse and P. J. Groenen, eds. Proc. of Int. Symp. on Aroma Res., Pudoc, Wageningen, The Netherlands.

DRAWERT, F., TRESSL, R., STANDT, G. and KOPPLER, H. 1973. Gaschromatographisch-massenspektometrische Differenzierung von Erdbeerarten. Z. Naturforsch. *28C*, 488.

ESKIN, N. A. M. 1979. Plant Pigments, Flavors and Texture, p. 43. Academic Press, New York, NY.

FREEMAN, G. G. 1979. Factors affecting flavour during growth, storage and processing of vegetables. *In* Progress in Flavor Research, p. 225, D. G. Land and H. E. Nursten, eds. Appl. Sci. Pub., London.

FREEMAN, G. G. and MOSSADEGHI, N. 1970. Effect of sulphate nutrition of flavour components of onion (*Allium apa*). J. Sci. Food Agric. *21*, 610.

FREEMAN, G. G. and MOSSADEGHI, N. 1971. Influence of sulphate nutrition on the flavour component of garlic (*Allium sativum*) and wild onion (*Allium vineale*). J. Sci. Food Agric. *22*, 330.

FREEMAN, G. G. and MOSSADEGHI, N. 1972a. Influence of sulphate nutrition on flavour components of three cruciferous plants: Radish (*Raphanus sativus*), cabbage (*Brassica capitata*) and white mustard (*Sinapis alba*). J. Sci. Food Agric. *23*, 387.

FREEMAN, G. G. and MOSSADEGHI, N. 1972b. Studies on sulphur nutrition and flavour production in watercress (*Rorippa nasturtium-aquaticum* (L) Hayek). J. Hort. Sci. *47*, 375.

FREEMAN, G. G. and WHENHAM, R. J. 1975. A survey of volatile components of some *Allium* species in terms of S-alk(en)yl-L-cysteine sulphoxide present as flavor precursors. J. Sci. Food Agric. *26*, 1869.

GALLIARD, F. and MATTHEW, J. A. 1977. Lipoxygenase-mediated cleavage of fatty acids to carbonyl fragments in tomato fruits. Phytochemistry. *16*, 339.

GALLIARD, T., MATTHEW, J. A., WRIGHT, A. J. and FISHWICK, M. J. 1977. The enzymatic breakdown of lipids to volatile and nonvolatile carbonyl fragments in disrupted tomato fruits. J. Sci. Food Agric. *28*, 863.

GEORGE-NASCIMENTO, C. and CORI, O. 1971. Terpene biosynthesis from geranyl and neryl pyrosphosphates by enzymes from orange flavedo. Phytochem. *10*, 1083.

GRANROTH, B. 1970. Biosynthesis and decomposition of cysteine derivatives in onion and other *Allium* species. Ann. Acad. Sci. Fenn. Ser. A2. *154*, 40.

JADHAV, S., SINGH, B. and SALUNKHE, D. K. 1972. Metabolism of unsaturated fatty acids in tomato fruit: Linoleic and linolenic acid as precursors of hexanol. Plant Cell Physiol. *13*, 449.

JENNINGS, W. G. 1967. Peaches and pears. *In* Chemistry and Physiology of Flavors, p. 419, H. W. Schultz, E. A. Day and L. M. Libbey, eds. AVI Publishing, Westport, CT.

JENSEN, K. A., CONTI, J. and KJAER, A. 1953. Isothiocyanates: II. Volatile isothiocyanates in seeds and roots of various Brassicas. Acta. Chem. Scand. *7*, 1267.

KAZENIAC, S. J. and HALL, R. M. 1970. Flavor chemistry of tomato volatiles. J. Food Sci. *35*, 519.

MAARSE, H. 1984. Volatile Compounds in Food, Quantitative Data, Vols. 1–3. Division for Nutrition and Food Research, TNO, Zeist, Netherlands.

MAGA, J. A. 1975. The role of sulfur compounds in food flavours: Part I: Thiazoles. Crit. Rev. Food Sci. Nutr. *6*, 153.

MACLEOD, A. J. and MACLEOD, G. 1970. Flavour volatiles of some cooked vegetables. J. Food Sci. *35*, 734.

MAGA, J. A. 1976a. The role of sulfur compounds in food flavour: Part III. Thiols. Crit. Rev. Food Sci. Nutr. *7*, 147.

MAGA, J. A. 1976b. Lactones in foods. Crit. Rev. Food Sci. Nutr. *8*, 1.

MATIKKALA, E. J. and VIRTANEN, A. I. 1967. On the quantitative determination of the amino acids and γ-glutamyl peptides of onions. Acta. Chem. Scand. *21*, 2891.

PAILLARD, N. 1967. Analyse des produits volatils émis par quelques variétés de pommes. Fruits *22*, 141.

PAILLARD, N. 1979. Biosynthesis des produits volatils de la pomme: formation des alcools et des esters a patir des acid gras. Phytochem. *18*, 1665.

PAILLARD, N. 1981. Factors influencing flavour formation in fruits. *In* Flavor '81, p. 479, P. Schreier, ed. W. de Gruyter Co., Berlin.

POTTY, V. H. 1969. Occurrence and properties of enzymes associated with mevalonic acid synthesis in the orange. J. Food Sci. *34*, 23.

POTTY, V. H. and BRUEMMER, J. H. 1970a. Mevalonate activating enzymes in the orange. Phytochem. *9*, 99.

POTTY, V. H. and BRUEMMER, J. H. 1970b. Oxidation of geraniol by an enzyme system from orange. Phytochem. *9*, 1003.

POTTY, V. H. and BRUEMMER, J. H. 1970c. Formation of isoprenoid pyrophosphates from mevalonate by orange enzymes. Phytochem. *9*, 1229.

SAGHIR, A. R. and MANN, L. K. 1965. The lachrymatory factor in the American wild *Alliums*. Suomen Kemistihehti. *B38*, 78.

SALUNKHE, D. K. and DO, J. Y. 1976. Biogenesis of aroma constituents of fruits and vegetables. CRC Crit. Rev. Food Sci. and Nutr., p. 161.

SCHREIER, P. 1984. Chromatographic Studies of Biogenesis of Plant Volatiles. Huethig Publishing, Berlin.

SCHREIER, P. and LORENZ, G. 1981. Formation of 'green-grassy' notes in disrupted plant tissues: Characterization of the tomato enzyme systems. *In* Flavour '81, p. 495, P. Schreier, W. ed. de Gruyter, Berlin.

SCHORMULLER, J. and GROSCH, W. 1965. Üntersuchungen uber Aromastoffe von Lebensmitteln. 2: Über das Vorkommen weiterer Carbonylverbindungen in der Tomate. Z. Lebensm.-Unters.-Forsch. *126*, 38.

SCHUTTE, L. 1973. Precursors of sulfur-containing flavour compounds. CRC Crit. Rev. Food Technol. *4*, 457.

SCHWIMMER, S. and AUSTIN, S. J. 1971. γ-Glutamyl transpeptidase of sprouted onion. J. Food Sci. *36*, 807.

SHAH, B. M., SALUNKHE, D. K. and OLSON, L. E. 1969. Effect of the ripening process on chemistry of tomato volatiles. J. Am. Soc. Hort. Sci. *94*, 171.

SHANKARANARAYANA, M. L., RAGHAVAN, B., ABRAHAM, K. O. and NATARAJAN, C. P. 1982. Sulfur compounds in foods. *In* Food Flavours Part A. Introduction, p. 169. I. D. Morton and A. J. MacLeod, eds. Developments in Food Science, Elsevier, London.

SIMON, P. W. and PETERSON, C. E. 1979. Genetic and environmental components of carrot culinary and nutritive value. Acta. Hort. *93*, 271.

SIMON, P. W., PETERSON, C. E., and LINDSAY, R. C. 1980a. Genetic and environmental influences on carrot flavor. J. Amer. Soc. Hort. Sci. *105*(3), 416.

SIMON, P. W., LINDSAY, R. C., and PETERSON, C. E. 1980b. Analysis of carrot volatiles collected on porous polymer traps. J. Agric. Food Chem. *28*(3), 549.

SOMOGYI, L. P., CHILDERS, N. F. and CHANG, S. S. 1964. Proc. Amer. Soc. Hort. Sci. *84*, 51.

STEVENS, M. A. 1970. Inheritance and flavor contribution of 2-isobutylthiazole, methyl salicylate and eugenol in tomatoes. J. Am. Soc. Hort. Sci. *95*, 9.

STONE, E. J., HALL, R. M. and KAZENIAC, S. J. 1975. Formation of aldehydes and alcohols in tomato fruit from U-^{14}C-labelled linolenic and linoleic acids. J. Food Sci. *40*, 1138.

TANG, C. S. and JENNINGS, W. G. 1968. Lactonic compounds of apricot. J. Agric. Food Chem. *16*, 252–254.

TORLINE, P. and BALLSCHMIETER, H. M. B. 1973. Volatile constituents from guava. I. A. comparison of extraction methods. Lebensm, Wiss. Technol. *6*, 32.

TRESSL, R. and DRAWERT, F. 1973. Biogenesis of banana volatiles. J. Agric. Food Chem. *21*, 560.

TRESSL, R. and JENNINGS, W. G. 1972. Production of volatile components in the ripening banana. J. Agric. Food Chem. *20*, 189.

TRESSL, R., DRAWERT, F., HEIMANN, W., and EMBERGER, R. 1970. Über die Biogenesis von Aromastoffen bei Pflanzen und Früchten. VIII. Zur Biogenese der im Bananen fefundenen bromastoffe. Z. Lebensm.-Unters-Forsch. *144*, 4.

TRESSL, R., HOLZER, M. and APETZ, M. 1975. Biogenesis of volatiles in fruit and vegetables. *In* Aroma Research, p. 41. H. Maarse and P. J. Groenen, eds. Proc. of Int. Symp. on Aroma Res., Pudoc, Wageningen, Netherlands.

VIRTANEN, A. I. 1962. On enzymatic and chemical reactions in crushed plants. Acta. Biochem. Biophys. Suppl. 1, 200.

WINTER, M. and KLOTI, R. 1972. Uber das Aroma der gelben Passionfrucht. Helv. Chim. Acta. *55*, 1916.

YASUMOTO, K., IWAMI, K., and MITSUDA, H. 1971. Enzyme catalyzed evolution of lenthionine from lentinic acid. Agric. Biol. Chem. *35*, 2070.

YU, M. H., and SPENCER, M. 1969. Conversion of L-leucine to certain keto acids by a tomato enzyme preparation. Phytochem. *8*, 1173.

YU, M. H. and SPENCER, M. 1970. α-Alanine aminotransferase from tomato fruit. Phytochem. *9*, 341.

YU, M. H., OLSON, L. E., and SALUNKHE, D. K. 1967. Precursors of volatile components in tomato fruits: I. Compositional changes during development. Phytochem. *6*, 1457.

YU, M. H., OLSON, L. E., and SALUNKHE, D. K. 1968a. Precursors of volatile components in tomato fruit: II. Enzymatic production of carbonyl compounds. Phytochem. *7*, 555.

YU, M. H., OLSON, L. E., and SALUNKHE, D. K. 1968b. Precursors of volatile components in tomato fruit: III. Enzymatic reaction products. Phytochem. *7*, 561.

YU, M. H., SALUNKHE, D. K. and OLSON, L. E. 1968c. Production of 3-methyl butanal from L-leucine by tomato extract. Plant Cell Physiol. *9*, 633.

3

Changes in Food Flavor Due to Processing

The predominant flavor of some food products develops during processing or home preparation of the food. Some examples of foods where the flavor develops during processing or home preparation are fermented foods, meats, chocolate, coffee, baked goods and deep fat fried products. The primary routes of flavor formation in these foods are metabolic pathways of microorganisms, nonenzymatic browning and thermal oxidations in fats. This chapter will present an overview of these means of flavor production.

A. Nonenzymatic Browning

Nonenzymatic browning (NEB) or Maillard browning is responsible for some of the most pleasant flavors enjoyed by man. There is no question that freshly baked bread, a nice steak, a freshly brewed cup of coffee or piece of chocolate is appreciated by the consumer. Yet none of these flavors existed in the product until the food processor (or cook) heated the product to develop the flavor. Hurrell (1982) has presented an excellent review of the Maillard reaction in food flavor formation. While the following will present an overview of this reaction, detail may be found in several other publications including Ohloff and Flament (1979), Vernin (1982), Olsson *et al.* (1981), and Shibamoto (1983).

1. General Overview of Nonenzymatic Browning

Hodge (1953) was the first to present an outline of the overall reactions involved in NEB. Generally speaking, NEB is a reaction between carbonyls and amines. The carbonyls in foods most often are reducing sugars, while the amines come from either amino acids or proteins. The major end products of NEB are melanoidins and other nonvolatile (and insignificant to flavor) compounds. However, a number of intermediates and minor products of the reaction are formed which are significant to flavor (Fig. 3.1). Reineccius *et al.* (1972) have shown that while 1.3 g (per 100 g beans) of reducing sugars and amino acids (total) were reacted via NEB during the roasting of cocoa beans, only 0.9 mg of pyrazines were formed. Thus, about 0.2% of the reactants were transformed into pyrazines while the remainder of the reactants went into other products. Yet these pyrazines are exceptionally important to the flavor of chocolate. It is typically these minor products which make the largest contribution to flavor.

2. Factors Influencing Browning Rate

While we will go through some of the pathways leading to the generation of different classes of browning flavors, it should be noted that many factors influence the rate of browning and the particular flavor pathway predominating in the reaction. The influence of system conditions on browning flavor development is obvious if one considers the sensory quality of roasted vs. stewed meats. A stewed meat lacks the roasted notes characteristic of the roasted product. This is partially because the stewed product has a water activity of approximately 1.0 and never exceeds a temperature of 100°C. The roasted meat, however, dries on the surface so water activity is substantially less than 1.0. Also, since the surface dries surface temperature will exceed 100°C. The lower water activity and higher surface temperatures favor the production of flavor compounds giving the meat roasted notes.

The influence of temperature on flavor formation may be understood better by recalling that each particular pathway of flavor formation has its own activation energy. To illustrate this influence, consider the plot shown in Fig. 3.2 (please note that this plot does not contain valid data but is only for illustrating an idea). In this plot, thiophene has the highest activation energy, followed by pyrazine and pyrrole. Activation energy is equal to the slope of the line divided by R (natural gas constant). This plot shows that thiophene would have the largest rate constant (or highest rate of production) at high temperatures followed by pyrazine and then pyrrole. As temperature is decreased to typical storage temperatures, the pyrrole would be produced at the highest rate and now thiophene would have the lowest rate. Since each pyrazine, thiazole, pyrrole, etc., would have its own activation energy, one could put several hundred lines on this figure and realize how important temperature is in influencing the formation of flavor compounds via NEB.

The vast majority of research on the factors which influence browning has related to either the disappearance of starting materials (e.g., lysine/

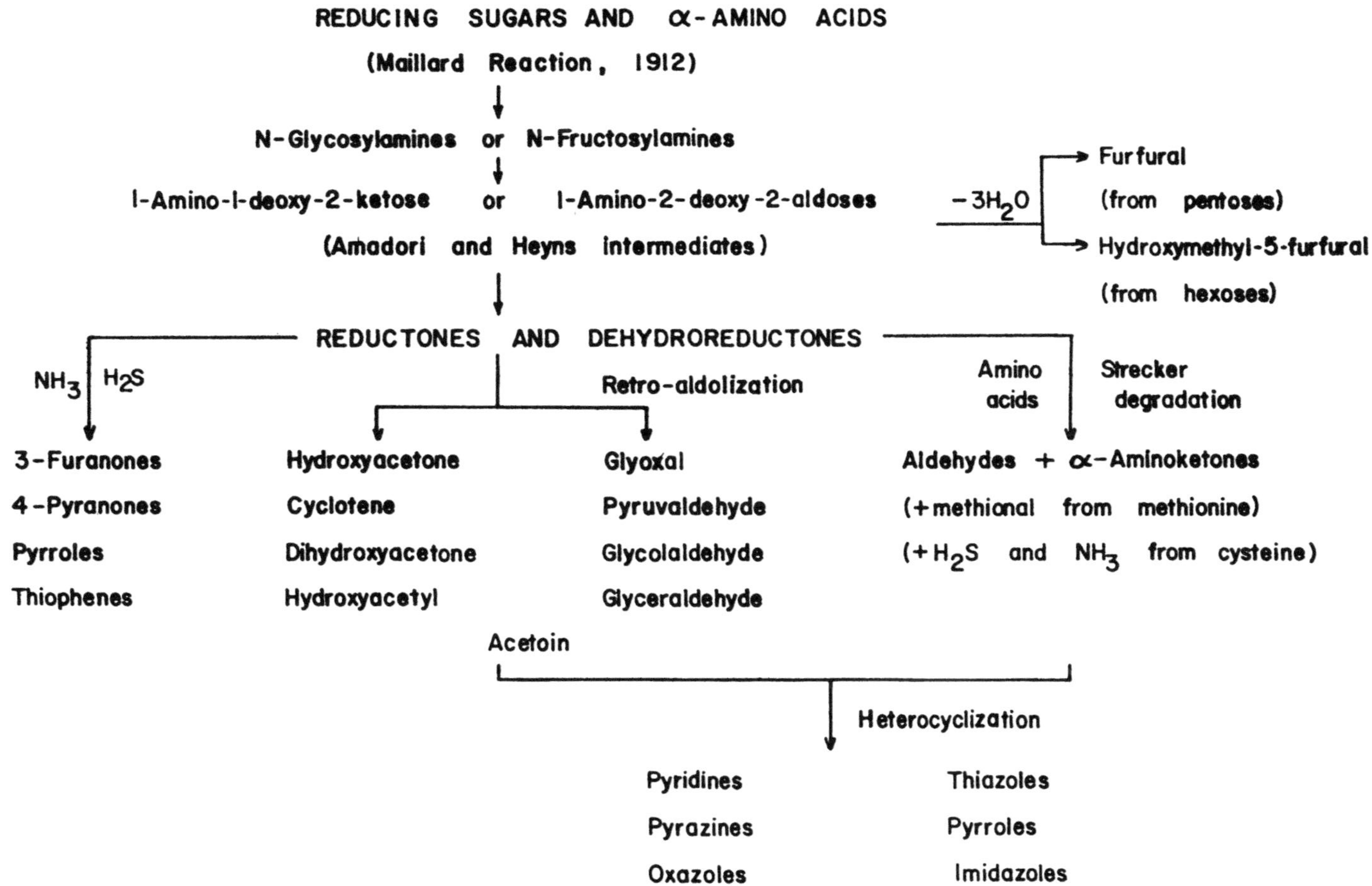

Fig. 3.1. Formation of flavor compounds via nonenzymatic browning.
Vernin and Parkanyi 1982

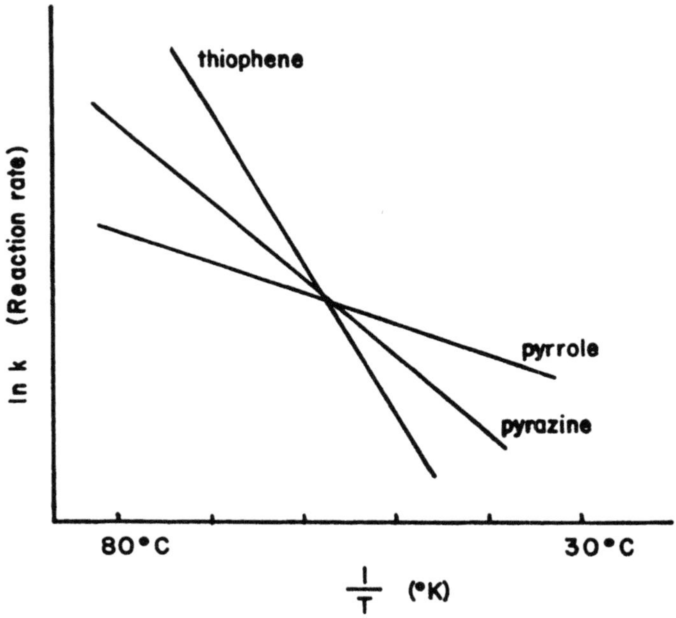

Fig. 3.2. Influence of temperature on rate of formation of various flavor compounds.

glucose reactions) or the production of brown pigments. The limited amount of work studying factors which influence flavor formation has largely focused on the flavor compounds formed via the heating of model systems (e.g., Tressl *et al.* 1985; Shu *et al.* 1985 a,b,c,d). While system composition (e.g., sugar and amino acid type) and thermal processing have been varied in some of these studies, there have been very few methodical studies conducted in this area. Most studies have related overall sensory properties to variations in reaction mixture composition or process (Table 3.1).

The influence of various parameters on the production of flavor via browning has been most thoroughly studied for the pyrazines. The factors which have been studied include source of nitrogen, source of carbon, pH, time and temperature of reaction, water activity and oxidizing and reducing agents.

Nitrogen source has been shown to influence the production of pyrazines via browning. Generally the same pyrazines were obtained irrespective of the amino acid used as nitrogen source but a different quantitative pattern was observed (Newell *et al.* 1967; Koehler *et al.* 1969; Wang and Odell 1972; Leahy 1985). Leahy (1985) found lysine to produce the greatest amount of pyrazines when heated with sugars followed by asparagine and cystine.

Table 3.1. Aromas Generated by Amino Acids with Sugars[a]

	Glycine	Glutamic Acid	Lysine	Methionine	Phenylalanine
Glucose	Burnt candy	Chicken tray	Burnt fried potatoes	Cabbage	Caramel
Fructose	Beef broth	Chicken	Fried potato	Bean soup	Dirty dog
Maltose	Beef broth	Baked ham	Stale potato	Harsh, horseradish	Sweet
Sucrose	Beef broth	Charred meat	Boiled meat	Overcooked cabbage	Chocolate

[a] From El'Ode *et al.* (1966).

Type of sugar has also been reported to influence the total yield and distribution of pyrazines formed during browning. Koehler *et al.* (1969) found about three times as much total pyrazines formed from the reaction of ammonium chloride with fructose as compared to glucose. Shibamoto and Bernhard (1977) reported pentoses to give the greatest yields of pyrazines during browning followed by the hexoses. Contrary to the results of Koehler *et al.* (1969), mannose, glucose and fructose were reported to be approximately equivalent. The work of Leahy (1985) supports the high reactivity of pentoses (ribose) but is contrary to the work of both Koehler *et al.* (1969) and Shibamoto and Bernhard (1977) in finding glucose to yield more pyrazines during browning than fructose. While the literature is not in agreement on which sugars yield the most pyrazines, there is little question in that sugar type does influence both yield and quantity of pyrazines.

The pH has been found to have a very strong influence on pyrazine production (Koehler and Odell 1970; Shibamoto and Bernhard 1976, 1977; Leahy 1985). Leahy (1985) reported nearly a 500-fold increase in total pyrazine production between a model system heated at pH 9.0 vs. the same system at pH 5.0 (13 ppm pyrazines vs. 24 ppb).

Temperature has also been found to strongly influence pyrazine formation. While Koehler and Odell (1970) reported that essentially no pyrazines were formed at temperatures less than 100°C, Shibamoto and Bernhard (1976) and Leahy (1985) both report pyrazine production at temperatures as low as 70°C. In all cases, pyrazine production increased rapidly with temperature. A kinetic study by Leahy (1985) reported an activation energy of approximately 35 kcal/mole for pyrazine and 2-methyl pyrazine and about 43 kcal/mole for the dimethyl pyrazines. These high activation energies indicate a very strong temperature dependence of pyrazine formation and explain why pyrazines (roasted, nutty, toasted notes) do not occur during storage but only during high temperature treatment.

The influence of water activity on pyrazine formation via browning has been studied by Leahy (1985). Data were presented that show increasing pyrazine production with water activity up to about 0.74.

Oxidation/reduction effects were investigated by Shibamoto and Bernhard (1977). They studied the effects of metal ions, oxygen, antioxidants and sodium hydroxide on the formation of pyrazines in glucose–ammonia systems. Their results indicated that oxygen does not catalyze pyrazine formation. Propyl gallate significantly inhibited formation of pyrazines for short reaction times (2 hr) at 100°C, but not for long reaction times (18 hr) as compared to the control. BHT inhibited production slightly with long reaction times. However, reaction under an oxygen blanket gave slightly depressed yields. They concluded that antioxidants may reduce yields for reasons other than reducing the amount of oxygen available to the system. Adding $CuCl_2$ and $ZnCl_2$ also decreased yields of pyrazines while increasing browning, as was noted visually.

It is of interest to make a brief comparison of what has been presented on flavor (pyrazine) production via browning to what is known to influence the nonflavor browning reactions. Browning is known to be strongly influenced by both pH and temperature (Wolf *et al.* 1982; Lea and Hannan 1949; Maillard 1916). Kinetic studies have indicated an activation energy range from about 15.5 kcal/mole for pigment production in a glycine/glucose model system (Stamp and Labuza 1983) to 33 kcal/mole for a solid intermediate moisture model food system (Warmbier *et al.* 1976). Pigment production is less temperature dependent than is pyrazine production.

Similar to what was observed for pyrazine production, both amino acid and sugar type have an influence on nonflavor browning reactions. Lewis and Lea (1950) found the pentoses to most reactive followed by the hexoses.

Water activity has been found to have a pronounced influence on the rate on nonflavor browning reactions (Labuza 1980). A maximum reaction rate is generally observed at a water activity of 0.6–0.7.

Overall, the conditions that enhance the browning reaction as measured by either the consumption of reactants or the generation of pigments, also enhance the generation of flavor. The exception is that the various reactions (e.g., color formation, reactant consumption, and flavor formation) have different activation energies. This leads to different flavor notes being formed when a product is heated at different temperatures.

3. Formation of Flavor Compounds

Pokorny (1980) made note that of the more than 3000 volatile flavor compounds identified in foods, about two-thirds are products of NEB or subsequent reactions. The most abundant flavor compounds formed via NEB are aliphatic aldehydes, ketones, diketones and lower fatty acids. However, heterocyclic compounds containing oxygen, nitrogen, sulfur or combinations of these atoms are much more numerous and significant to the flavor of browned foods (Vernin and Vernin 1982). The remainder of this chapter will provide examples of flavor compounds formed via NEB, proposed mechanisms of formation, as well as information on sensory properties. Quite obviously, this section cannot be comprehensive but only a sampling of the diverse number of compounds formed via NEB.

a. Carbonyls

The major pathway leading to the formation of carbonyls is the Strecker degradation. This reaction occurs between dicarbonyls and free amino acids as is shown below:

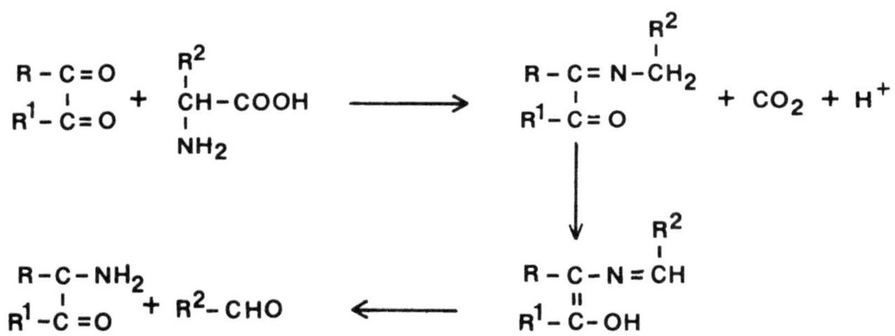

The dicarbonyls consist of any compounds that have vicinal carbonyls or carbonyl groups separated by one double bond or several conjugated double bonds (Pokorny 1980). While these carbonyls typically are intermediates in NEB, they may also be normal constituents of the food (e.g., ascorbic acid), be end products of enzymatic browning (e.g., quinones) or be products of lipid oxidation. The reactivity depends upon the carbonyl moiety. For example, methyl glyoxal and glyceraldehyde are very reactive while dehydroascorbic acid and glyoxal are moderate in reactivity.

The end products of the Strecker degradation are CO_2, an amine and the corresponding aldehyde of each deaminated and decarboxylated amino acid. At one time these aldehydes were considered to be quite important to the flavor of heated food products. This was primarily because they are the most abundant volatiles formed via NEB. It is now realized that the heterocyclic volatiles are more important. The Strecker aldehydes are often monitored and in foods since they are present in quantity and can serve as indicators of NEB (Sullivan *et al.* 1974).

Strecker aldehydes may also be formed via free radical mechanisms

(Nilov and Ogorodnik 1967). The oxidation of amino acids by hydrogen peroxide or lipid peroxides yields CO_2, ammonia and the corresponding Strecker aldehydes. While this is a viable pathway for the production of Strecker aldehydes, the first pathway typically predominates.

The aldehydes formed in this reaction may undergo additional reactions. Pokorny (1980) has outlined aldol condensations leading to dimeric unsaturated aldehydes and, eventually, high molecular weight polymers. Takken *et al.* (1976) have demonstrated the reaction of aldehydes with alpha-dicarbonyls, ammonia and hydrogen sulfide. A host of thiazoles, dithiolanes, thiazolines and hydroxythiazolines were found as products of these reactants. An additional fate of the Strecker aldehydes is that they may end up being bound onto various food constituents (e.g., protein). The binding may be sufficiently strong that the aldehydes may not have adequate vapor pressure to contribute to food odor.

One must remember that few products of NEB are final products. Many of the primary products of NEB undergo secondary reactions and so on to yield a large number of flavor compounds.

b. Pyrazines

The most studied class of nitrogen heterocyclics formed via NEB is the pyrazine group. Currently more than 100 different pyrazines have been identified in various food products. Their distribution and occurrence has been reviewed quite thoroughly by Maga (1982). The sensory properties of the pyrazines are quite diverse. The alkyl pyrazines (I) generally possess roasted nutlike notes (Ohloff and Flament 1979). The methoxy pyrazines (II) often possess earthy vegetable notes. The 2-isobutyl-3-methoxy pyrazine has a freshly cut green pepper flavor with a sensory threshold of 0.002 ppb in water. The 2-isopropyl-3-methoxy pyrazine has an earthy note often used to flavor mashed potatoes, while the acetyl pyrazines (III) typically have a popcorn character, and 2-acetonyl pyrazine (IV) has a toasted or burned note. Initially 2-(-2-furyl) pyrazine was found in coffee. Later it was found in cigarette smoke and other roasted foodstuffs. Several bicyclic pyrazines have been identified in foods (VI, VII, VIII). The sensory properties of these pyrazines have been described as being burnt, roasted, grilled and/or animal in character. Several mechanisms have been proposed for the formation of various pyrazines (Vernin and Parkanyi 1982). The alkyl pyrazines are probably formed from the reaction of α-diketones with amino acids to form α-amino ketones (Strecker degradation). These α-amino ketones may condense with other α-amino ketones to yield a heterocyclic compound. This heterocyclic may undergo oxidation to form the triunsaturated pyrazines. Note that the alkyl substitution is dictated by the

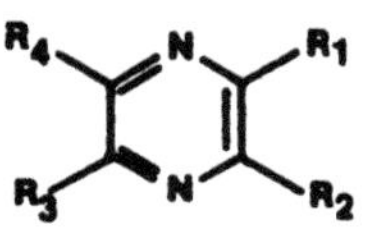

I.

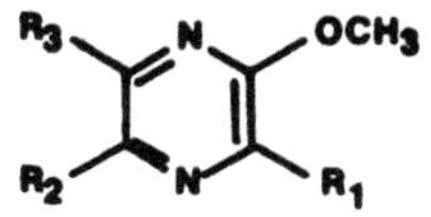

II.

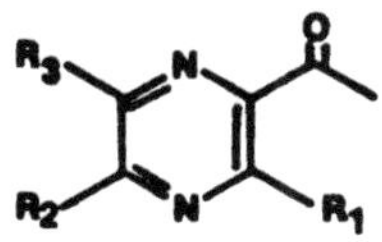

III.

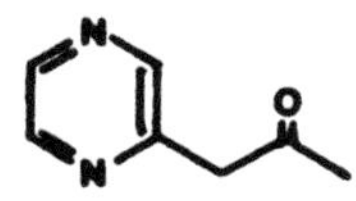

IV.

dicarbonyl fragment. The amino acid contributes only the amine to the pyrazine (Koehler *et al.* 1969; Fig. 3.3A).

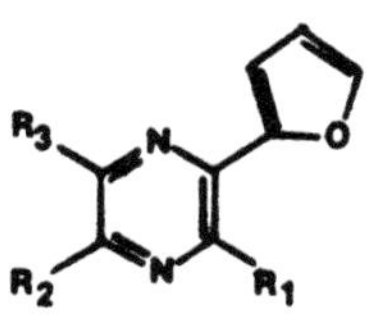

V.

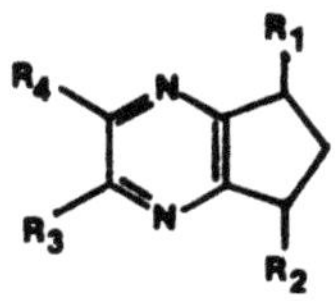

VI.

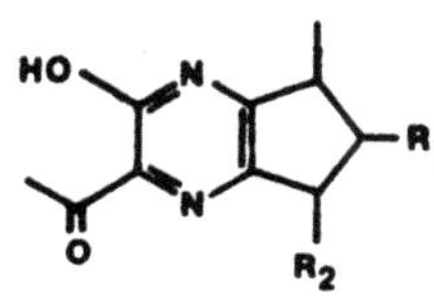

VII.

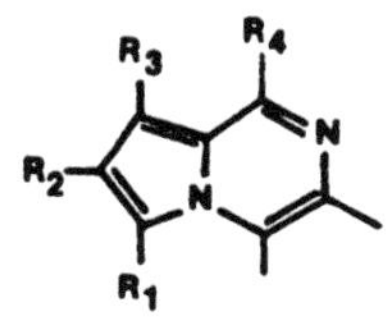

VIII.

Rizzi (1972) has provided a second mechanism for the generation of alkyl pyrazines via NEB. He pointed out that amino acetone is a product of the Strecker degradation which may condense with a second amino acetone molecule to form an acyclic imino ketone. This ketone may cyclize to yield water plus a dihydropyrazine. Thermal rearrangements could then produce the pyrazine. Additional aldol condensations and reactions with amino acids could result in the formation of several different pyrazines via this mechanism.

Mechanisms B, C, and D (Fig. 3.3) illustrate how additional ring structures may be added to the base pyrazine. Pyrazines with larger alkyl or alkenyl groups require an alternate pathway. Shibamoto *et al.* (1979) have proposed a pathway involving addition of an aldehyde onto a dihydropyrazine carbanion (Fig. 3.3E). Reduction or dehydration of this addition product would yield a larger or unsaturated side chain.

c. Pyrroles

Pyrroles are nitrogen-containing heterocyclic compounds, also quite widely distributed in foods (Maga 1981a). The 2-formyl pyrrole (IX) and 2-acetyl pyrrole (X) are the two most abundant and widely occurring pyrroles in foods (Ohloff and Flament 1979). The 2-formyl pyrrole has a sweet cornlike odor while the 2-acetyl pyrrole has a carmellike odor. The 1-acetonyl pyrrole (XI) has been found to have a cookie or mushroom aroma while the pyrrole lactones (XII) have a spicy peppery flavor.

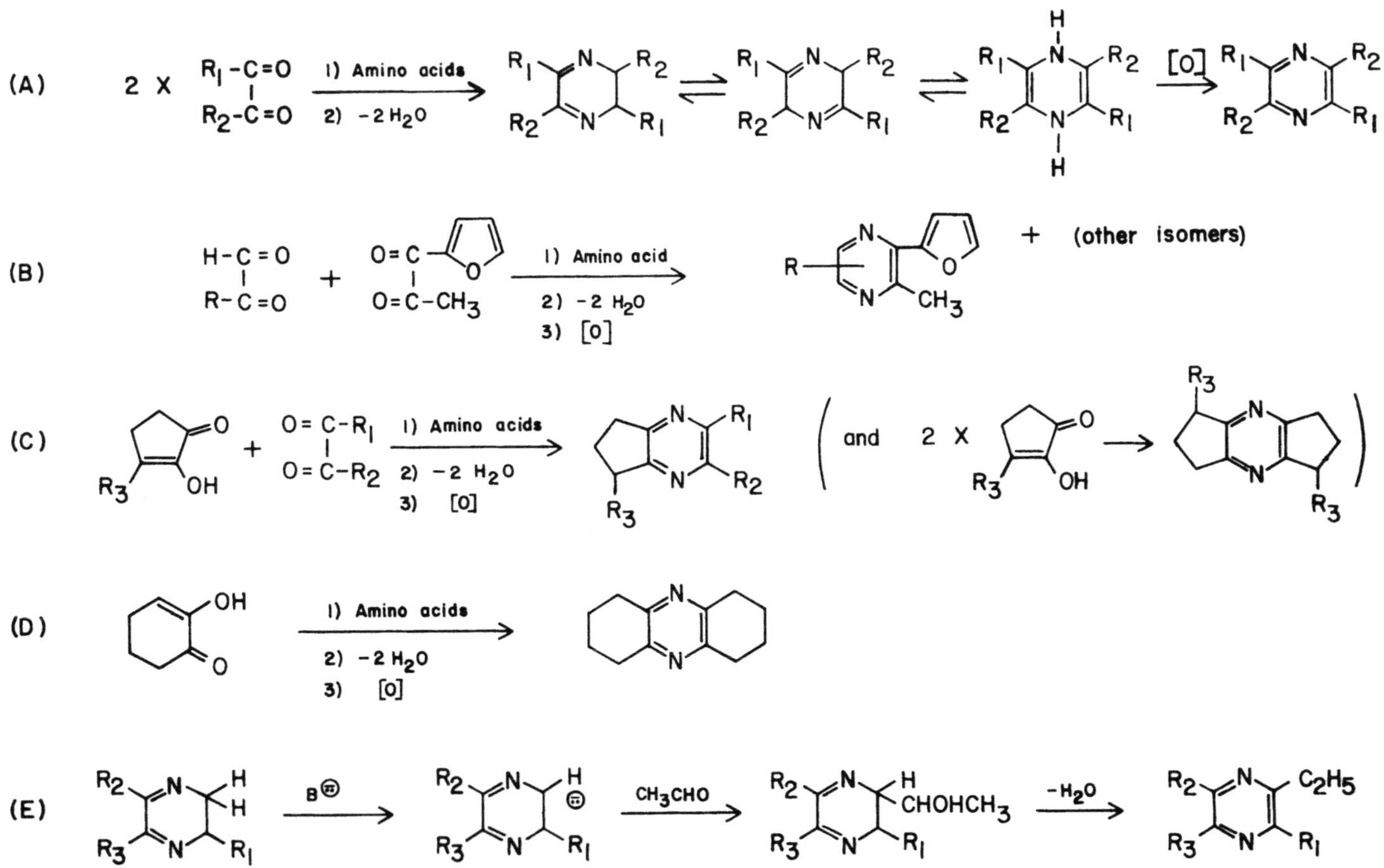

Fig. 3.3. Mechanisms proposed for the formation of pyrazines.
Vernin and Parkanyi 1982

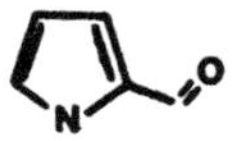
IX.

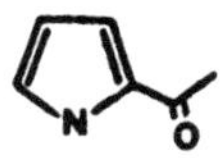
X.

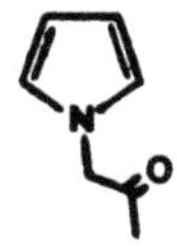
XI.

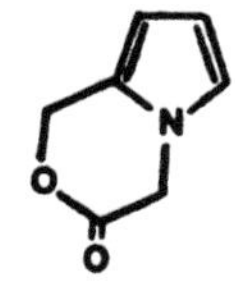
XII.

Hodge *et al.* (1972) originally proposed a mechanism for the formation of pyrroles, which is also supported by Tressl *et al.* (1985a). This basically is the participation of proline and hydroxy proline in the Strecker degradation to yield pyrroles. If the food system does not contain either proline or hydroxy proline, then a sugar of at least five or more carbons is required for pyrrole formation (Shibamoto 1983). Rizzi (1974) proposed an alternate pathway for the formation of acylalkyl pyrroles involving the reaction of an amino acid with the corresponding furan (see Fig. 3.4). The amino acid reacts with the number 5 carbon on the furan ring to open the furan ring. The ring closes via dehydration with N in the ring structure. Depending upon the amino acid involved and subsequent rearrangements, a variety of pyrroles may arise.

d. Pyridines

Pyridine compounds appear to be less widely distributed in browned foods than the pyrroles and pyrazines (Maga 1981b). While they possess a wide range in odor properties, green notes are quite common (Pittet and Hruza 1974). The 3-methyl pyridine (XIII) has a green odor as do many of the singly substituted pyrazines. The 3-methyl-4-ethyl pyridine (XIV) has been characterized as being sweet and nutty. Pyridine (XV) is judged as being pungent and offensive while 2-acetyl-pyridine (XVI) is tobaccolike in odor.

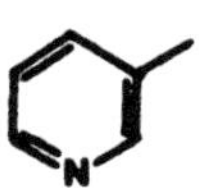
XIII.

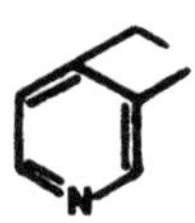
XIV.

XV.

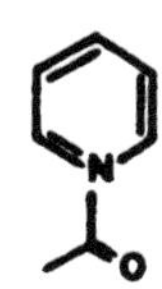
XVI.

The contribution of the pyridines to browning flavor is dependent on the individual pyridine formed and its concentration in the food. At low concentrations, the pyridines typically contribute very pleasant notes. However, they generally become harsh and offensive at higher concentrations.

The mechanisms of pyridine formation are poorly characterized. Vernin and Parkanyi (1982) have proposed three mechanisms which are outlined in Fig. 3.5. The first two mechanisms are dependent upon aldol condensations to yield unsaturated aldehydes with side groups. Reaction then with ammonia or an amino acid (pathway A and B, respectively) and ring closure would yield a nitrogen-containing heterocyclic. The oxidation of this heterocyclic would result in the formation of a pyridine. Pathway C involves the reaction of dialdehydes with ammonia followed by dehydration to produce pyridines.

e. Miscellaneous Nitrogen Heterocyclics

Tressl *et al.* (1985a,b,c) have identified an additional 59 nitrogen-containing heterocyclic compounds which, like the pyrroles, are believed to arise from the reaction of proline and hydroxy proline with dicarbonyls via the Strecker degradation. These compounds fall into the classes of pyrrolines, pyrrolidines, pyrrolizines, piperidines and azepines. Odors associated with these compounds are most often cereal or roasted in nature. For example, 2-acetyl-l-pyrroline (XVII) is characterized as being popcornlike (Tressl *et al.* 1985a). 1-Acetonyl pyrrolidine (XVIII) has a cereal aroma while the furan-substituted pyrrolidines are considered to have sesamelike notes (Tressl *et al.* 1985c). While the piperidines [e.g., 2-acetyl piperidine (XIX)] have been characterized as having weak woody odor and very bitter taste, they are readily transformed into the corresponding tetrahydro pyridines which have bready, crackerlike aromas (Tressl *et al.* 1985b). The 2-acetyl pyrrolizine (XX) has a smoky roasted aroma. The azepine derivatives generally are exceptionally bitter with little odor (Tressl *et al.* 1985a).

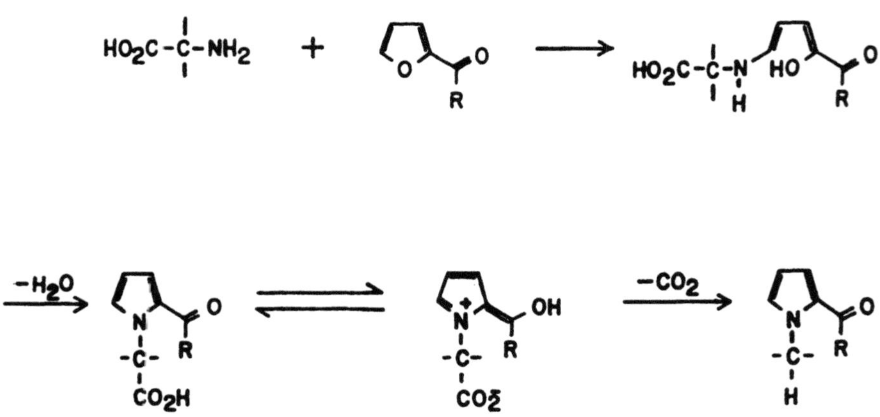

Fig. 3.4. Mechanism proposed for the formation of acylalkyl pyrroles.
Reprinted with the permission of Rizzi (1974). Copyright 1974. American Chemical Society

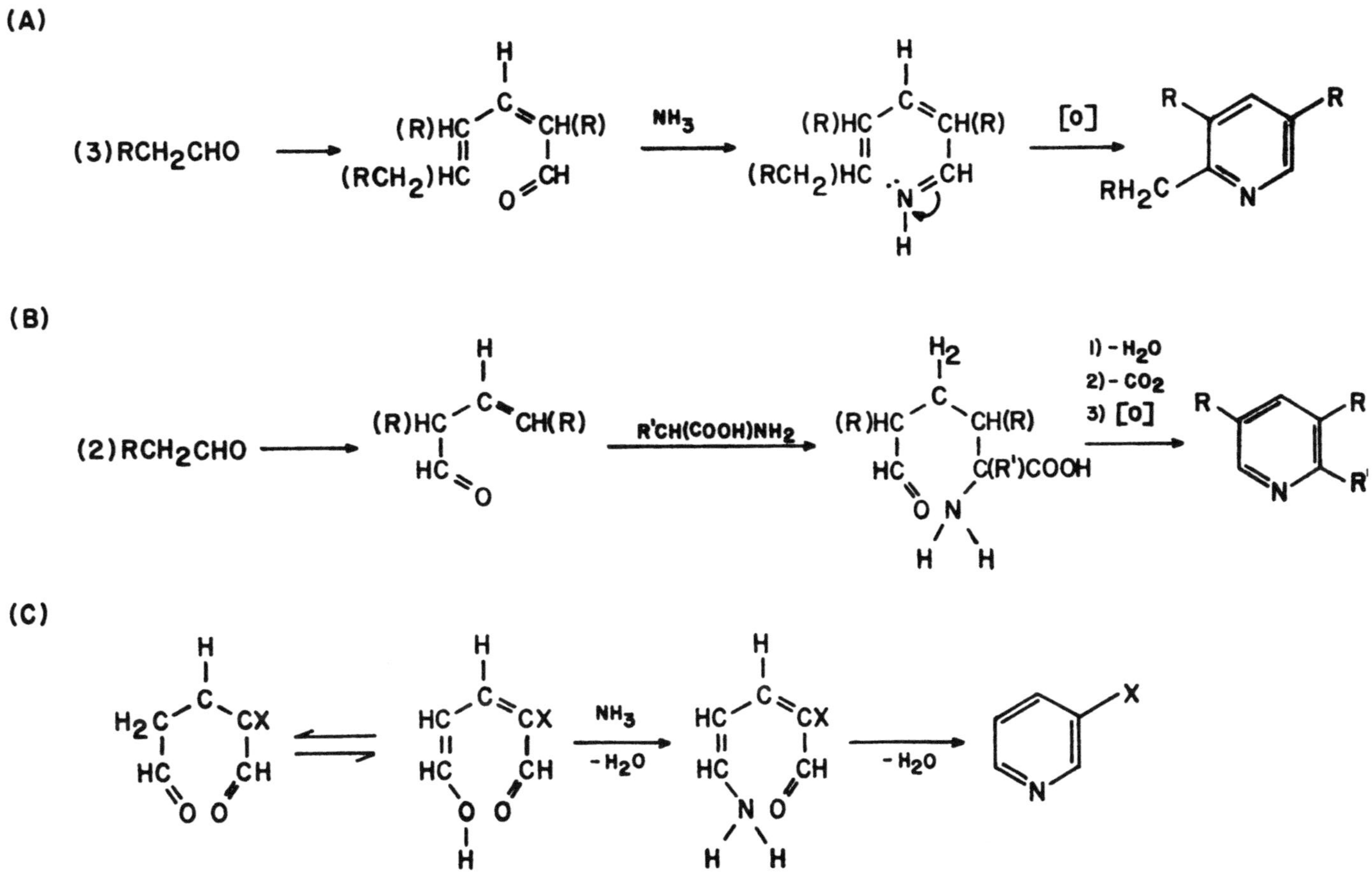

Fig. 3.5. Formation of pyridines from aldehydes (or 1,5-dicarbonyl compounds) and ammonia (or α-amino acids).
Vernin and Parkanyi 1982

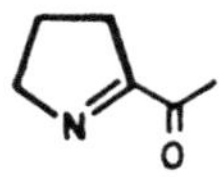

XVII.

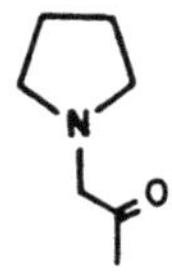

XVIII.

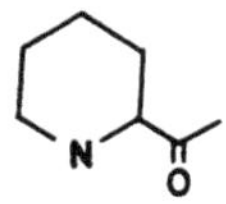

XIX.

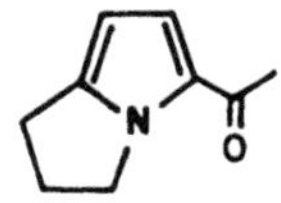

XX.

Reaction schemes have been presented for the formation of these heterocyclics (Tressl *et al.* 1985a). As is shown in Figure 3.6A, pyruvaldehyde and L-proline form an iminium carboxylate intermediate which is transformed by decarboxylatioin to a reactive ylide or iminium ion. These two intermediates then may undergo hydration to yield 1-pyrroline and acetol or pyrrolidine and pyruvaldehyde. Reduction would yield 1-acetonyl pyrrolidine.

The pyrrolizines may arise from proline/dicarbonyl reactions as shown in Fig. 3.6B. This involves aldol-type reactions via the iminium ion intermediate. Further reaction by Michael-addition or aldol condensation and dehydration would yield the pyrrolizines.

f. Furanones and Pyranones

The furanones and pyranones are oxygen-containing heterocyclic compounds associated with both caramelized and NEB flavors. The odor characters most common to this group of compounds would be caramellike, sweet, fruity, butterscotch, nutty or burnt. They predominate both in proportion and absolute amount in condensates of carbohydrates that are subjected to browning reactions (Ohloff and Flament 1979). It is of interest that about 70 furan derivatives have been identified in coffee aroma, 25 in the volatiles of bread and 37 in caramel (Ohloff and Flament 1979).

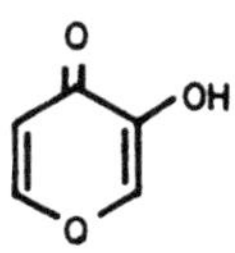

XXI.

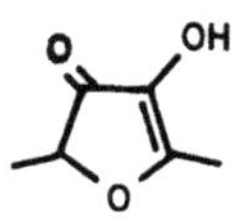

XXII.

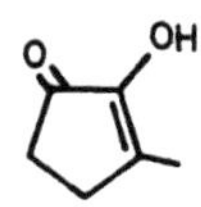

XXIII.

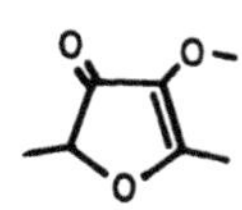

XXIV.

The caramel character is associated with a planar, contiguous C-alkylenol carbonyl group (alkyl–C=C(OH)–C:O) in the molecule (Pittett *et al.* 1970). Maltol (XXI) was one of the first compounds in this class to be identified in foods. Ethyl maltol (2-ethyl-3-hydroxy-4(4*H*)-pyranone) also has a caramel odor but is approximately four to six times stronger in flavor strength than maltol. Furaneol (4-hydroxy-2,5-dimethyl-

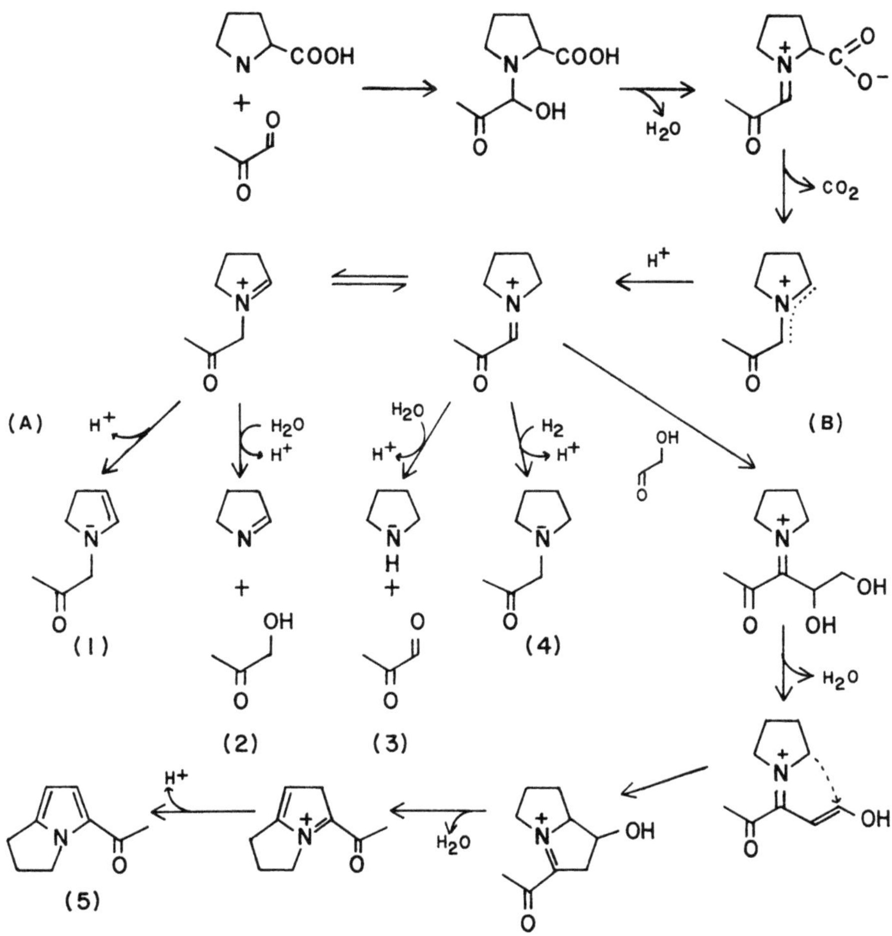

Fig. 3.6. Strecker degradation of proline to yield (1) 1-acetonyl-2-pyrroline; (2) 1-pyrroline; (3) pyrrolidine; (4) 1-acetonyl-pyrrolidine; (5) 5-acetyl-2,3-dihydro-1*H*-pyrrolizine.
Tressl et al. 1985a

3(2*H*)furanone, XXII) is the tradename of a compound used extensively in the flavor industry and, like maltol and ethyl maltol, is a flavor enhancer for sweet products. Furaneol itself has a "burnt pineapple" odor.

The five carbon analogs of maltol (XXIII) and furaneol possess odor properties similar to their oxygen-containing counterparts. Cyclotene (tradename for compound XXIII) has a very characteristic sweet maple character. The related compound, 3-ethyl-2-hydroxy-2-cyclopenten-1-one, also finds extensive use in the flavor industry for similar applications (caramel, nut, maple and butterscotch flavorings).

Compounds of this class that do not meet the structural requirements for being caramel in character may have quite different sensory properties. For example, 4-methoxy-2,5-dimethyl-3(2*H*)-furanone (XXIV) has an odor similar to sherry. The *n*-butyl ether version of this furanone has a jasminelike odor.

This group of compounds does not contain nitrogen. Therefore, the mechanism of formation generally involves the cyclization of nonnitrogen-containing browning intermediates (Fig. 3.7). These intermediates may be products of major browning pathways involving sugar dehydration or the Strecker degradation (Reynolds 1970; Hodge *et al.* 1972).

g. Sulfur Heterocyclics

Several different types of heterocyclic compounds containing sulfur are produced via NEB. These include thiophenes, dithioles, dithianes, dithiins, trithiolanes, trithanes, tetrathianes, thiazoles, thiazolines and

(A)

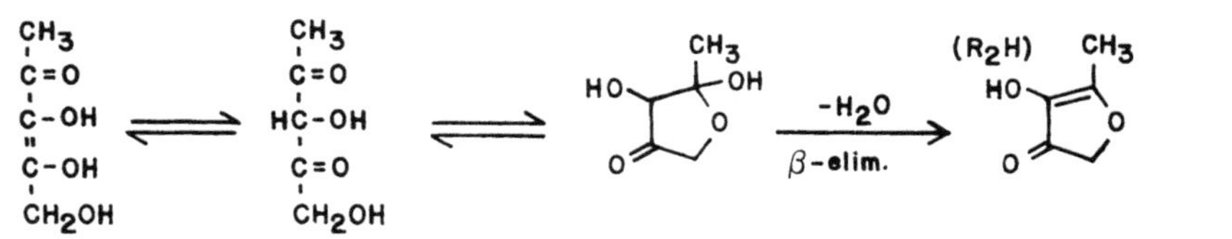

(B)

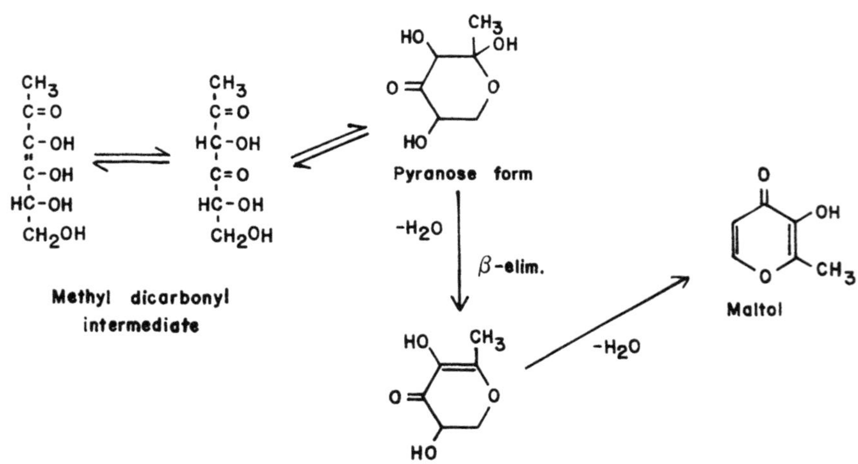

Fig. 3.7. Mechanisms proposed for the formation of furans and furanones: (A) Formation of a dihydrofuranone. (B) Formation of maltol.
Hodge et al. 1972

thiazolidines. The major heterocyclic sulfur compounds produced via NEB are thiazoles and thiophenes.

Thiazoles and pyrazines have somewhat similar sensory properties. Pittet and Hruza (1974) and Ho and Jim (1985) have reported that the alkylthiazoles give green, nutty, roasted, vegetable or meaty notes. Trimethyl thiazole (XXV) is reported to have a cocoa, nutty character. The 2-isobutyl thiazole (XXVI) is one of the best known as having the strong green odor of tomato leaf. This compound is considered to be important to tomato flavor. The 2,4-dimethyl-5-vinyl thiazole (XXVII) has a nutlike odor. The 2-acetyl thiazole (XXVIII) is characterized as having a nutty, cereal and popcorn flavor (Ohloff and Flament 1979).

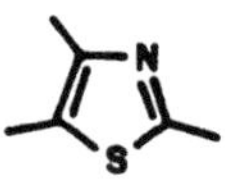

XXV.

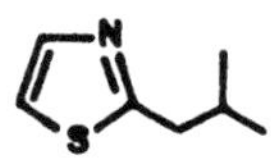

XXVI.

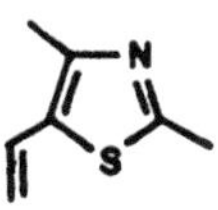

XXVII.

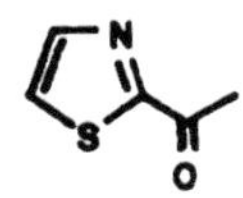

XXVIII.

Thiophenes have only recently been identified as food flavors. With the exception of cranberries, thiophenes have only been found in cooked or roasted food products (Ohloff and Flament 1979). The sensory properties of many of the thiophenes have been summarized by Ohloff and Flament (1979). Thiophene (XXIX) has a pungent character while 2,4-dimethyl thiophene is well known for its importance to the flavor of fried onions. The 2-acetyl-3-methyl thiophene (XXX) has a honeylike sensory quality at 0.25 g/100 L concentration but is nutty and starchy at 0.11 g/100 L concentration. Praline and woody notes are characteristic of compound XXXI. The 5-methyl thiophene-2-carboxaldehyde (XXXII) has a burnt coffee note (Shibamoto 1983).

XXIX.

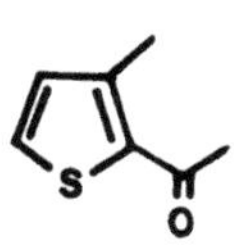

XXX.

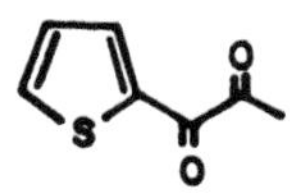

XXXI.

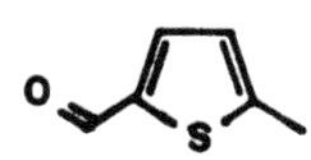

XXXII.

The sensory properties of some of the other sulfur-containing heterocyclic compounds are presented in Table 3.2. These compounds, like the thiazoles and thiophenes, typically are formed via the reaction of sulfur-containing amino acids with intermediates of the browning reaction. An alternative mechanism involves the initial formation of H_2S from the sulfur-containing amino acids and then the reaction of the H_2S with browning intermediates (Sakaguchi and Shibamoto 1978). Some of

Table 3.2. Sensory Properties of Sulfur-Containing Heterocyclic Componds via NEB

Structure	Name	Sensory Properties
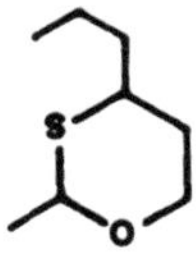	2-Methyl-4-propyl-1,3-oxathiane	Fruity, green, slightly burnt
	3*H*-1,2-dithiole	Cooked asparagus
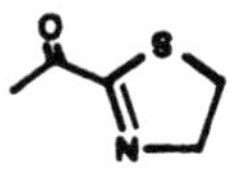	2-Acetyl-2-thiazoline	Freshly baked bread crust
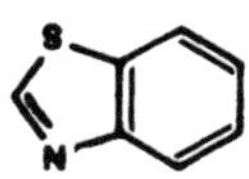	Benzothiazole	Quinolinelike, rubbery odor
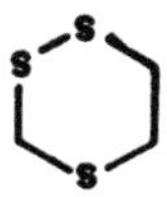	1, 2, 4 Trithiolane	Sulfurous
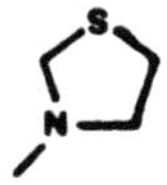	*N*-Methyl-thiazolidine	Green coconut

the intermediates and pathways involved in the formation of the sulphur heterocyclics are presented in Fig. 3.8.

It is suggested that the reader go to Shankaranarayana *et al.* (1982), Vernin and Vernin (1982), Shibamoto (1983), and Ohloff and Flament (1979) for excellent reviews on the role of sulfur compounds in food flavor.

h. Oxazoles and Oxazolines

With the exception of the report by Stoffelsma and Pypker (1968), oxazoles and oxazolines have been found only in food systems that have undergone NEB (Maga 1978, 1981c).

The oxazoles typically have green, sweet, floral or vegetablelike aromas (Ho and Tuorto 1981; Jin *et al.* 1984).

As an example, 4-methyl-5-propyl oxazole (XXXIII) has been charac-

(A)

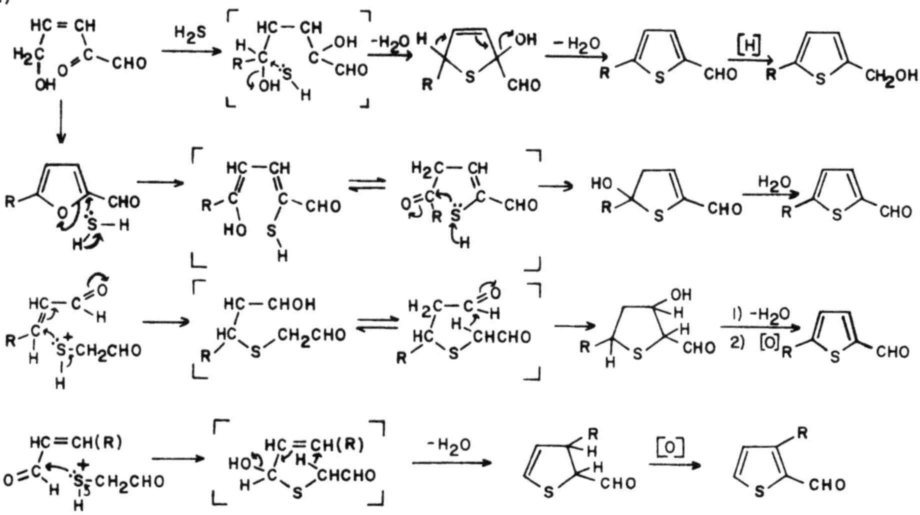

(B)

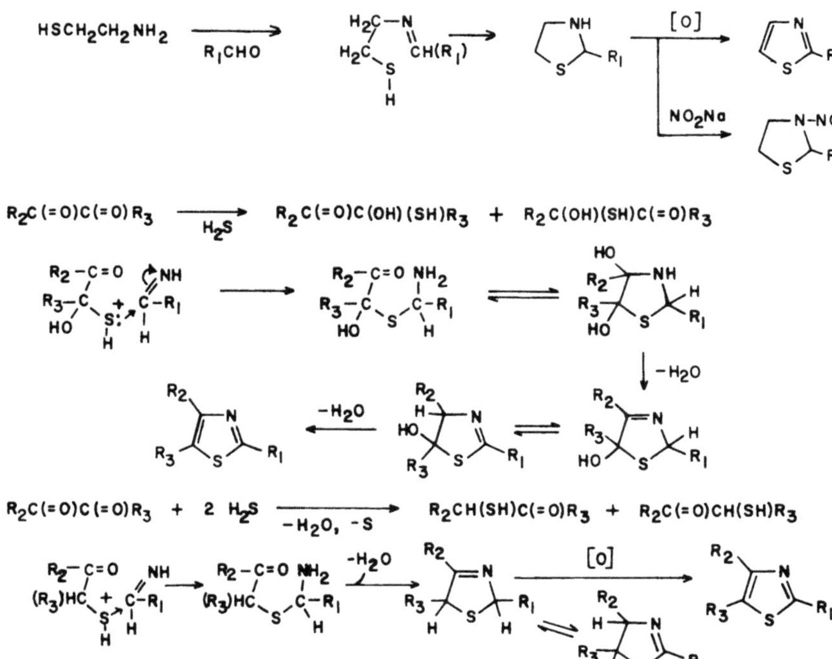

Fig. 3.8. Pathways proposed for the formation of various sulfur compounds via nonenzymatic browning; (A) Formation of 2-formyl thiophenes. (B) Formation of thiazoles, thiazolines and thiazolidines. *Continued.*

(C)

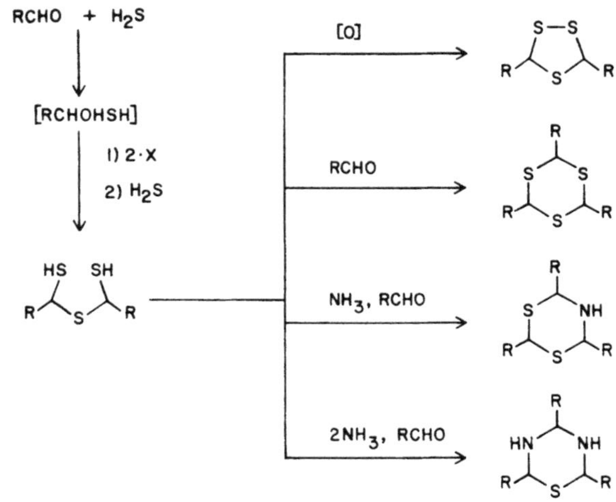

Fig. 3.8. Continued. (C) Formation of sulfur-containing heterocyclics.
Vernin and Parkanyi 1982

terized as having a green vegetable aroma. However, oxazoles which have a 4 or 5 carbon length chain on carbon-5 of the oxazole ring and no alkyl group on carbon-2 or 4 have distinct bacon-fatty notes [e.g., 5-butyl oxazole (XXXXIV)]. When a methyl or ethyl group is substituted on carbon-2 (e.g., 2-ethyl-5-butyl oxazole), the fatty aroma is reduced and sweet-floral aromas become more characteristic. The sweet-floral character is further enhanced by additional methyl or ethyl substitution on carbon-4 (Jin *et al.* 1984). The oxazolines tend to have a wide variety of sensory properties. The 2-isopropyl-4,5,5-trimethyl-3-oxazoline (XXXV) has a rumlike note while the 2-isopropyl-4,5-diethyl-3-oxazoline (XXXVI) has a typical cocoa aroma.

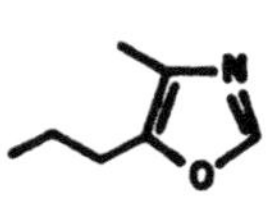
XXXIII.

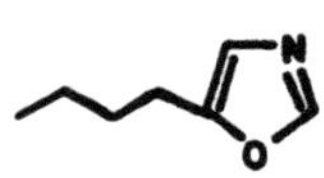
XXXIV.

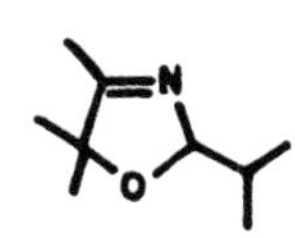
XXXV.

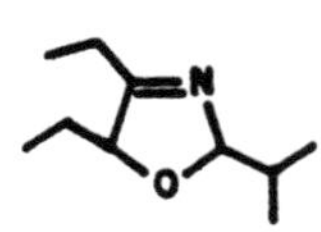
XXXVI.

Ho and Hartman (1982) have proposed mechanisms for the formation of 2,4,5-trimethyl oxazole, 4,5-dimethyl oxazole and 2,4,5-trimethyl-3-oxazoline. These mechanisms are presented in Fig. 3.9.

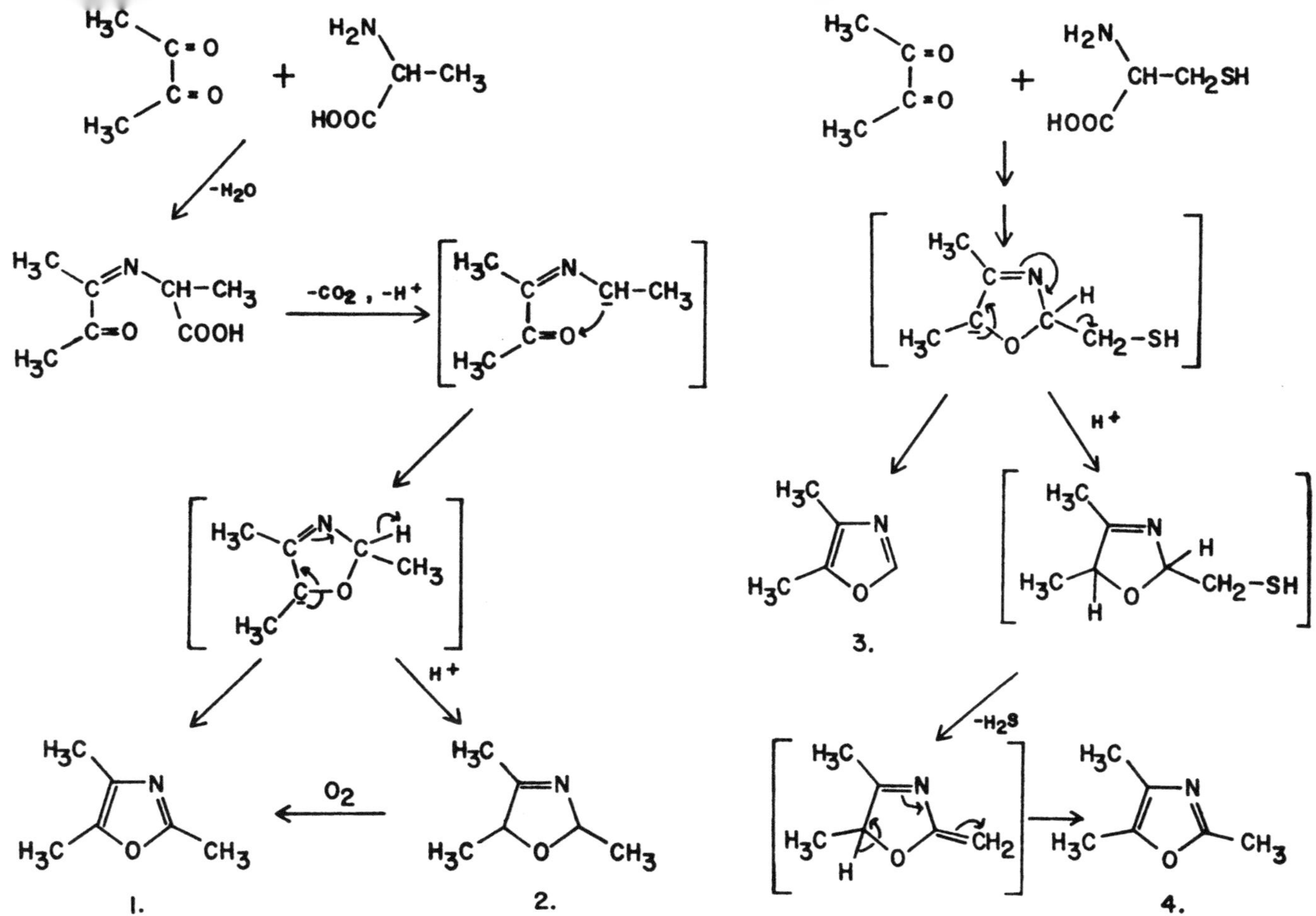

Fig. 3.9. Formation of oxazoles and oxazolines: (1) and (4) 2,4,5-Trimethyl oxazole. (2) 2,4,5-Trimethyl-3-oxazoline. (3) 4,5-Dimethyl oxazole.
Reprinted with the permission of Ho and Hartman (1982). Copyright 1982. American Chemical Society

5. Summary

The browning reaction is exceptionally important to the development of flavor in many foods. Hopefully the reader has gained an appreciation for the complexity of flavor formation and the large number of volatile compounds formed via NEB. While structures, organoleptic properties and mechanisms of formation for several classes of compounds and many individual flavor compounds have been provided, this is only just a start in understanding and controlling the formation of flavor via NEB. A great deal of research remains to be done.

B. Flavors from Lipids

Lipids may undergo changes during the processing of foods which make a flavor contribution. Later in this text (see Chapter 4) we shall discuss lipid oxidation and lipolysis as undesirable flavors imparted by changes in the lipid fraction of a food. This chapter discusses lipid changes that impart desirable flavors.

1. Deep Fat Fried Flavor

Chang *et al.* (1978) have pointed out the significance of deep fat frying in their review of deep fat fried flavor by noting that over 1.1 billion lb of fat is used in the United States annually to process french fried potatoes, doughnuts and potato chips. They also note that the major part of the 10 billion lb of fat consumed by Americans yearly is used for fried foods. While part of the desirability of deep fat fried foods lies in the physical properties imparted to the food by this process, the characteristic "french fried" flavor is quite popular and unique to this process.

The mechanism of flavor development in heated oils is essentially that of lipid oxidation. Since lipid oxidation is so important to the development of off-flavors in foods, a detailed presentation of the mechanism of lipid oxidation is postponed until the following chapter on off-flavors in foods. As Ohnishi and Shibamoto (1984) have outlined (Fig. 3.10), thermally induced oxidation involves hydrogen radical abstraction, the addition of molecular oxygen to form the peroxide radical, formation of the hydroperoxide and then decomposition to form volatile flavor compounds. The products of thermally induced oxidations are different from the typical lipid oxidation products at room temperature. The reason for these differences goes back to kinetic considerations as was illustrated in the previous section on NEB. Each chemical reaction has its own

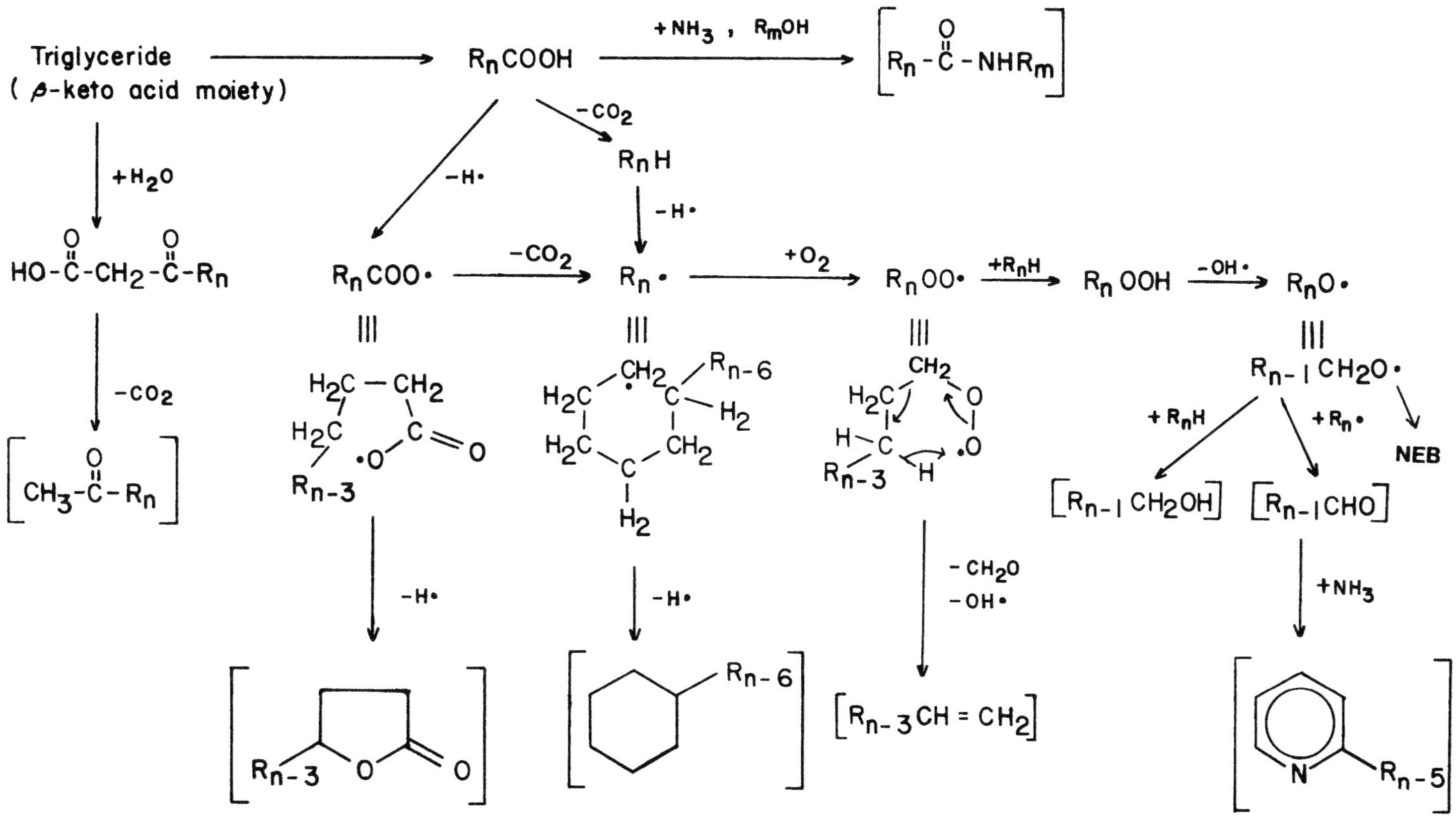

Fig. 3.10. Proposed mechanisms for the formation of volatiles from heated animal fats.
Reprinted with the permission of Ohnishi and Shibamoto (1984). Copyright 1984. American Chemical Society

unique activation energy and, therefore, rate at a given process temperature. Thus, the reactions occurring in the oil and the volatiles formed are dependent on the temperature of processing.

A second distinguishing factor of thermal oxidations is their more random nature than typical room temperature oxidations. The very high temperatures make more sites available on the fatty acid for oxidation to occur. Thus a wider range in end products (volatile flavor components) may occur. So even though the same chemical mechanisms are involved in flavor formation in deep fat fried foods, the flavor developed is unique to this process.

Chang *et al.* (1978) and Nawar *et al.* (1978) have identified many of the volatiles formed during deep fat frying and some of the variables influencing their formation. The list of volatiles identified is composed primarily of acids, alcohols, aldehydes, hydrocarbons, ketones, lactones, esters, aromatics and a few miscellaneous compounds (e.g., pentyl furan and 1,4-dioxane). Chang *et al.* (1978) reported the identities of 220 volatiles.

Both Chang *et al.* (1978) and Nawar *et al.* (1978) studied the volatiles produced by different types of oil upon heating. The study reported by Chang *et al.* (1978) was primarily on qualitative differences while Nawar *et al.* (1978) focused on quantitative differences. The qualitative study noted numerous differences between the volatiles produced by heated corn oil vs. hydrogenated cottonseed oil. This is expected since the oils differ greatly in fatty acid composition. The quantitative study by Nawar *et al.* (1978) indicated that corn and soybean oils produced similar volatiles but coconut oil was quite different. The coconut oil produced less of the decadienals and unsaturated aldehydes and more saturated aldehydes relative to the corn and soybean oils. Methyl ketones and gamma and delta lactones were also present in greater quantities in the coconut oil.

The quantitative effects of heating time and introduction of moisture during deep fat frying were also reported by Nawar *et al.* (1978). In general, the production of individual volatile components increased with heating time up to 48 h of heating. The introduction of moisture during heating of the oil resulted in a large reduction in volatiles in the oil. This apparent "protective" effect of water may be due to the steam stripping of volatiles during frying, the displacement of oxygen by the steam or some combination of both (Nawar *et al.* 1978).

Of the host of volatiles produced during deep fat frying, Chang *et al.* (1978) have shown that the gamma lactones with unsaturation at the 2 or 3 position are of particular significance to deep fat fried flavor. May *et al.* (1978) have found that 2.5 ppm 4-hydroxy-2-nonenoic acid in cottonseed oil imparts nutty, buttery and fried fat notes to the oil.

2. Lactones

Lactones may be formed in foods via microbial action, extensive lipid oxidation or biosynthetic routes (Maga 1976). It appears that due to inborn errors in metabolism, hydroxy acids may be synthesized and incorporated into the triglycerides (Dimick *et al.* 1969). These hydroxy acids are relatively unstable and are readily hydrolyzed from the triglyceride in the presence of water with heat. Hydrolysis of the hydroxy acid results in its cyclization to a lactone. Since the hydroxy acid has no flavor as long as it is bound to the triglyceride, thermal processing frees the acid and a flavorful lactone is produced. This process is one of the simpler ways in which flavor may be formed during the processing of foods.

3. Secondary Reactions

It is well established that fat plays a major role in the formation of meat flavors (Yamato *et al.* 1970). The lean portion of the meat supplies the meaty brothy character and the fat provides the species character (Wasserman 1972). This occurrence is used in the manufacture of reaction products (meat flavors containing no meat). Meat reaction products contain approximately the same sugars and amino acids for the basic meat flavors but contain different fats to give pork, beef or chicken notes.

The characteristic meat aromas which arise from animal fats during thermal processing are due both to the volatiles formed as thermal degradation products of the lipids and to secondary reactions of these degradation products. As Chang *et al.* (1978), Nawar *et al.* (1978) and Ohnishi and Shibamoto (1984) have shown, a large number of carbonyls are produced via lipid degradation. These carbonyls may readily participate in nonenzymatic browning with the small quantities of amino acids, ammonia and protein present in animal fats. Heterocyclic compounds from these secondary reactions produce much of the characteristic meat flavor (Fig. 3.10).

C. Flavors Formed via Fermentation

Fermentation is a process which provides us with a variety of new flavors to enjoy. We daily come in contact with a variety of food products produced via fermentation. Some examples include soy sauce, cheese, yogurt, bread, beer, wine, fermented fish products and sausages.

Flavor may be developed from the primary metabolism of the micro-

Table 3.3. Reactions Catalyzed by Microbes[a]

Oxidation	*Acylation*
Oxygenation of CH or C=C	*Transglycosidation*
Dehydrogenation of CHOH	*Methylation*
Dehydrogenation of CH–CH	*Condensation*
Reduction	*Cleavage of C–C*
Hydrogenation of C=O	*Decarboxylation*
Hydrogenation of C=C	*Dehydration*
	Amination/Deamination
Isomerization	*Halogenation*
Esterification/Hydrolysis	*Phosphorylation*

[a] From Tressl *et al.* (1978).

organism or from residual enzymatic activity once the microbial cell has lysed. Primary metabolism is responsible for much of the flavor of alcoholic beverages while residual enzymatic activity is essential for the development of aged cheese flavor.

A large number of different classes of flavor compounds can be produced by microorganisms. Tressl *et al.* (1978) have summarized some of the reactions which microorganisms may accomplish (Table 3.3). This section will present some metabolic pathways leading to the development of flavor in fermented food products. The book by Margalith (1981) is recommended for greater detail.

1. Esters

Esters are quite important to the flavor of both natural foods (e.g., fresh fruit) and fermented foods. Of the fermented foods, esters probably are most important to the flavor of some of the alcoholic beverages. Engan (1974) pointed out that 82 different esters had been identified in beer at that time. Most of the esters found in beer are formed via primary fermentation. They have been linked to the metabolism of lipids by yeast. Lipid metabolism provides a large number of acids and alcohols which may undergo esterification to yield a variety of esters. While pure chemical reactions can lead to ester formation, this reaction is too slow to account for the esters in most foods. Nordstrom (1965) has demonstrated that esterification is controlled by intracellular enzymes.

2. Acids

Acid production via microorganisms is important to the flavor of many fermented foods. Certainly the importance of acids to both fermented dairy products and pickled products cannot be overemphasized.

The acid of greatest importance to the flavor of fermented dairy products is lactic acid (Dumont and Adda 1979). Lactic acid is an optically active acid existing as the D, L or optically inactive mixture depending upon the microorganism involved in its synthesis. There appears to be no difference in flavor between the optical isomers. Lactic acid is characterized as being odorless and having a sour milk taste.

The organisms most commonly associated with lactic acid formation are classified as being either homofermentative (e.g., *Lactobacillus bulgaricus,* which produces primarily lactic acid, >85% of metabolic end products) or heterofermentative (e.g., *Leuconostoc* sp., which produces lactic acid, acetic acid, ethanol, carbon dioxide and other metabolites). The homofermentative organisms employ the Embden-Meyerhof pathway for lactose fermentation while the heterofermentation organisms metabolize lactose via the hexose–monophosphate pathway (Margalith 1981). Lactic acid will reach concentrations of up to 2% or greater in cultures of *L. bulgaricus* or *L. acidophilus* (Fig. 3.11).

Lactic acid is also formed in wines during fermentation from L-malic acid. Malic acid is a very tart acid present in grapes which needs to be converted to lactic acid to "smooth out" the harsh acidity. This "malolactic" fermentation is the result of bacterial (most commonly *Leuconostoc* sp.) action during the last stages of fermentation.

Numerous other acids are produced during fermentation. Most microorganisms have active lipase systems which attack triglycerides to yield glycerol, monoglycerides, diglycerides and free fatty acids. These free fatty acids can make a substantial contribution to the flavor of various cheeses and sausages. Secondary reactions of the free fatty acids (e.g., oxidation) can again yield a host of new flavor compounds.

Acids may also come from the deamination of amino acids. The products are various aliphatic (linear and branched chain) and aromatic acids.

3. Carbonyls

Carbonyl flavor compounds make a particularly significant contribution to the flavor of fermented dairy products. Diacetyl is one of the most important carbonyls to the flavor of these products. It has a buttery, nutlike flavor. Diacetyl is produced via the fermentation of citrate in dairy products. The most important citric acid fermenters are *Leuconostoc citrovorum, L. creamoris, L. dextranicum, Streptococcus lactis* subspecies *diacetylactis, S. thermophilus* and certain strains of *Proprionibacterium shermani* (Kempler 1983; Vedamuthu 1979). The metabolic pathway leading to the synthesis of diacetyl is somewhat controversial. Kempler and McKay (1981) have presented a summary of the two best accepted pathways (Fig. 3.12). Both pathways involve the degradation of citrate to

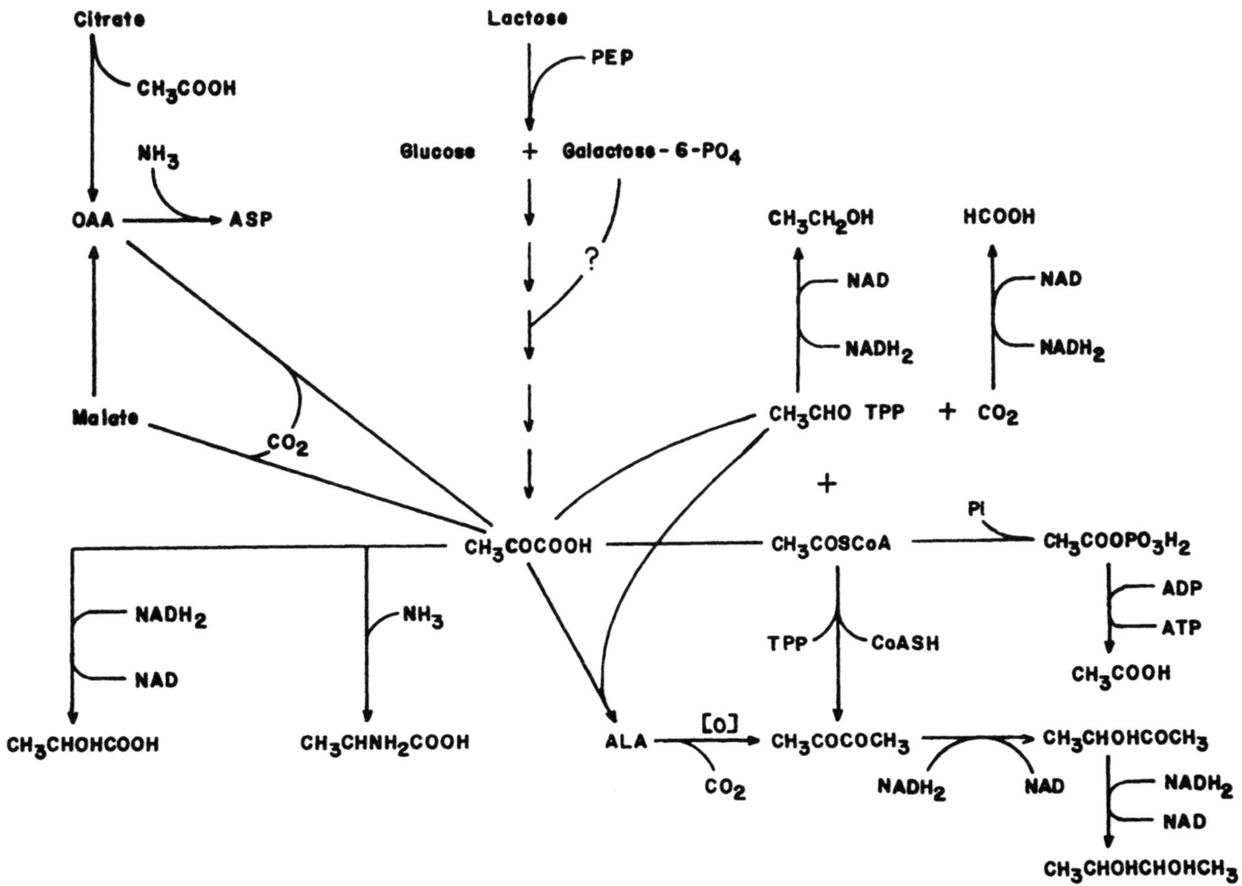

Fig. 3.11. Formation of various flavor compounds via fermentation of lactose and citrate.
Reprinted with the permission of Sandine and Elliker (1970). Copyright 1970. American Chemical Society

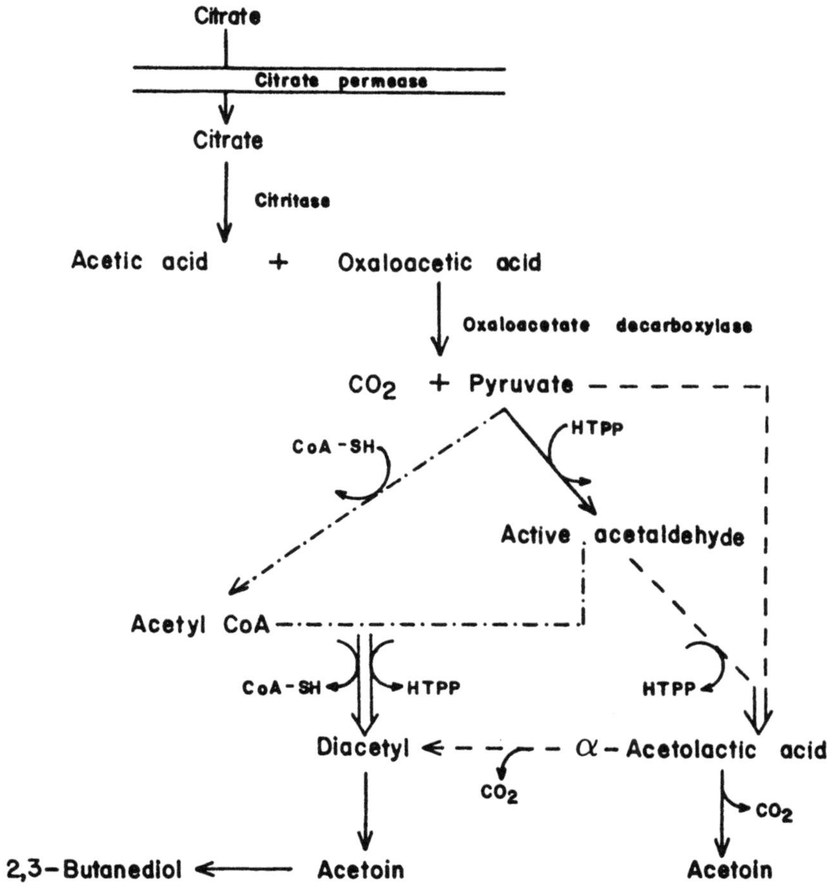

Fig. 3.12. Formation of diacetyl during fermentation in dairy cultures.
Kempler and McKay 1981

acetate and oxaloacetate which is then decarboxylated to form pyruvate. Apparently diacetyl is relatively nontoxic to the bacteria cell so excess pyruvate is channeled into diacetyl (Harvey and Collins 1963). The pathway proposed by Speckman and Collins (1968) involves the formation of "active acetaldehyde" from pyruvate and thiamine pyrophosphate which then condenses with acetyl CoA. This product is enzymatically cleaved to yield diacetyl. DeMan (1959) and Seitz *et al.* (1963) propose that the "active acetaldehyde" condenses with a molecule of pyruvate to form α-acetolactate via acetolactate synthase. The α-acetolactate would then be oxidized to form diacetyl.

Unfortunately diacetyl is not stable in most cultured food products.

The microorganisms which synthesized the diacetyl also contain diacetyl reductase which reduces diacetyl to flavorless acetoin and 2,3-butanediol (Fig. 3.12). Thus fermented proudcts such as buttermilk which depend on diacetyl for flavor have an optimum or peak flavor. As the product is stored, the diacetyl is reduced and flavor strength and quality decrease.

A second dairy product where carbonyls are considered as flavor impact compounds is yogurt. Here acetaldehyde is credited with providing the characteristic pungent green character. Acetaldehyde is a metabolic end product of *L. bulgaricus* and/or *S. thermophilus* during the fermentation of milk to yogurt (Margalith 1981).

Carbonyls are quite important to the flavor of cheeses. Lamparsky and Klimes (1981) have noted that 28 aldehydes and methyl ketones have been identified in the volatile fraction of cheddar cheese. Carbonyls (methyl ketones) may arise in fermented products initially via lipase activity of the microorgainsm. Dairy products contain a significant quantity of β-keto acids which are readily hydrolyzed from the triglyceride by microbial lipases and then decarboxylated to form the odd carbon number methyl ketones.

Methyl ketones and aldehydes may also be formed via microbially induced lipid oxidations (Haymon and Acton 1978). The oxidation may be initiated by microbial lipases, hydrogen peroxide produced by microorganisms (Tjaberg *et al.* 1969) and/or lipoxidaselike activity.

Alford *et al.* (1971) have reported on the action of various microorganisms on lard oxidation. They found 10 of the 28 microorganisms studied destroyed the low levels of peroxide in fresh lard. Fourteen of the 28 had no effect on peroxides in the lard while five strains of streptomyces increased peroxides by threefold. *Pseudomonas ovalis* increased peroxides by eightfold and *Micrococcus freudenreichii* increased peroxides fourteenfold. The extent of oxidation observed in the lard was found to be very dependent upon the microflora present on the lard.

A final mechanism for the formation of carbonys via microorganisms involves the transamination and decarboxylation of free amino acids. Morgan (1976) has demonstrated the production of 3-methyl butanal by *S. lactis* var. *maltigenes* (Fig. 3.13). MacLeod and Morgan (1958) have demonstrated the production of 2-methyl butanal, methional and phenylacetaldehyde from leucine, methionine and phenylalanine, respectively, by the same microorganism.

Carbonyl compounds typically are significant to the flavor of most fermented food products. As has been shown, these carbonyls may arise due to carbohydrate or citrate metabolism, lipid oxidation or amino acid degradation.

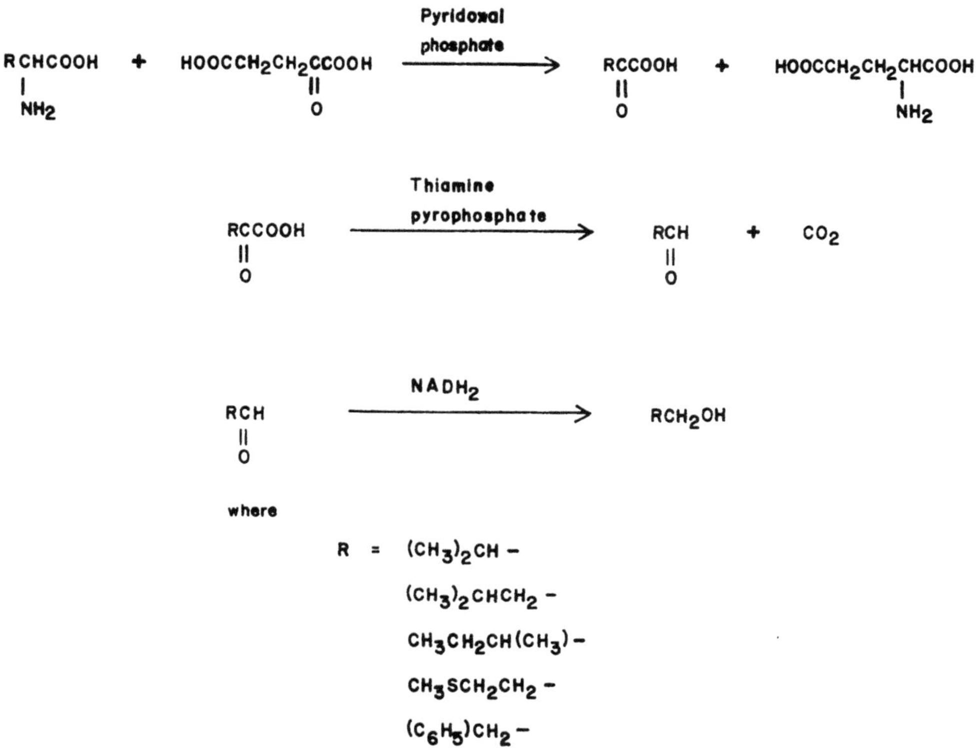

Fig. 3.13. Metabolic pathway leading to the formation of aldehydes and alcohols. *Morgan 1976*

4. Alcohols

Alcohols typically make a minor contribution to flavor unless present in relatively high concentrations (ppm) or they are unsaturated (e.g., 1-octen-3-ol). Alcohols may arise via primary metabolic activity of a microorganism (e.g., ethanol) or by reduction of a carbonyl to the corresponding alcohol.

Reazin *et al.* (1970) have outlined metabolic pathways leading to the major congeners during grain fermentation (Fig. 3.14). This figure illustrates that the fusel alcohols (alcohols larger in size than ethanol) can be formed from either carbohydrate or amino acid metabolism.

Alcohol production from amino acids may occur either by transamination, decarboxylation and reduction or by oxidative deamination followed by decarboxylation and reduction. In either pathway, the product is always the amino acid minus the amine group and one carbon atom.

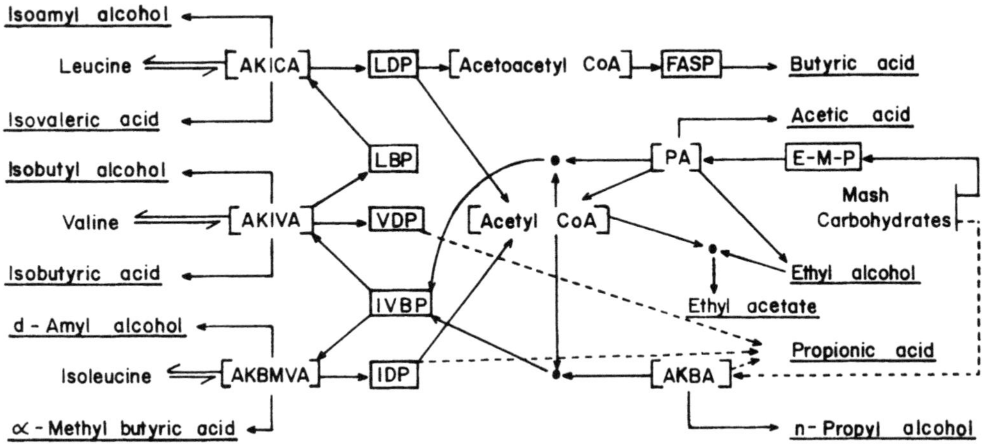

Fig. 3.14. Proposed metabolic pathways for the synthesis of whiskey congeners:—, synthetic and degradative pathways;---, hypothetical pathways; LBP, leucine biosynthesis; IVBP, isoleucine–valine biosynthesis; FASP, fatty acid biosynthesis; LDP, leucine degradation; VDP, valine degradation; IDP, isoleucine degradation; E-M-P, Embden–Meyerhof–Parnas fermentation pathway.
Reprinted with the permission of Reazin et al. (1970). Copyright 1970. American Chemical Society

The mechanism of fusel oil production from carbohydrates follows the typical Embden–Meyerhof–Parnas pathway to pyruvic acid. Pyruvic acid may be reacted with a second pyruvic acid to yield acetolactic acid. Acetolactic acid enters the valine–isoleucine biosynthesis pathway and is converted to α-keto isovaleric acid. The α-keto isovaleric acid may then be reduced to isobutyl alcohol or isobutyric acid. The interconnection of pathways as shown in Fig. 3.14 results in the formation of numerous alcohols. Note that acetyl CoA may also enter into these pathways. The acetyl CoA may come from fatty acid metabolism which ties fatty acids also into alcohol production.

The alternate metabolic pathway yielding alcohols involves the reduction of the corresponding carbonyl (Fig. 3.15). Lamparsky and Klimes (1981) demonstrated this mechanism by finding the equivalent alcohols (as aldehydes) in cheddar cheese. Secondary alcohols corresponding to the methyl ketones found in blue cheeses have also been reported.

5. Terpenes

Mono and sesquiterpenes are most commonly associated with the flavor of citrus products or spices. They are hydrocarbons based upon

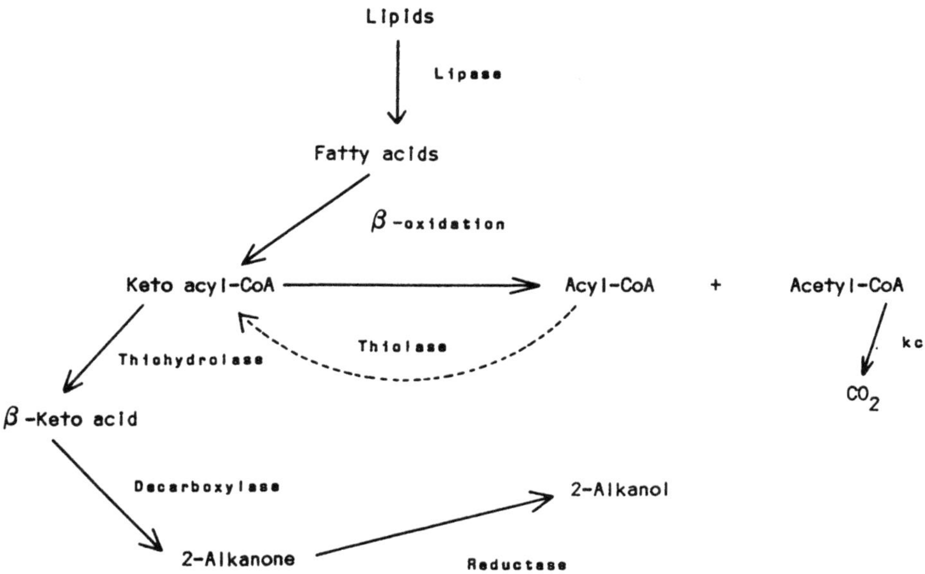

Fig. 3.15. Metabolic pathways leading to the formation of methyl ketones and the corresponding 2-alkanols.
Kinsella and Hwang 1976

the five carbon isoprene unit (2-methyl-1,3-butadiene) with structures that may be open chain, closed chain, saturated or unsaturated.

Currently the production of terpenes by microorganisms is primarily of academic interest. There are no fermented food products commercially available which derive their characteristic flavor from terpenes. The interest in terpene synthesis by microorganisms is due to the potential for the commercial production of specific terpenes via fermentation (Kempler 1983; Tressl *et al.* 1978; Jourdain *et al.* 1985; Kieslich *et al.* 1985).

Collins and Halim (1970) identified geraniol, citronellol, linalool, neral, citral, geranyl acetate and citronellyl acetate as metabolic products of *Ceratocystis variospora*. Later, nerol, citronellol and geraniol were found to be produced by *Trametes odorata* (Halim and Collins 1971; Collins and Halim 1972).

The biosynthetic pathways leading to terpene formation in microorganisms is not well characterized. It is assumed that a pathway similar to that found in plants also operates in microbial syntheses. (See Fig. 2.9, Chapter 2.) This basically involves the conversion of mevalonate (MVA) into isopentyl pyrophosphate and 3,3-dimethyl allyl pyrophosphate which condense to form geranyl pyrophosphate (Banthorpe *et al.* 1972). Lanza and Palmer (1977) have provided evidence via radioactive labeling that this pathway occurs in the synthesis of geraniol in *Ceratocystis moniliformis*.

6. Lactones

Lactones make a significant contribution to the flavor of several fermented food products, particularly dairy products and alcoholic beverages. Tressl *et al.* (1978) have summarized some of the previous literature on lactone synthesis by microorganisms in fermented alcoholic beverages. As can be seen in Fig. 3.16, lactones may be formed from amino acids. Pathway A shows the oxidative deamination of glutamic acid to yield 2-oxoglutarate. The 2-oxoglutarate is decarboxylated to produce 4-oxobutyrate which is then reduced to the 4-hydroxy butyrate. The cyclization of 4-hydroxy butyrate yields γ-butyrolactone, alkoxy- and acyl lactones.

Pathway B illustrates the conversion of 2-oxobutyric acid into 3-hydroxy-4,5-dimethyl-2(5*H*)-furanone. This lactone has burnt aroma and has been identified in aged sake (Takahashi *et al.* 1976).

7. Pyrazines

Pyrazines are generally associated with nonenzymatic browning. However, pyrazines have also been found as a result of microbial action. *Bacillus subtilis* was the first organism found to produce a pyrazine (tetramethyl pyrazine; Kosuge and Kamiya 1962). Since that time, several pyrazines (2-acetyl, dimethyl, 2-methoxy-3-ethyl, 2,5 or 2,6-diethyl-3-methyl, trimethyl, tetramethyl and trimethyl pyrazine) have been identified in aged cheese (Sloot and Harkes 1975; Lin 1976; Liardon *et al.* 1982). While some of these pyrazines may be formed via NEB during aging of the cheese, it is likely that they were formed via microbial action.

Adachi *et al.* (1964) postulated that tetramethyl pyrazine was formed by *B. subtilis* from 2 moles of acetoin and 2 moles of ammonia (Fig. 3.17). Acetoin is an intermediate in the biosynthesis of leucine, valine and pantothenate.

Pyrazines have also been linked to the generation of off-flavors in foods due to microbial action. *Pseudomonas taetrolens* has been shown to produce a musty-potato aroma (2,5-dimethyl pyrazine and 2-methoxy-3-isopropyl pyrazine) in milk and other refrigerated foods (Morgan 1976). A similar flavor defect has been found in cheese containing *Ps. mucidolens*.

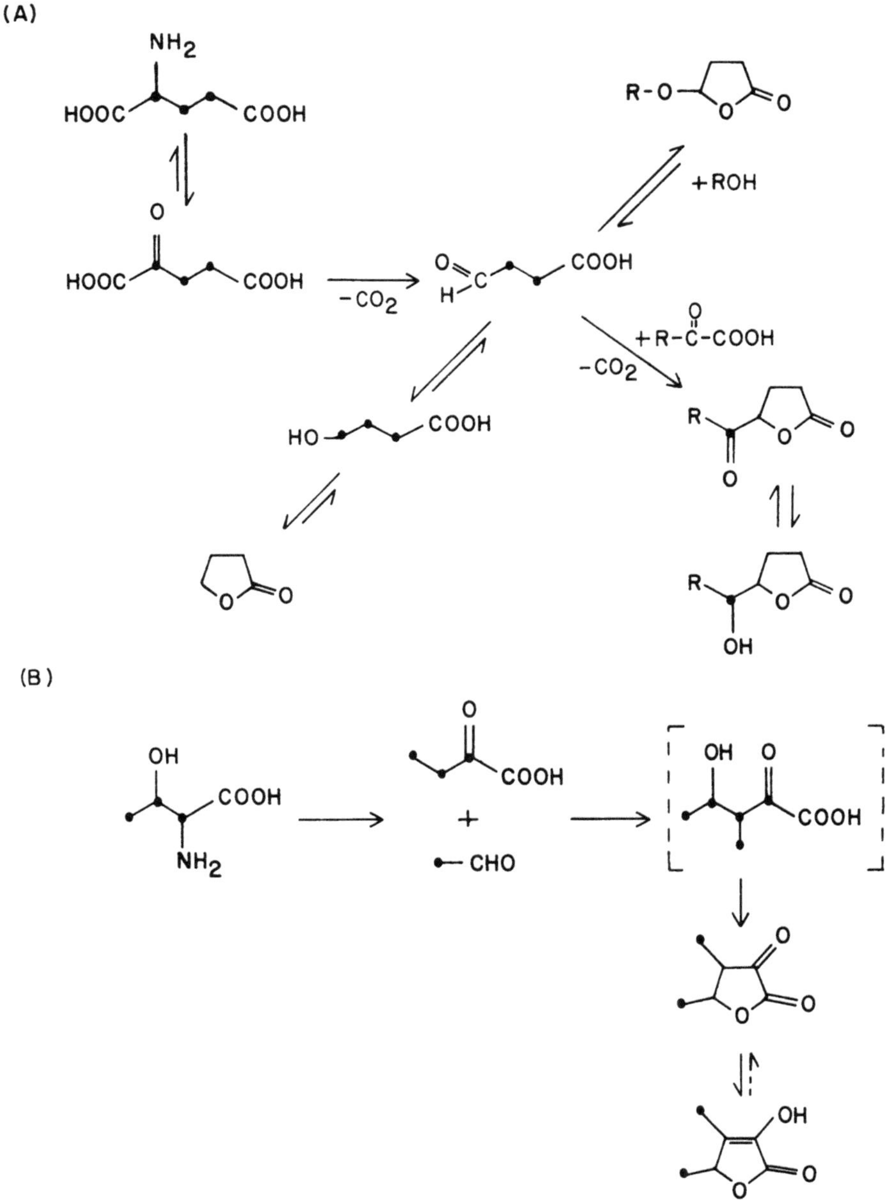

Fig. 3.16. Formation of lactones by microorganisms.
Tressl et al. 1978

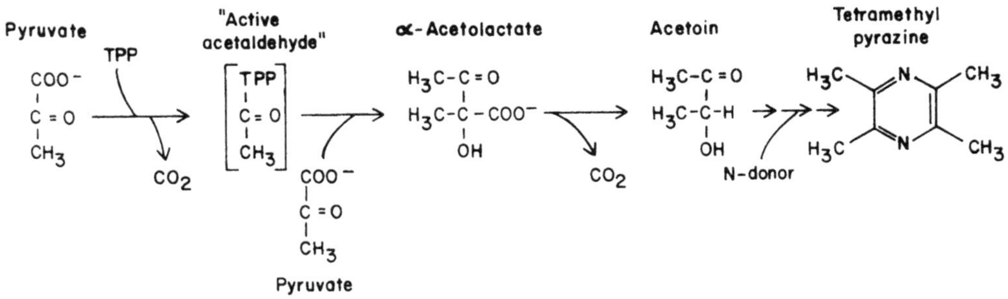

Fig. 3.17. Metabolic pathway proposed for the formation of pyrazines by microorganisms.

A mutant of *Ps. taetrolens* has recently been isolated which is of interest for the commercial production of 2-methoxy-3-isopropyl pyrazine (McIver and Reineccius 1986). The parent culture was found to produce about 40 ppb of this pyrazine while a mutant produced by chemical mutation produced about 15,000 ppb. Since the sensory threshold of this compound is about 0.001 ppb, 15,000 is a sufficient concentration to be of commercial significance.

8. Conclusion

Some of the mechanisms leading to the formation of several classes of flavor compounds via microorganisms are outlined in this chapter. At this point, two additional ideas should be mentioned. First, we have only considered the formation of flavor via metabolic activity of the living microorganism. Microorganisms could be viewed as a small packet of enzymes, which upon death and lysis of the cell, may continue to function. A good example of the significance of bacterial enzymes to flavor is in aged cheeses. The primary metabolic acitivity of the microflora in the cheese occurs during the early stages of cheese making and aging. Yet there is little characteristic flavor at this point. Only during aging do we see protein degradation via microbial proteases to yield the soft texture, free amino acids (meaty, brothy flavor background), peptides (bitterness) and free sulfur compounds so important to the flavor of aged cheeses. Changes in lipids during aging of the cheese produce free fatty acids, oxidized fatty acids (e.g., methyl ketones) and a host of other flavor compounds. These changes during aging are occurring when microbial activity is decreasing and enzymes from lysed cells are increasing. One must remember that microbial growth in a food product can have a pronounced effect both on the development of desirable flavor and off-flavor even though no viable bacterial cells exist in the food.

The second idea that should be brought out in closing this section of the text is that the flavor system in a fermenting product is very dynamic (Schreier 1979; Haymon and Acton 1978). One microorganism may produce a metabolite which initially accumulates in the product and may make a significant contribution to flavor. A second organism may then reach significant metabolic activity (population) to start metabolizing this first metabolite. Therefore, the flavor of the product will change. Microorganisms may also alter flavor compounds produced in the product via other mechanisms, e.g., lipid oxidation or nonenzymatic browning. Thus lipid oxidation or NEB may result in a different flavor in a fermented product than in a nonfermented product.

References

ADACHI, T., KAMIYA, H. and KOSUGE, T. 1964. Studies on the metabolic products of *Bacillus subtilis* IV. Determination and mechanism of formation of tetramethyl pyrazine. J. Pharm. Soc. Japan *84*, 545.

ALFORD, J. A., SMITH, J. L. and LILLY, H. D. 1971. Relation of microbial activity to changes in lipids of foods. J. Appl. Bacteriol. *34*, 133.

BANTHORPE, D. V., LE PATOUREL, G. N. J. and FRANCIS, J. J. O. 1972. Biosynthesis of geraniol and nerol and their α-D-glucosides in *Pelargonium graveolens* and *Rosa dilecta*. Biochem. J. *130*, 1045.

CHANG, S. S., PETERSON, R. J. and HO, C. T. 1978. Chemistry of deep fat fried flavor. *In* Lipids as a Source of Flavors, p. 18, M. K. Supran, ed. ACS Symp. Ser. 75, Washington DC.

COLLINS, R. P. and HALIM, A. F. 1970. Production of monoterpenes by the filamentous fungus *Cerafocystis variospora*. Lloydia *33*, 581.

COLLINS, R. P. and HALIM, A. F. 1972. An analysis of the odorous constituents produced by various species of *Phillinus*. Can. J. Microbiol. *18*, 65.

DEMAN, J. C. 1959. The formation of diacetyl in cultures of *Betacoccus cremoris*. Neth. Milk Dairy J. *10*, 38.

DIMICK, P. S., WALKER, N. J. and PATTON, S. 1969. Lipid metabolism. Evidence of a β-oxidation pathway for saturated Fatty Acids. Biochem. J. *111*, 395.

DUMONT, J. P. and ADDA, J. 1979. Flavour formation in dairy products. *In* Progress in Flavour Research, p. 245, D. G. Land and H. E. Nursten, eds. Applied Science Publ., London.

EL'ODE, K. E., DORNSEIFER, T. P., KEITH, E. S. and POWERS, J. J. 1966. Effects of pH and temperature on the carbonyls and aromas produced in heated amino acid–sugar mixutres. J. Food Sci. *31*, 351.

ENGAN, S. 1974. Esters in beer. Brewers Digest, *49*, 40.

HALIM, A. F. and COLLINS, R. P. 1971 An analysis of the odorous constituents of *Trametes odorata*. Lloydia *34*, 451.

HARVEY, R. J. and COLLINS, E. B. 1963. Roles of citrate and acetoin in the metabolism of *Streptococcus diacetilactis*. J. Bacteriol. *86*, 1301.

HAYMON, L. W. and ACTON, J. C. 1978. Flavors from lipids by microbiological action.

In Lipids as a Source of Flavor, p. 94, M. K. Supran, ed. ACS Symp. Ser. 75, Washington DC.
HO, C. T. and HARTMAN, G. J. 1982. Formation of oxazolines and oxazoles in Strecker degradation of DL-alanine and L-cysteine with 2,3-butanedione. J. Agric. Food Chem. *30*, 793.
HO, C. T. and JIN. Q, Z. 1985. Aroma properties of some alkyl thiazoles. Perf. Flavorist *9*(6), 15.
HO, C. T. and TUORTO, R. M. 1981. Mass spectra and sensory properties of some 4,5-dialkyl-oxazoles. J. Agric. Food Chem. *29*, 1306.
HODGE, J. E. 1953. Chemistry of browning reactions in model systems. J. Agric. Food Chem. *1*, 928.
HODGE, J. E., MILLS, F. D. and FISHER, B. E. 1972. Compounds of browned flavor derived from sugar–amine reactions. Cereal Sci. Today *17*(2), 34.
HURRELL, R. F. 1982. Maillard reactions in flavour. *In* Food Flavours, Part A: Introduction, p. 399, I. D. Morton and A. J. MacLeod, eds. Elsevier, New York.
JIN, Q. Z., HARTMAN, G. J. and HO, C. T. 1984. Aroma properties of some oxazoles. Perfumer Flavorist *9*(4), 25.
JOURDAIN, N., GOLI, T., JALLAGEAS, J. C., CROZET, C., GHOMMIDH, C., NAVARRO, J. and CROZET, J. 1985. Aroma components production by immobilized cells. *In* Topics in Flavour Research, R. G. Berger, S. Nitz and P. Schreier eds. H. Eichhorn, Marzling-Hangenham.
KEMPLER, G. M. 1983. Production of flavor compounds by microorganisms. Adv. Appl. Micro., *29*, 29.
KEMPLER, G. M. and McKAY, L. L. 1981. Biochemistry and genetics of citrate utilization in *Streptococcus lactic* sp. *diacetylactis*. J. Dairy Sci. *64*, 1527.
KIESLICH, K., ABRAHAM, W. R., STUMPF, B. and WASHAUSEN, P. 1985. Microbial transformation of terpenoids. *In* Topics in Flavour Research, R. G. Berger, S. Nitz and P. Schreier, eds. H. Eichorn, Marzling-Hangenham.
KINSELLA, J. E. and HWANG, D. H. 1976. Enzymes of *Penicillium roqueforti* involved in the biosynthesis of cheese flavor. CRC Crit. Rev. Food Sci. Nutr. *8*, 191.
KOEHLER, P. E. and ODELL, G. V. 1970. Factors affecting the formation of pyrazine compounds in sugar–amine reactions. J. Agric. Food Chem. *18*(5), 895.
KOEHLER, P. E., MASON, M. E. and NEWELL, J. A. 1969. Formation of pyrazine compounds in sugar–amino acid model systems. J. Agric. Food Chem. *17*(2), 393.
KOSUGE, T. and KAMIYA, H. 1962. Discovery of a pyrazine in a natural product: Tetramethyl pyrazine from cultures of a strain of *Bacillus subtilis*. Nature. *193*, 776.
LABUZA, T. P. 1980. Effect of water activity on the reaction kinetics of food deterioration. Food Technol. *34*(2), 36.
LAMPARSKY, D. and KLIMES, I. 1981. Cheddar cheese flavour—its formation in the light of new analytical results. *In* Flavour '81, p. 557, P. Schreier, ed. deGruyter, Berlin.
LANZA, E. and PALMER, J. K. 1977. Phytochem. *16*, 1555.
LEA, C. H. and HANNAN, R. S. 1949. The effect of activity of water, of pH and of temperature on the primary reaction between casein and glucose. Biochem. Biophys. Acta. *3*, 313.
LEAHY, M. M. 1985. The effects of pH, types of sugar and amino acid and water activity on the kinetics of the formation of alkyl pyrazines. Ph.D. Thesis, Feb., University of Minnesota, Department of Food Science and Nutrition, St. Paul, Minnesota.
LEWIS, V. M. and LEA, C. H. 1950. A note of the relative rates of reaction of several reducing sugars and sugar derivatives with casein. Biochem. Biophys. Acta. *4*, 532.
LIARDON, R., BOSSET, J. O. and BLANC, B. 1982. The aroma composition of Swiss Gruyere cheese I. The alkaline volatile components. Lebensm.-Wiss. U.-Technol. *15*, 143.

LIN, S. S. 1976. Alkyl pyrazines in processed American cheese. J. Agric. Food Chem. *24*, 1252.

MACLEOD, P. and MORGAN, M. E. 1958 Differences in the ability of lactic streptococci to form aldehydes from certain amino acids. J. Dairy Sci. *41*, 908.

MAGA, J. A. 1976. Lactones in Foods. Crit. Rev. Food Sci. Nutr. *8*, 1.

MAGA, J. A. 1978. Oxazoles and oxazolines in foods. J. Agric. Food Chem. *26*, 1049.

MAGA, J. A. 1981a. Pyrroles in foods. J Agric. Food. Chem. *29*, 691.

MAGA, J. A. 1981b. Pyridines in Foods. J. Agric. Food Chem. *29*, 895.

MAGA, J. A. 1981c. The chemistry of oxazoles and oxazolines in foods. CRC Crit. Rev. Food Sci. Nutr. *14*, 295.

MAGA, J. A. 1982. Pyrazines in flavour. *In* Food Flavours Part A: Introduction, p. 283, I. D. Morton and A. J. MacLeod, eds. Devel. Food Sci., Elsevier, Amsterdam.

MAILLARD, L. C. 1916. Syntheses des matieres humique par action des acides amines sur les sucre reducteurs. Ann. Chim. *6*, 258.

MARGALITH, P. Z. 1981. Flavor Microbiology. Charles C. Thomas, Springfield.

MAY, W. A., PETERSON, R. J. and CHANG, S. S. 1978. Synthesis of some unsaturated lactones and their relationship to deep-fat fried flavor. J. Food Sci. *43*, 1248.

McIVER, R. L. and REINECCIUS, G. A. 1986. Synthesis of 2-alkyl-3-methoxy pyrazines by *Pseudomonas perolens*. *In* Biogeneration of Aromas, T. Parliment, ed. Amer. Chem. Soc., Washington DC.

MORGAN, M. E. 1976. The chemistry of some microbiologically induced flavor defects in milk and dairy foods. Biotechnol. Bioeng. *18*, 953.

NAWAR, W. W., BRADLEY, S. J., LOMANNO, S. S., RICHARDSON, G. C. and WHITEMAN, R. C. 1978. Volatiles from frying fats: a Comparative study. *In* Lipids as a Source of Flavor, p. 42, M. K. Supran, ed. ACS Symp. Ser., Washington DC.

NEWELL, J. A., MASON, M. E. and MATLOCK, R. S. 1967. Precursors of typical and atypical roasted peanut flavor. J. Agric. Food Chem. *15*(5), 767.

NILOV, V. I. and OGORODNIK, S. J. 1967. Kharchove Prom. Nauk.-Virob. ZB. No. *1*(31), 40; Ref. Zh. Khim. H. II 18R 327. Cited by Pokorny (1980).

NORDSTROM, K. 1965. Formation of volatile esters by brewer's yeast. Brewers Digest *40* (Nov), 60.

OHLOFF, G. and FLAMENT, I. 1979. The role of heteroatomic substances in the aroma compounds of foodstuffs. Fortschr. Chem. Org. Naturst. *36*, 231.

OHNISHI, S. and SHIBAMOTO, T. 1984. Volatile compounds from heated beef fat and beef fat with glycine. J Agric. Food Chem. *32*(5), 987.

OLSSON, K., PERNEMALM, P. A. and THEANDER, O. 1981. Reaction products and mechanisms in some simple model systems. Prog. Food Nutr. Sci. *5*,45.

PITTET, A. O. and HRUZA, D. E. 1974. Comparative study of flavor properties of thiazole derivatives. J. Agric. Food Chem. *22*, 264.

PITTET, A. O., RITTERSBACHER, P. and MURALIDHARA, R. 1970. Flavour properties of compounds related to maltol and isomaltol. J. Agric. Food Chem. *18*, 929.

POKORNY, J. 1980. Effect of browning reactions on the formation of flavour substances. Die Nahrung *24*(2), 115.

REAZIN, G., SCALES, H. and ANDREASEN, A. 1970. Mechanism of major congener formation in alcoholic grain formations. J. Agric. Food Chem. *18*, 585.

REINECCIUS, G. A., KEENEY, P. G. and WEISBERGER, E. 1972. Factors affecting the concentratiion of pyrazines in cocoa beans. J. Agric. Food Chem. *20*, 202.

REYNOLDS, T. M. 1970. Flavours from nonenzymatic browning reactions. Food Technol. Aust. *22*(11), 610.

RIZZI, G. P. 1972. A mechanistic study of alkylpyrazine formation in model systems. J. Agric. Food Chem. *20*(5), 1081.

RIZZI, G. P. 1974. Formation of *N*-alkyl-2-acyl-pyrroles and aliphatic aldimines in model nonenzymatic browning reactions. J. Agric. Food Chem. *22*, 279.

SAKAGUCHI, M. and SHIBAMOTO, T. 1978. Formation of sulfur-containing compounds from the reaction of D-glucose and hydrogen sulfide. J. Agric. Food Chem. *26*, 1260.

SANDINE, W. E. and ELLIKER, P. R. 1970. Microbially induced flavors and fermented foods. Flavor in dairy products. J. Agric. Food Chem. *18*(4), 557.

SCHREIER, P. 1979. The role of microorganisms in flavour formation. *In* Progress in Flavour Research, p. 175, D. G. Land and H. E. Nursten, eds. Applied Sci. Publ., London.

SEITZ, E. W., SANDINE, W. E., ELLIKER, P. R. and DAY, E. A. 1963. Studies on diacetyl biosynthesis by *Streptococcus diacetilactis*. Can. J. Microbiol. *9*, 431.

SHANKARANARAYANA, M. L., RAGHAVEN, B., ABRAHAM, K. O. and NATARAJAN, C. P. 1982. Sulfur compounds in foods. *In* Food Flavours: Part A. Introduction, p. 169, I. D. Morton and A. J. MacLeod, eds. Elsevier, Amsterdam.

SHIBAMOTO, T. 1983. Heterocyclic compounds in browning and browning/nitrite model systems: Occurrence, formation mechanisms, flavor characteristics and mutagenic activity. *In* Instrumental Analysis of Foods: Recent Progress, p. 229, G. Charalambous and G. Inglett, eds. Academic Press, New York.

SHIBAMOTO, T. and BERNHARD, R. A. 1976. Effect of time, temperature and reactant ratio on pyrazine formation in model systems. J. Agric. Food Chem. *24*, 847.

SHIBAMOTO, T. and BERNHARD, R. A. 1977. Investigation of pyrazine formation pathways in glucose–ammonia model systems. J. Agric. Food Chem. *25*(1), 143.

SHIBAMOTO, T., AKIYAMA, T., SAKAGUCHI, M., ENOMOTO, Y. and MASUDA, H. 1979. A study of pyrazine formation. J. Agric. Food Chem. *27*, 1027.

SHU, C. K., HAGEDORN, M. L., MOOKHERJEE, B. D. and HO, C. T. 1985a. Volatile components of the thermal degradation of cystine in water. J. Agric. Food Chem. *33*(3), 438.

SHU, C. K., HAGEDORN, M. L., MOOKHERJEE, B. D. and HO, C. T. 1985b. pH effect on the volatile components in the thermal degradation of cystine. J. Agric. Food Chem. *33*(3), 442.

SHU, C. K., MOOKHERJEE, B. D. and HO, C. T. 1985c. Volatile components of the thermal degradation of 2,5-dimethyl-4-hydroxy-3-(2*H*)-furanone. J. Agric. Food Chem. *33*(3), 446.

SHU, C. K., MOOKHERJEE, B. D., BONDAROVICH, H. A. and HAGEDORN, M. L. 1985d. Characterization of bacon odor and other flavor components from the reaction of isovaleraldehyde and ammonium sulfide. J. Agric. Food Chem. *33*(1), 130.

SLOOT, D., and HARKES, P. D. 1975. Volatile trace components in Gouda cheese. J. Agric. Food Chem. *23*, 356.

SPECKMAN, R. A. and COLLINS, E. B. 1968. Diacetyl biosynthesis in *Streptococcus diacetilactis* and *Leuconostoc citrovorum*. J. Bacteriol. *94*, 174.

STAMP, J. A. and LABUZA, T. P. 1983. Kinetics of the Maillard reaction between aspartame and glucose in solution at high temperatures. J. Food Sci. *48*, 543.

STOFFELSMA, J. and PYPKER, J. 1968. J. Rccl. Trav. Chim. Pays-Bas. *87*, 241. Cited by Ho and Tuorto (1981).

SULLIVAN, J. F., KONSTANCE, R. P., CALHOUN, M. J., TALLEY, F. B., CORDING, JR., J. and PANASIUK, O. 1974. Flavor and storage stability of explosion-puffed potatoes: Nonenzymatic browning. J. Food Sci. *39*, 58.

TAKAHASHI, K., TADENUMA, M. and SATO, S. 1976. 3-Hydroxy-4,5-dimethyl-2(5*H*)-furanone, a burnt flavoring compound from aged sake. Agric. Biol. Chem. *40*, 325.

TAKKEN, H. J., VAN DER LINDE, L. M., DE VALOIS, P. J., VAN DART, H. M. and

BOELENS, M. 1976. Reaction products of α-dicarbonyl compounds, aldehydes, hydrogen sulfide and ammonia. *In* Phenolic Sulfur and Nitrogen Compounds in Food Flavors, p. 114, G. Charalambous and I. Katz, eds. ACS Symp. Ser., Washington DC.

TJABERG, T. B., HAUGAM, M. and NURMI, E. 1969. Studies on discoloration of Norwegian salami sausages. *In* Proceedings 15th European Meeting of Meat Research Workers, p. 138. Helsinki, Finland.

TRESSL, R., APETZ, M., ARRIETA, R. and GRUNEWALD, K. G. 1978. Formation of lactones and terpenoids by microorganisms. *In* Flavor of Foods and Beverages, p. 145, G. Charalambous and G. E. Inglett, eds. Academic Press, New York.

TRESSL, R., HELAK, B. and MARTIN, N. 1985a. Formation of flavor components from L-proline. *In* Topics in Flavour Research, R. G. Berger, S. Nitz and P. Schreier, eds. H. Eichhorn, Marzling-Hangenham.

TRESSL, R., REWICKI, D., HELAK, B., KAMPERSCHROER, H. and MARTIN, N. 1985b. Formation of 2,3-dihydro-1*H*-pyrrolizines as proline specific Maillard products. J. Agric. Food Chem. *33*(5), 919.

TRESSL, R., REWICKI, D., HELAK, B. and KAMPERSCHROER, H. 1985c. Formation of pyrrolidines and piperidines on heating L-proline with reducing sugars. J. Agric. Food Chem. *33*(5), 924.

VEDAMUTHU, E. R. 1979. Microbiologically induced desirable flavors in the fermented foods of the west. Dev. Ind. Microbiol. *20*, 187.

VERNIN, G. 1982. The Chemistry of Heterocyclic Flavouring and Aroma Compounds. Ellis Horwood, Chichester.

VERNIN, G. and PARKANYI, C. 1982. Mechanisms of formation of heterocyclic compounds in Maillard and pyrolyses reactions. *In* The Chemistry of Heterocyclic Flavouring and Aroma Compounds, p. 151, G. Vernin, ed. Ellis Horwood, Chichester.

VERNIN, G. and VERNIN, G. 1982. Heterocyclic compounds in foods: Occurrence and organoleptic properties. *In* The Chemistry of Heterocyclic Flavouring and Aroma Compounds, p. 72, G. Vernin, ed. Ellis Horwood, Chichester.

WANG, P. and ODELL, G. V. 1972. Characterization of some volatile constituents of roasted peanuts. J. Agric. Food Chem. *20*, 206.

WARMBIER, H., SCHNICKLES, R. A. and LABUZA, T. P. 1976. Nonenzymatic browning in intermediate moisture model systems, effect of glucose to lysine ratio. J. Food Sci. *41*, 981.

WASSERMAN, A. 1972. Thermally produced flavor compounds in the aroma of meat and poultry. J. Agric. Food Chem. *20*(4), 737.

WOLF, J. C., THOMPSON, D. R. and REINECCIUS, G. A. 1982. Predicting available lysine losses during heat processing. J. Food Proc. Eng. *6*, 201.

YAMATO, T., KURATA, T., KATO, H. and FUJIMAKI, M. 1970. Volatile carbonyl compounds from heated beef fat. Agric. Biol. Chem. *34*, 88.

4

Off-Flavors in Foods

The major cause of consumer complaints to the food industry is off-flavors (Whitfield and Shaw 1985). These off-flavors may be due to incidental contamination of the food from environmental sources (e.g., air, water or packaging materials) or may arise in the food itself (e.g., lipid oxidation, nonenzymatic browning or enzymatic action). This chapter discusses some of the more commonly occurring off-flavors.

A. Environmental Contamination

1. Airborne Sources

The air can be a very effective means of conveying an off-odor to a food product. The food products susceptible to this off-odor are generally limited to baked goods which are baked and then cooled open to the air environment. While the off-odor may enter the product during cooling, it may also be a problem once packaged. The packaging on these products typically presents little barrier to the transmission of volatile organic compounds.

Often the source of the off-odor can be several miles from the food production or storage facility and yet be responsible for tainting the food. Goldenberg and Matheson (1975) have provided a summary of off-flavors (OF) they have observed in the products of Marks and Spencer due to airborne contamination (Table 4.1). It is of interest that in one case, the source of contamination was as far away as 20 miles from the food production facility.

Airborne contamination often is exceedingly difficult to identify as the source of product contamination. This is partly due to the fact that the OF cannot be reproduced in the product at will for studying. This is contrasted with an OF arising from lipid oxidation, for example, during storage of food. The oxidized OF can be reproduced and subjected to analysis and eventual correction. The airborne OF may occur randomly,

Table 4.1. Off-Flavors Due to Airborne Contamination[a]

Food	Contaminant	Method of Transmission
Biscuits	Herbicides (Chlorophenols)	Carried by wind from nearby herbicide factories (3–5 miles).
Biscuits	Tar	Tar boiler gave off smell through gaps left in wall.
Chocolates	"Aniseed taste"	"Aniseed-type" odors airborne from adjacent factory.
Cakes	"Catty off-flavor"	Airborne contaminant produced by interaction of ketones and sulfur compounds in chemical factory in another town 20 miles away.
Wrapped bakery and confectionery products	Diesel oil fumes	May occur in depots, warehouses and loading bays. Diesel oil fumes can penetrate some types of wrapping materials and be absorbed by foods, especially "fatty" foods. May also be caused by forklift trucks driven by diesel oil.
Wrapped bakery and confectionery products	Petrol vapors	May occur in depots, warehouses and loading bays, where petrol has been spilled.
Cakes, biscuits, meat pies, chocolate confectionary		Off-flavors due to paints used in stock rooms and production areas.
Biscuits	Oxidized oil	Oil used for spraying savory biscuits and other products overheated through failure of heater thermostat. Oil oxidized in metal tank with "fishy" off-flavor. Products uneatable.

[a] From Goldenberg and Matheson (1975).

depending on the production schedule of the offending source of OF, wind direction and speed, production schedule of the food processing plant, etc. The offending chemicals also typically have a very low sensory threshold so they are difficult to isolate and identify by established analytical techniques. An example is the sensory threshold of 6-chloro-*o*-

cresol in biscuits (5×10^{-5} ppm). This compound was found to contribute a disinfectant taint (Goldenberg and Matheson 1975).

An examination of the compounds contributed to foods by airborne contamination in Table 4.1 shows that a variety of sources may be involved in the production of OF. The most common source appears to be a chemical manufacturing facility. The chemical plant may be producing offensive odorants (e.g., chlorophenols) or may be contributing compounds which undergo secondary reactions in foods to produce the OF. An example of OF contributed directly to the food product are the chlorophenols (Griffiths and Land 1972). An OF coming from secondary reactions in food products would be the reaction of mesityloxide (Saxby 1982) or pentoxone (Patterson 1968a, 1969) with sulfur compounds in the food to yield mercaptans with very low sensory thresholds (0.01 ppb).

Chloroanisoles have been found to be responsible for musty OF in foods (Saxby 1982; Curtis *et al.* 1974; Engel *et al.* 1966; Bemelmans and Noever deBrauw 1974; Buser *et al.* 1982; Maarse *et al.* 1985; Whitfield and Shaw 1985). The chloroanisoles generally have been transferred to the food product via air contamination. The chloroanisoles in the food may arise from direct exposure to the chloroanisoles or the exposure to chlorinated phenols which later get transformed (i.e., methylated) to the corresponding anisoles via microbial action. An interesting example of the latter occurrence is the finding of 2,3,4,6-tetrachloroanisole in eggs and broiler meat. It appears that some of the timber in Britain is treated with chlorophenols as antisapstain agents. These chlorophenols were converted by microorganisms in the chicken litter to the corresponding chloroanisoles (Curtis *et al.* 1974). The chloroanisoles are somewhat volatile and appeared to enter the chicken via air contamination. The chicken and eggs assumed a musty taint from the chloroanisoles.

Other sources of airborne contamination include unburned fuel in cooking ovens and emissions from forklift trucks and transport vehicles. These contaminants typically have a reasonably high sensory threshold so that contamination must be quite severe in order to cause an objectionable OF.

2. Waterborne Sources

Large amounts of water are typically used in most food processing operations. If this water contains significant quantities of OF components, the potential exists for the OF to be transferred into the food being processed. However, this is generally not a problem unless the water becomes a significant component of the food product. This may arise by reconstitution of the food with water prior to processing (e.g., spray drying of cheese or reconstitution of concentrated juices for retail sale) or from direct steam injection as a means of product heating. A second means by which OF may enter our food supply from tainted water is via absorption of OF by fish and shellfish.

Goldenberg and Matheson (1975) have summarized some of the OF problems they have encountered from water contamination in Table 4.2. It is interesting that chlorophenols again show up as the major source of problems.

There has been a substantial amount of work on OF in fish and shellfish due to water contamination (Reineccius 1979). Fish are very susceptible to absorbing chemicals from their environment resulting in an OF. The OF may come from water pollution or bacterial growth in the water.

Water pollution has been linked with various OF in fish. Vale *et al.* (1970) and Shipton *et al.* (1970) have reported a "kerosene" taint in mullet. The tainted mullet were found to contain high levels of hydrocarbons from polluted waters. Bemelmans and denBraber (1983) reported an iodinelike taint in herring from the Baltic Sea. They traced this OF to O-bromophenol. While the source of the O-bromophenol was not determined, it is possible that it came from industrial water pollution.

The ease with which chemicals are absorbed into fish (particularly

Table 4.2. "Off-Flavors" Due to Waterborne Contaminants[a]

Food	Contaminant	Method of Transmission
Canned fruits and canned vegetables	Chlorophenols	Chlorinated water used for cooling lithographed cans subsequently used as boiler feed water. Chlorine reacted with phenol-based lithographic print. Resulting chlorophenols volatized in steam, subsequently used for heating brines and syrups by direct steam injection.
Canned products	Compounds used in boiler-water treatments	Can cause "off-flavors" if live steam is used for direct steam-injection for heating.
Spray-dried cheese powder	Disinfectant?	Source of "off-flavor" water borne. Problem solved by use of activated carbon water filter.

[a] From Goldenberg and Matheson (1975).

fresh water fish) is a major problem facing aquaculture. While water pollution is not typically a problem, the growth of bacteria or tainted algae in the pond does present OF problems. Lovell and Sackey (1973) reported that catfish developed an objectionable "earthy-musty" taint within 2 days holding in tanks containing odor-causing algae. This taint is associated with the presence of *Anabaena, Ceratium and Pediastrum* algae (Nichols and Lacewell 1971; Brewer 1972). The off-flavor component could be geosmin since this compound has been identified in "muddy" tasting fish (Yurkowski and Tabachek 1974). Magligalig *et al.* (1975) demonstrated how quickly water odors can be transferred to fish muscle. They found an exposure of 10 min in 15 ppm 2 pentanone or dimethyl sulfide resulted in significant OF. This very rapid absorption of odorants makes fish very susceptible to picking up OF from their environment.

3. Disinfectants, Pesticides, and Detergents

Disinfectants, pesticides, and detergents comprise a group of chemicals which occasionally contribute to OF in foods due to inadequate care in their application. That is, they are present only because someone was careless in their usage or handling. Again, the summary of OF observed at Marks and Spencer (Table 4.3) by Goldenberg and Matheson (1975) provides a useful illustration of how these chemicals may enter a food and produce an OF. They noted several incidences of OF in food due to detergent contamination which are not included in Table 4.3. It is of interest that the Campden Food Preservation Research Association has a taint test program for chemicals applied to agricultural crops (Arthey 1974). The intent of this program is to provide agricultural chemical manufacturers with data to indicate whether their existing or new chemicals will potentially contribute a taint to food products. This program could reduce the incidence of OF in foods due to chemical residues.

4. Packaging Sources

Food packaging materials are frequent contributors to OF found in food products. Virtually the only food packaging material which will not potentially contribute an OF to foods is glass. Since food is not sold in glass ampoules (i.e., sealed entirely in glass) but needs a closure of some type, a small potential exists for food contamination due to this closure even for glass packaging.

Some occurrences of food contamination by packaging materials and resulting in OF are presented in Table 4.4. The contamination seldom comes from the major packaging component. Most often the OF is due

Table 4.3. Off-Flavors Due to Disinfectants, Pesticides and Detergents[a]

Food	Contaminant	Method of Transmission
Cake	Chlorophenol	Due to sultanas, contaminated on board ship by chlorophenol. On a previous voyage, holds had been used for cargo of herbicide containing 0.5% chlorophenol as impurity. Hold washed out but timber still impregnated. Finally timber parts had to be destroyed.
Spray-dried milk powder	Chlorophenol	"Off-flavor" of chlorophenol type: waterborne; use of activated carbon eliminated off-flavor.
Canned peas	Chlorophenol	Peas transported to cannery in open steel troughs on newly tarred road. Peas contaminated by tar-covered stones flying up from road. Peas then blanched in chlorinated water and chlorophenol formed.
Canned raspberries	Pesticide	"Catty" off-flavor may have been due to mesityl oxide in ketonic solvent.
Pork	Disinfectant	Preventative spraying with disinfectant in slaughterhouse against spread of swine vesicular disease.
Tomatoes	Pesticide	"Off-flavor" of chlorophenol type.
Smoked bacon	Disinfectant	Hypochlorite used as sterilizing agent; excess used and inadequately washed off. Chlorine combined with phenolic compounds present in smoked bacon to give chlorophenol. Bacon uneatable. Sterilizing agents containing chlorine should not be used where smoked meats or fish are handled, unless special precautions are taken.

[a] From Goldenberg and Matheson (1975).

Table 4.4. Off-Flavors in Foods Due to Packaging Materials

Food	Packaging Material	Source and Method of Transmission
	Paper	
Cakes, biscuits, chocolate[a]	Chipboard	Off-flavor due to either preservatives added to repulped waste paper or compounds in print in waste paper.
Doughnuts[b]	Coated paper divider plus cellophane overwrap	Inadequate removal of hydrocarbon solvents used to apply divider coating.
Cold cereals[c]	Glassine liner coupons	Residual solvents. Residual solvents in inks.
Chocolate cakes[a] Yogurt containers[b] Confectionery[a]	Coated paper—print on one side	Inadequate "curing" of inks used in printing.
	Plastic	
Orange and lemon drinks[a]	PVC bottle	Mercaptide stabilizer used in PVC—split off in processing of beverage.
Chocolate and lemon[d] cookies	Polystyrene trays wrapped in printed cellophane	Residual styrene monomer.
TV dinners[b]	Polystyrene trays	Residual styrene monomer.

[a] Goldenberg and Matheson 1975. [b] G. A. Reineccius 1983, unpublished.

to trace amounts of residual solvents in the packaging material. This may be related to the production of the packaging material (e.g., styrene monomer), trace contaminants in the packaging material (e.g., ethyl benzene in plastics or residual inks in recycled paper), solvents used in inks to print on the food package or solvents in adhesives. In some cases the contaminant responsible for the OF is a normal constituent of the packaging system but has not been adequately removed by the processing. This typically occurs due to "pushing" production too fast. The alternative is that the solvents may, on occasion, be contaminated with higher boiling constituents which are not removed by established process conditions.

A quick review of Table 4.4 indicates that virtually all food packaging materials (except glass) may contribute an off-flavor to food products.

Food	Packaging Material	Source and Method of Transmission
	Laminates	
Fruity soft drinks[d]	Polyester/aluminum foil/ polyethylene laminate	Residual toluene in laminate adhesive—faulty drying process.
	Metal–Paper	
Refrigerated doughs[b]	Paper tube with metal caps	Residual hydrocarbon solvents in triglycerides used to lubricate metal caps to facilitate packaging operation.
	Metal	
Canned pork products[a]	Lacquered can	"Catty" OF due to use of lacquer to cover sideseam to prevent blackening. Solvent used contained mesityl oxide which reacted with sulfur compounds in food to cause OF.
Refrigerated milk[b] beverage	Lacquered can	Trace contaminant (isophorone) in solvents used in lacquer which did not "flash off."
Beer[e]	Can	Lubricants used in can manufacture.

[c] Heydanek 1977; Heydanek *et al.* 1979. [d] Passey 1983. [e] Hardwick 1985.

However, OF due to packaging materials should be controllable by an adequate quality control program. The OF arising from packaging materials are nearly always due to volatile constituents occurring in the packaging material at reasonably high concentrations. High concentrations are necessary since the weight of packaging material to product is typically quite low. Therefore, the volatile OF constituents usually are in the packaging material at sufficiently high levels to be monitored by established quality control methods (Heydanek 1978).

B. Off-Flavors Due to Genetics or Diet

It is obvious to each of us that genetics and diet have a strong influence on the flavor of both plants and animals. We appreciate the differences in flavor found between the numerous varieties of apples, grapes, etc. We do not, however, accept differences in meat flavor due to either genetic or diet differences between animals. We have come to expect all meats (of the same species) to taste the same. If a steak tastes different, we consider it to have an off-flavor.

1. Genetics

Two animal species in which genetics plays an important role in producing an off-flavor are swine and ovine. Swine products (pork) occasionally contain an OF characterized as being perspiration or onionlike (Lerche 1936). This OF occurs primarily in sexually mature uncastrated hogs. It is of interest that 46% of the male population and 92% of the female population can smell this off-odor (Griffiths and Patterson 1970). The OF is caused by 5-α-androst-16-ene-2-one (Patterson, 1968b) and 3-hydroxy-5-α-androst-16-ene (Beery *et al.* 1969).

Like swine, the mature ovine animal is also considered to have an OF. This OF is called mutton flavor and is characterized as being "sweaty-sour" in nature (Wong 1975). The source of this OF has been traced to the synthesis of 4-methyl octanoic and 4-methyl nonanoic acids in the mature animal (Wong *et al.* 1975).

2. Diet

Animal diet can have a pronounced influence on the flavor quality of meats, especially in the nonruminant species. Earlier in this chapter, mention is made of musty taints in broilers due to the inhalation of chloroanisoles. While the wood litter was primarily responsible for chloroanisoles in the broiler, Bemelmans and Noever deBrauw (1974) also found chloroanisoles in the poultry feed. Thus the diet may also contribute musty OF to the poultry meat. Off-flavor may also enter poultry meat via the diet from the consumption of significant levels (>2%) of highly unsaturated fats (Crawford *et al.* 1975). For example, turkeys fed 2% tuna oil develop a "fishy" OF when cooked (Crawford and Kretsch 1976a,b). The fishy character was found to be due to oxidation of the highly unsaturated fatty acids during cooking of the turkey. Poultry very readily deposit dietary fat. Goldenberg and Matheson (1975) reported similar OF problems with the feeding of fish oil to hogs. They recommended limiting fish oil to less than 1% of the animal diet.

Off-flavors entering fish via the diet have already been partially discussed in this chapter. Mention was made of "muddy" taints being the result of the fish eating certain algae (*Anabaena, Ceratium* and *Pediestrum*).

Petroleumlike OF have been found in fish consuming *Limacina helicina* as a major part of their diet (Motohiro 1962). Earthy taints have been associated with the consumption of certain *Actinomyces* species (Thaysen 1936). In addition to OF arising in fish from certain algae in their diet, OF may also occur if the fish are fed turkey livers or cereal. Maligalig *et al.* (1973) found turkey liver notes in pond-reared catfish after 19 days of feeding livers and cereal notes in the fish following 33 days on a cereal diet. It appears that the maintenance of flavor in fish farming is a challenging task.

There are numerous studies showing how diet can produce OF in lamb (Reineccius 1979). Park *et al.* (1972a) reported on a "sickly" and "nauseating" aroma in lamb which had been fed rape. They postulated that the natural glucosinolates of the rape were metabolized to aliphatic sulfides, disulfides or mercaptans which caused the OF. Park *et al.* (1972b) also noted that feeding vetch had an influence on lamb flavor. The vetch-fed lambs had a sweeter and stronger meat flavor.

C. Off-Flavors Due to Chemical Changes in the Food

Foods contain a number of constituents which may either react with other constituents or simply degrade into OF components. The major reactions leading to OF include lipid oxidation, nonenzymatic browning, enzymatic changes and photocatalyzed reactions. This chapter will present some examples of OF arising in foods from these reactions.

1. Lipid Oxidation

Lipid oxidation is with few exceptions (e.g., American milk chocolate and deep fat fried flavor) a detrimental reaction. This reaction is probably the most common one leading to OF in foods during storage. While dry foods are most susceptible to OF from lipid oxidation, nearly all foods are susceptible to deterioration via this reaction. Lillard (1978) and Grosch (1982) have recently presented reviews of lipid oxidation as it relates to OF production. The review of Grosch (1982) is an exceptionally well done offering and is highly recommended to the reader for detail beyond what is presented here.

Lipid oxidation (LO) typically involves the reaction of molecular oxygen with unsaturated fatty acids via a free radical mechanism. The mechanism may be presented as follows:

Initiation

$$R_1H \longrightarrow R_1\cdot + H\cdot$$
$$R_1H + O_2 \longrightarrow R_1OO\cdot + H\cdot$$

Propagation

$$R\cdot + O_2 \longrightarrow R_1OO\cdot$$
$$R_1OO\cdot + R_2H \longrightarrow R_1OOH + R_2\cdot$$

Termination

$$ROO\cdot + R\cdot \longrightarrow ROOR$$
$$R\cdot + R\cdot \longrightarrow RR$$
$$ROO\cdot + ROO\cdot \longrightarrow ROOR + O_2$$

where RH is unsaturated lipid; R·, lipid radical; and ROO·, lipid peroxy radical.

One can see that this is a self-propagating reaction, i.e., once started it continues itself. It is interesting that an initial radical will produce about 100 hydroperoxides before its termination occurs (Bolland and Gee 1946).

Off-flavors arise from LO via secondary reaction products. Lipid hydroperoxides are very unstable and break down to produce short chain volatile flavor compounds. Keeney (1962) has shown some of the general reactions leading to flavor production:

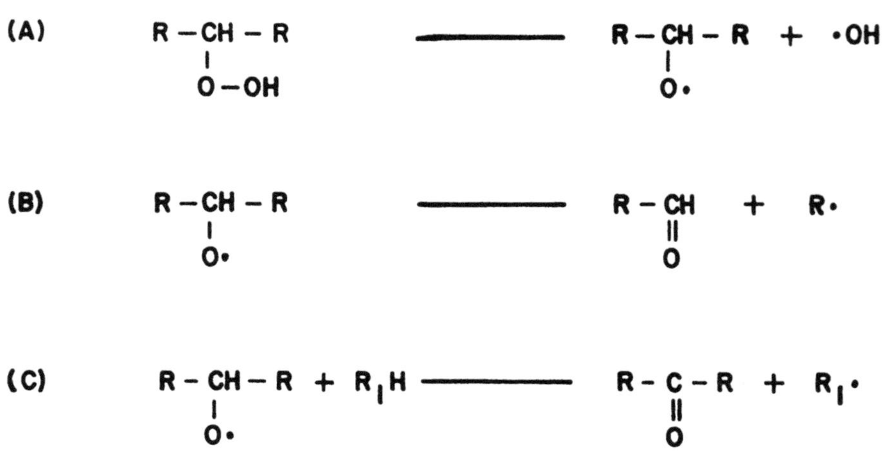

(D)

(E)

Steps B–E result in the production of short chain saturated or unsaturated aldehydes ketones and alcohols. These primary reaction products can undergo further oxidation if unsaturated, or secondary reactions to yield a host of off-flavor volatiles (see Fig. 4.1). These final products are generally aldehydes, ketones, acids, alcohols, hydrocarbons, lactones or esters. The unsaturated aldehydes and ketones have the lowest sensory thresholds and are, therefore, most often credited with being responsible for oxidized flavors. The metallic taint in oxidized butter, for example, has been attributed to oct-1-ene-3-one (Stark and Forss 1962).

An insight into the factors that enhance lipid oxidation will help one understand why different food products develop oxidized OF. The number and type of double bonds in a fatty acid have a strong influence on rate of LO. Fatty acids containing three or more double bonds are extremely prone to oxidation. In some food applications (e.g., dehydrated coffee whiteners), the linolenic acid must be selectively hydrogenated in order to permit an acceptable shelf life. The presence of any linolenic acid in the coffee whitener would limit shelf life to probably less than 3 months.

Trace metals, particularly copper, cobalt and iron, greatly increase the rate of LO and influence the direction of peroxide decomposition (Labuza *et al.* 1966). These metals function both to reduce the induction period and increase reaction rate by decomposing the hydroperoxides. As little as 0.3 ppm iron or 0.01 ppm copper will result in oxidized OF in soybean oil (Dutton *et al.* 1948). The iron may also come from heme in meats. This aspect of metal catalysis will be discussed in connection with warmed-over OF in meats.

There is substantial literature indicating a strong relationship between water activity and rate of LO (Labuza 1971). This relationship is shown in Fig. 4.2. One can see that very dry products or intermediate moisture products are very susceptible to LO. Dry foods, such as dry milk, potatoes, potato chips and dry cereals, are particularly susceptible to LO because they have substantial area for oxygen exposure, are permeable to oxygen and may be at water activities (less than the monolayer) such that oxidation rate is high.

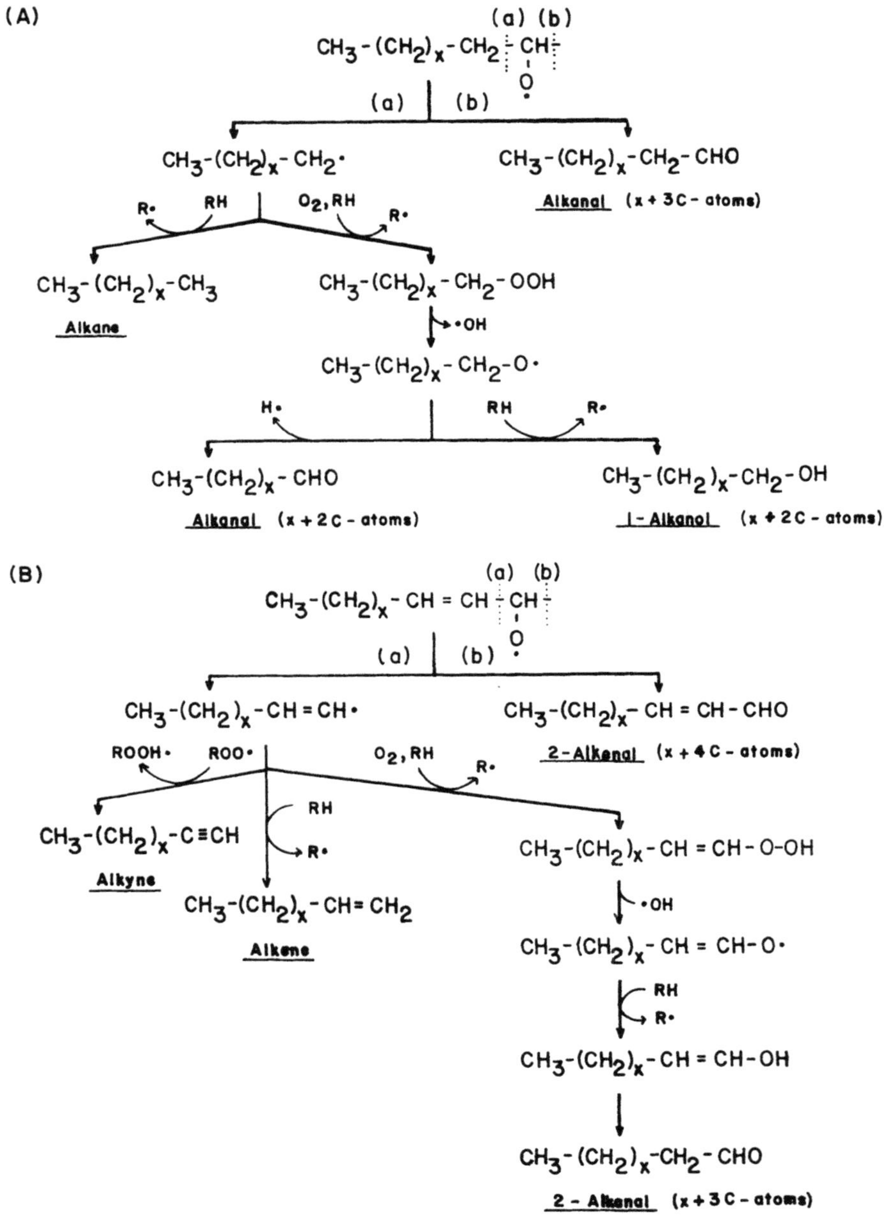

Fig. 4.1. Mechanism for the formation of volatile flavor compounds via lipid oxidation. (A) R′ is saturated. (B) R′ contains an ene system.

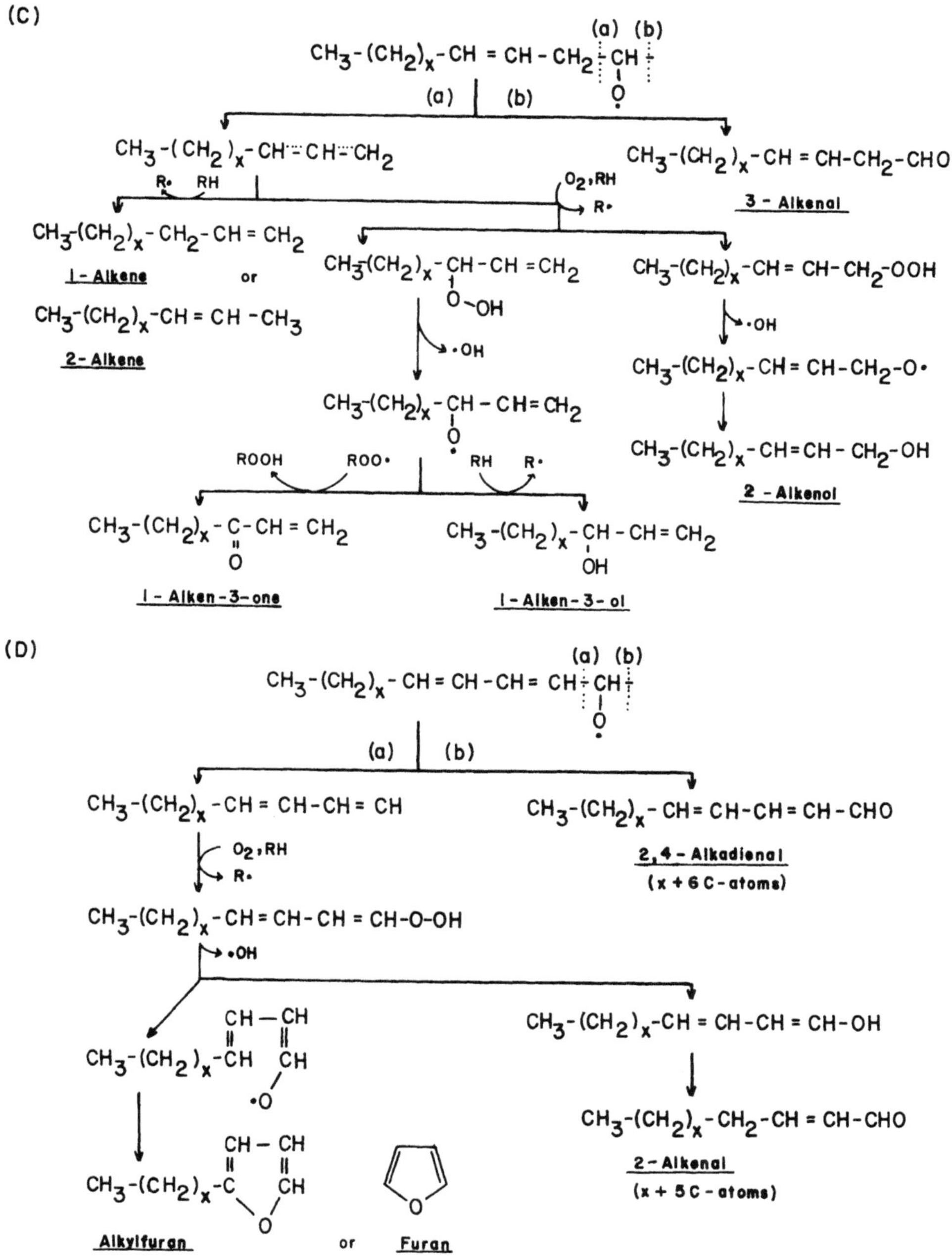

Fig. 4.1. *Continued.* (C) R′ contains an allylic system. (D) R′ contains a diene system.
Grosch 1982

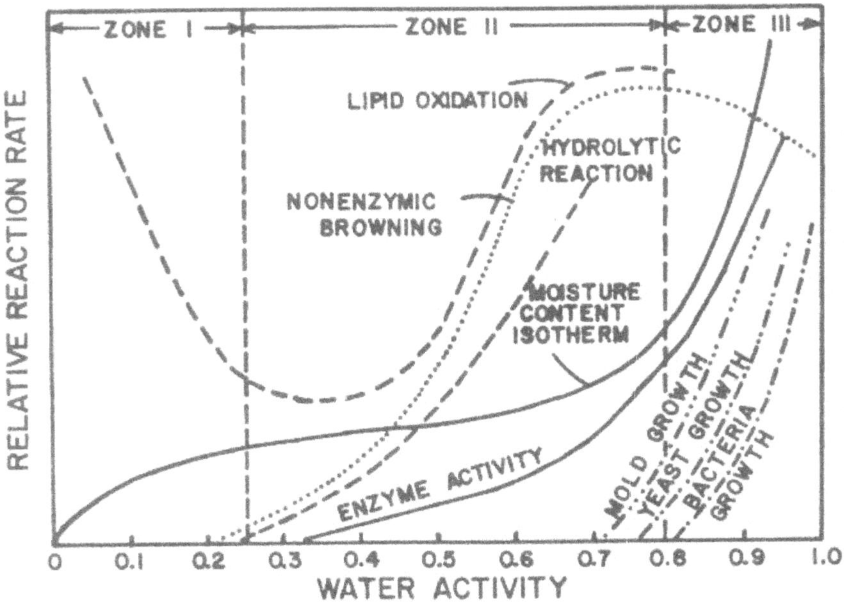

Fig. 4.2. Rate of lipid oxidation as influenced by water activity.
Labuza 1971

Lipid oxidation rate is temperature dependent. While LO may be initiated by enzymes (e.g., lipoxygenase), it most often is a "pure" chemical reaction responding to reaction temperature. The higher the temperature, the faster the rate of LO.

As discussed later, light is an effective initiator of LO. The shorter wavelengths are higher in energy and therefore, are more effective in enhancing the rate of LO.

A final factor determining rate of LO is the presence of antioxidants. Most of the vegetable oils contain some tocopherols which are quite effective antioxidants. The amount of natural (or added synthetic) antioxidant will influence rate of LO.

If one considers the factors that influence rate of LO, one can see that very dry products high in polyunsaturated fatty acids are most likely to develop OF due to LO. However, one must also recognize that virtually all foods contain sufficient lipid to develop OF via LO. It is only a question of whether some other mode of product failure occurs before LO becomes significant. It is a question of time.

While we usually think of LO in terms of dry milk, breakfast cereals, and snacks, occasionally, we do find fish and meats developing OF due to oxidation. Fish are very susceptible due to their very highly unsaturated fatty acids. Mendenhall (1972) reported significant oxidation in fresh mullet and bluefish after 3 and 5 days storage, respectively. However, most fresh fish spoil via microbiological means before oxidation becomes significant. Frozen fish have a sufficiently long shelf life that oxidation becomes the major mode of deterioration. While other meats are not nearly as susceptible to oxidation (nor bacterial spoilage), one occasionally finds oxidized frozen meats, particularly from foreign countries where the meat has been stored for several months during shipping, etc.

It should be mentioned that foods containing carotenoids also undergo oxidative deterioration. Carotenoids contain two end groups (often cyclic) joined by a series of conjugated trans double bonds. The oxidation of carotenoids yields highly unsaturated compounds plus cyclic derivatives of the carotenoids. Ayers *et al.* (1964) were the first to identify alpha-ionone, beta-ionone and beta-ionone-5,6-epoxide as OF components of dehydrated carrots. The ionones have a floral character and are typically used in the formulation of raspberry flavorings. If dehydrated yellow or red vegetables (containing carotenoids) are to be used in a food product, an antioxidant must be used to provide adequate shelf life. Off-flavor has occurred due to oxidation of the carotenoids in products such as dry carrot cake mixes and foods containing dehydrated tomato products.

It is also interesting that frozen tomato products, e.g., pizza, will undergo substantial carotenoid deterioration during the average distribution and storage system. It is in fact carotenoid deterioration that typically limits the shelf life of frozen pizzas.

Warmed-over flavor (WOF) is an OF associated with precooked meats (Tims and Watts 1958). This OF is generally accepted as being due to the oxidation of intramuscular phospholipid fatty acids. Oxidation proceeds at a very rapid rate following heating of the meat due to denaturation of heme proteins and destruction of cellular structure thereby permitting intimate mixing of cellular constituents. While the reaction is probably catalyzed by both heme and nonheme iron, the nonheme iron is considered to be of greater importance. WOF can be inhibited by the same methods that are used for lipid oxidation. This includes the use of metal chelators, primary and secondary antioxidants, vacuum packaging, etc. This subject has been reviewed in detail by Pearson *et al.* (1977).

2. Nonenzymatic Browning

While nonenzymatic browning (NEB) is very important to the production of desirable flavors (as was shown in section A of the previous chapter), it is also a primary source of undesirable flavors in foods.

Whether one wishes to say that canned peas have an OF due to the thermal processing is a matter of semantics. Canned foods do not taste like their fresh counterparts. The primary reason for a flavor change is NEB which occurs during the canning process. Virtually any food product which undergoes substantial heat treatment develops a flavor from NEB.

It is somewhat interesting that the browning reaction which occurs during storage is undesirable while that which occurs during heating (e.g., toasting of cereal) is desirable. Why is it that a loaf of bread does not get a better flavor as it is stored? Browning is occurring in the bread during storage and browning produced much of the characteristic flavor of the bread. The reason, of course, goes back to reaction kinetics. The NEB reaction is exceedingly complex with a host of possible pathways. The particular pathways active during NEB depend on many factors including water activity, reactants present, temperature and pH. The primary difference between developing desirable flavors via NEB as contrasted to developing OF during storage is temperature. At storage temperatures, the reactions leading to toasted, roasted, nutty and meaty flavor compounds are not favored and, therefore, these notes disappear from the product during storage. Unfortunately the NEB pathways leading to stale, sour green sulfury notes can and do progress at typical storage temperatures. Therefore, OFs develop in foods during storage due to NEB. Like lipid oxidation, nearly all foods have sufficient reactants (reducing sugars and amino acids) for NEB to occur, therefore, NEB will result in an OF unless some other mode of deterioration occurs before NEB becomes significant.

The OF due to NEB are most typically labeled as being "stale" in character. Stale is one of the more nebulous terms used in sensory evaluation. Stale is a rather nondescript term denoting a lack of fresh character. Benzothiazole and *O*-aminoacetophenone are two compounds believed to be formed via NEB and which contribute to stale flavor. These compounds were found in stale, dry milk by Parks *et al.* (1964). The *O*-aminoacetophenone along with some furans were also found to be partially responsible for the "gluey" flavor of old casein (Ramshaw and Dunstone 1969).

While OFs due to NEB are quite common, few studies other than those noted have been successful in identifying the components responsible for NEB staleness.

3. Photo-Induced Off-Flavors

Some quite unique OF may be produced via photo-catalyzed reactions in foods. "Skunkiness" in beer and "burnt feathers" in dairy products are two of the more unique examples of OF in foods due to light exposure.

Light-induced (activated) OFs in milk are again becoming more common in the United States. The OF due to light exposure at one time was very common. Milk was bottled in clear bottles and left on home doorsteps exposed to light. Changes in packaging (plastic-coated paper) and distribution (supermarkets) nearly eliminated this problem for several years. A recent change in packaging to clear or translucent plastic packages and brightly illuminated diary display cases has resulted in the revival of this OF. Activated OFs in dairy products are actually due to two concurrent chemical reactions in the milk: lipid oxidation and protein (amino acid) degradation (Shipe *et al.* 1978). Amino acid degradation produces a "burnt feather" OF while lipid oxidation produces rather typical metallic notes. The "burnt feather" or activated flavor initially predominates (2–3 days), however, metallic notes dominate later and the milk would be judged as "typical oxidized" after a few days of exposure.

Amino acid degradation is catalyzed by light and riboflavin as is shown:

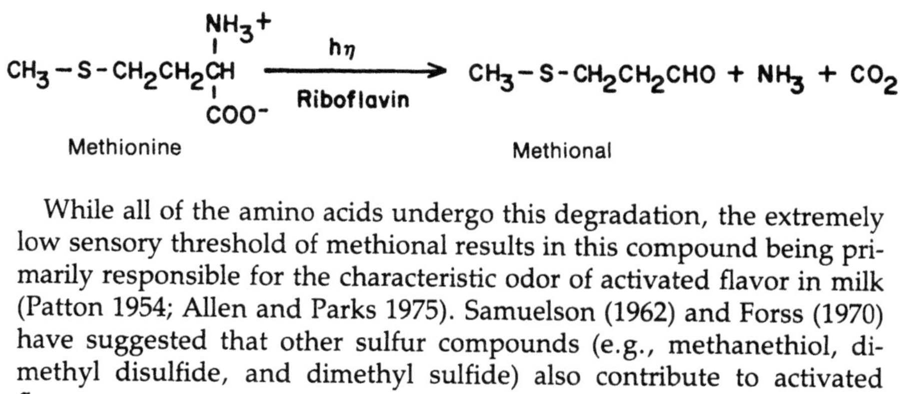

While all of the amino acids undergo this degradation, the extremely low sensory threshold of methional results in this compound being primarily responsible for the characteristic odor of activated flavor in milk (Patton 1954; Allen and Parks 1975). Samuelson (1962) and Forss (1970) have suggested that other sulfur compounds (e.g., methanethiol, dimethyl disulfide, and dimethyl sulfide) also contribute to activated flavor.

Off-flavors due to light exposure have also been reported in champagne (Charpentier and Maujean 1981) and beer (Kuroiwa and Hashimoto 1961). Tressl *et al.* (1980) have provided an excellent review of the "sunstruck" flavor in beer (as well as other OF in this beverage). Like dairy products, the photo-induced changes in beer flavor are complex involving photo-oxidation of lipids (later stages of the OF), photo-catalyzed decarboxylation and deamination of amino acids and hydrolysis of the isohumulones. Unlike dairy products, however, beer will take on an initial OF often characterized as being "skunky." This skunky OF is attributed to the photo-catalyzed degradation of the isohumulones

(Hashimoto 1976). Tressl *et al.* (1980) have provided a pathway illustrating the degradation of isohumulone (Fig. 4.3). Note that the 4-methyl-3-pentenoyl side chain of isohumulone is cleaved to yield compounds II (2,7-dimethyl-2,6-octadiene) and III (4-methyl-2-pentenal) as major end products. This pathway, in the presence of cysteine, has been shown to yield 3-methyl-2-butene-1-thiol (prenyl mercaptan) via free radical mechanisms. The prenyl mercaptan has a sensory threshold of 0.005 ppb in water and 0.05 ppb in beer. This mercaptan is responsible for the skunky OF. Control of this OF is typically accomplished by packaging the beer in light impermeable bottles or cans. The alternative is to use hops which have had the iso-α-acids (isohumulones) reduced to their saturated counterparts. Reduced α-acids will not form the objectionable thiols. They will, however, still undergo photo-catalyzed cleavage into "green" OF and result in a somewhat less objectionable OF (Tressl *et al.* 1980).

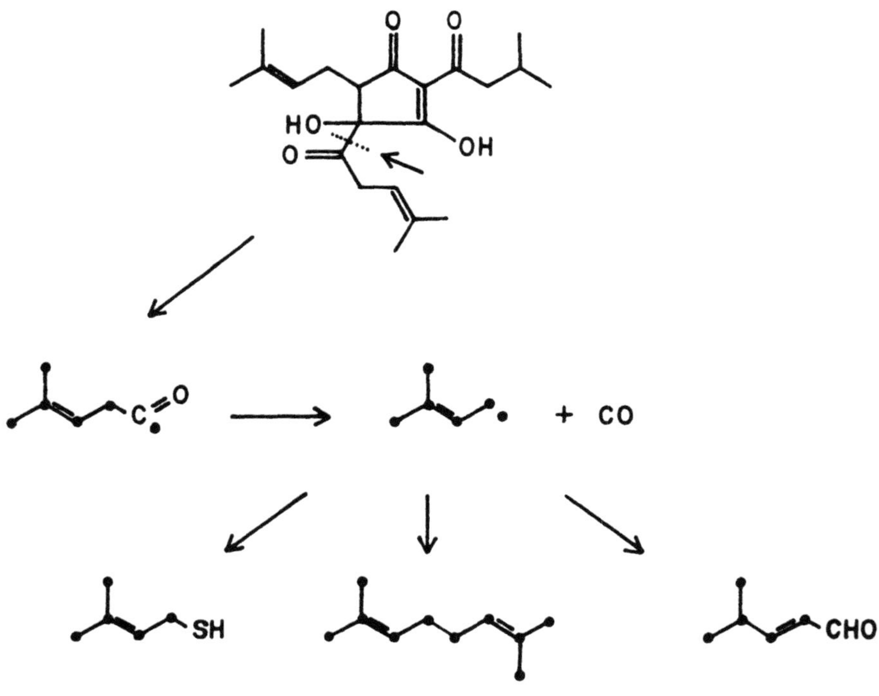

I II III

Fig. 4.3. Photodegradation of iso-α-humulones in a cysteine model system.
Tressl et al. 1980

4. Enzymatic Flavor Changes

One must remember that virtually all foods were once living organisms. These organisms depended on enzymes for metabolism and subsequent growth. Therefore, foods contain a variety of enzymes. While many are inactivated by processing or inactive due to inadequate conditions for functioning (e.g., low water activity, temperature, and pH), some foods do contain active enzymes which may cause OF to develop in the food product. The three enzymes most commonly associated with OF in foods are lipoxygenase, lipase and various proteases.

Lipoxygenase is an enzyme which will attack cis–cis double bonded, methylene interrupted fatty acids (most commonly linoleic and linolenic acids). This enzyme initiates lipid oxidation by abstracting a hydrogen radical from the fatty acids. Once lipid oxidation has been initiated by this enzyme, the process of oxidation follows the typical auto-oxidative process leading to OF which has been previously described in this chapter.

Lipoxygenase enzymes are quite common to plant tissues. The legumes (especially soybeans) contain substantial quantities of lipoxygenase enzymes. The beany flavor of soybeans is believed to be due to lipoxygenase activity in the bean once tissue damage has occurred (Rackis *et al.* 1972; Sessa and Rackis 1977; Kalbrener *et al.* 1974). The reversion flavor of soybean oil is also believed to be due to lipoxygenase activity in the oil. Hsieh *et al.* (1978) have shown 2-pentyl furan to be the oxidation product responsible for this OF.

Lipoxygenase also plays a role in the deterioration of frozen vegetables. Many vegetables must be blanched prior to frozen storage in order to inactivate lipoxygenase enzymes which otherwise result in oxidized OF (Whitfield and Shipton 1966).

Off-flavors may arise from lipase activity in foods.Lipolyzed flavor is due to the hydrolysis of fatty acids from the triglycerides. Since lipase is quite widely occurring in food products, one would expect to find lipolyzed OF more commonly than encountered. The major reason lipolyzed OF is not encountered more frequently is that most lipids in foods contain only mid- to long-chain fatty acids. These fatty acids (>12 carbons) are too large to make a significant contribution to either odor or taste. Therefore, the only foods where lipolysis produces significant OF are those which contain short chain fatty acids (e.g., dairy or coconut oil products).

The OF arising from lipase activity in dairy products is characterized as being "butyric," "goaty," and/or "bitter" in nature (Shipe 1980). There appear to be two lipases in dairy products: one associated with the fat globule membrane and the other with the proteins. While the point is controversial, it seems that these two lipases are identical but located in different parts of the milk. Milk lipases are highly specific for the primary positions of the triglyceride. Unfortunately, nature has also shown a preference to put the most sensorially significant fatty acids (i.e., butyric and caproic acids) in the primary positions.

The lipases of milk are rather unique in that they are inactive until activated by some treatment (the exception being rare that a cow will produce spontaneously rancid milk). Activation of the lipase enzyme system may be accomplished in several different ways. Occasionally, fluctuations in temperature such as cooling, warming and then cooling, will cause the lipase to become active. Mechanical agitation or foaming can also activate the lipase systems. This was a particular problem when pipeline milking systems were initially introduced. Homogenization of raw milk is especially effective in producing lipolyzed flavor. In a matter of seconds, the milk will become so acidic it will nearly burn the mouth. The milk fat globule membrane is inadequate to cover the new surface generated upon homogenization of the milk. This unprotected fat is particularly susceptible to lipase attack.

It is of interest that lipase activity can be influenced by the presence of certain chemicals (Chandran and Shahani 1964). Aureomycin, penicillin, streptomycin and terramycin were shown to provide 7.6–49.8% inhibition of lipase.

While dairy products may become lipolyzed due to indigenous enzymes, the lipases may also be from outside sources. Bacterial activity, spices and some fruits contain lipases which may attack the triglycerides of milk.

The other food product which is subject to lipolyzed OF is coconut. Coconut contains lauric acid (C_{12}) as a major fatty acid. Lauric acid tastes soapy when hydrolyzed from the triglyceride. While coconut does not contain a significant amount of lipases, coconut is often used in foods with other ingredients which do contribute lipases. It should be mentioned that lipases are reasonably heat stable enzymes. They will often survive a thermal treatment that one would expect to be adequate to denature the lipase.

One food product where "soapy" flavors are a problem is the pina colada beverage. Pina colada is an alcoholic drink consisting primarily of coconut and pineapple juice. The pineapple is an excellent source of quite heat stable lipases. If one encounters a "soapy" OF in a food product, the first step would be to search the ingredients for coconut oil and then a source of enzyme.

Bitterness in foods is an additional OF that may occur due to enzyme activity. In this case, we are considering bitterness arising from proteo-

Table 4.5. Bitter Peptides Identified in Protein Hydrolysates and Cheese Products

Soy globulin hydrolyzed with pepsin:
Arg–Leu, Gly–Leu, Leu–Lys, Phe–Leu, Gln–Tyr–Phe–Leu, Ser–Lys–Gly–Leu
Zein hydrolyzed with pepsin:
Ala–Ile–Ala, Gly–Ala–Leu, Leu–Val–Leu, Leu–Pro–Phe–Ser–Gln–Leu
Casein hydrolyzed with trypsin:
Gly–Pro–Phe–Pro–Ile–Ile–Val, Phe–Ala–Leu–Pro–Gln–Tyr–Leu–Lys
Cheddar cheese:
Pro–Phe–Pro–Gly–Ile–Pro
Pro–Phe–Pro–Gly–Pro–Ile–Pro–Asn–Ser
Leu–Val–Tyr–Pro–Phe–Pro–Gly–Pro–Ile–Pro

[a] From Arai (1980).

lytic activity on food proteins. Murry and Baker (1952) are often credited with being the first to determine that this bitterness was due to short chain peptides.

Bitter peptides have been found to produce OF in soy products (Fujimaki *et al.* 1968; Noguchi *et al.* 1970; Arai *et al.* 1970). Arai (1980) pointed out that most of these peptides have leucine at the C-terminal end. Otherwise there appears to be little similarity between peptide structures (Table 4.5). Zein (Wieser and Belitz 1975); casein (Matoba *et al.* 1969, 1970; Clegg *et al.* 1974; Huber and Klostermeyer 1974); and cheddar cheese (Hamilton and Hill 1974) have also been found to develop bitterness due to peptides.

Probably the largest amount of work on bitterness due to peptides has focused on aged cheeses (Dulley and Kitchen 1972; Harwalkar and Elliott 1971; Emmons *et al.* 1962; Harwalkar 1972; Richardson and Creamer 1973). Proteolytic activity in the cheese comes from the rennin, starter organisms and/or contaminating microorganisms. Rennin initially breaks a portion of the protein down into large peptides. These peptides then are further degraded into small peptides which may cause bitterness unless they are further metabolized by the starter organisms (Lawrence *et al.* 1978). Therefore starter failure or retardation via chemicals or contaminating organisms may end up causing bitterness in the cheese. The proper starter organisms and their growth is essential to minimize the possibility of bitterness during aging of cheese (Sullivan *et al.* 1973).

D. Microbial Off-Flavors

Microorganisms (MO) are a common source of OF in foods. If the product is not sterile (via heating or irradiation) or preserved via antimicrobial agents (e.g., humectants or chemical preservatives), microorganisms may grow in the product and produce OF.

Off-flavors may arise in food products via microbial sources in several different ways. The OF may be the result of a planned fermentation going wrong. The fermented product may then develop an OF. An example of this would be a starter culture failure in a cheese vat. Failure of the starter culture would result in poor acid production and the opportunity for the growth of other organisms. These contaminating organisms often result in bitterness and unclean OF. This type of OF development is possible in virtually all types of fermented foods. Only through very careful handling of the starter cultures and manufacture of the fermented products can this problem be minimized.

A second way in which microorganisms may produce OF in foods is somewhat indirect. Enzymes from lysed MO can catalyze reactions in foods leading to OF. Microbial enzymes are frequently quite stable to heat denaturation. Therefore a thermal process may kill the MO but leave active enzymes (e.g., lipases or proteases) which can produce OF (Shipe 1980; Tressl *et al.* 1980). One can appreciate that a food may develop an OF in this manner but yet show very low *viable* cell counts. The low viable cell count could mislead the investigator into believing that the OF was not due to MO.

The final means by which MO can produce OF in foods is the most common. That is from growth in the food and contamination with primary or secondary metabolic products. Fresh fish (Reineccius 1979) and dairy products (Shipe 1980) are two foods which are very susceptible to OF from microbial sources. Fresh fish is generally considered to have very little odor. However, if fish is stored above freezing temperatures, the fish develops a "fishy" OF initially which later becomes putrid and foul in character. Fish is particularly susceptible to microbial growth since the fish muscle has substantial soluble material and cells tend to lyse once the fish has been killed. The fish muscle provides a rather ideal growth medium for MO.

The "fishy" OF is primarily due to the generation of trimethylamine via bacterial action (Davies and Gill 1936; Stansby 1962). Trimethylamine is formed from trimethylamine oxide which is a natural constituent of fish muscle. This reduction is accomplished through bacterial enzymes and involves a coupled oxidation of lactic acid to acetic acid and CO_2 (Watson 1939). The latter stages of fish spoilage involve the production of various nitrogen- and sulfur-containing compounds. These compounds produce putrid sulfury notes in the fish.

It should be noted that different MO vary in their ability to produce OF in foods. The ability to produce an OF is dependent upon the metabolic compounds which are produced and excreted into the growth medium (food) by the MO. Work on OF in fish demonstrated this very nicely. Herbert and Shewan (1976) have shown that gram-negative bacteria of the genera *Pseudomonas, Achromobacter* and *Vibrio* are most often responsible for OF in fresh fish. Within these genera, only a few species can cause spoilage odors. Adams *et al.* (1964) found that typically less than 10% of the bacterial population present on fish can cause OF. This should make one cautious in concluding that high plate counts (or the converse) necessarily mean an OF came from microbial sources. It is the specific organisms and populations which are significant in determining the contribution to OF.

Shipe (1980) has reviewed some of the OFs which may occur in milk due to MO. Prior to wide usage of pasteurization and refrigeration, milk would most commonly spoil due to *Streptococcus lactis* growth. This would result in an acid sour OF. Pasteurization quite effectively kills *S. lactis* and their growth is not favored by storage at refrigeration temperatures so this organism (and OF character) is no longer a common means of spoilage.

Today the psychrotrophic bacteria are typically responsible for OF in milk. These MOs can multiply at or below 7°C so they will grow at typical refrigeration temperatures. Even though pasteurization destroys these MOs, they do produce substantial quantities of heat stable lipases and proteases which may produce OF later during storage. They also are common postpasteurization contaminants which will grow in the milk and produce OF due to metabolic activity. The OFs arising from psychrotrophic bacteria are often characterized as being unclean, foreign, fermented or bitter.

Two quite characteristic OFs found occasionally in milk are fruity and malty OFs. Fruity defects have been traced to the presence of *Pseudomonas fragi*. *Pseudomonas fragi* produces a lipase which initially hydrolyzes short chain fatty acids from the milk lipids and then esterifies these acids to their corresponding ethyl esters. Ethyl butyrate and ethyl caproate have been found to be the OF compounds responsible for this defect (Shipe *et al.* 1978).

Malty OF occurs occasionally in milk which has had some temperature abuse (storage at 10°C or above). The malty OF is caused by *Streptococcus lactic* subsp. *maltigenes* (Bradfield and Duthie 1966) or *Lactobacillus maltaromicus* (Morgan 1976). The malty character itself is due primarily to 3-methyl butanal production from leucine.

Additional examples of OF produced in foods from MO are provided by Goldenberg and Matheson (1975). A summary of OF they have encountered at Marks and Spencer is presented in Table 4.6. It is quite obvious from their tabulation that OF due to MO can occur in a wide variety of products.

Table 4.6. Off-Flavors Due to Microbial Spoilage

Food	Type of Spoilage	Comments
Cakes	Bacteria	"Off-flavor" due to "bad" egg; either use of poor quality egg or careless handling during or after thawing of frozen egg.
Cakes; angel cakes; apple tarts; marzipan	Mold	Mold due to contamination by flour dust in bakery.
Cakes	Lipolytic bacteria	"Soapy" off-flavor in coconut.
Apricot jam	Bacteria-producing butyric acid from sugar	Inadequate processing or inadequate cleaning or washing of equipment.
Fresh cream	Bacteria	Contamination by yoghurt.
Cream	Lipolytic bacteria	"Cheesy" off-flavor due to growth of lipolytic bacteria either in milk or during processing before pasteurization.
Cream	*Bacillus cereus*	Causes "bitter" flavors.
Chocolate tea cakes	Osmophilic yeasts	Caused fermentation and beery off-flavor.
Biscuits	Molds	Musty flour used and not examined before used.
Chicken	Molds	Wood shavings used as litter in chicken broiler-house made from wood preserved with chlorophenols. This metabolized to chloroanisoles and absorbed by chicken flesh.
Canned ham	Fecal streptococci	"Cheesy" off-flavors caused by fecal streptococci which survive heat process, especially in large 15-lb packs.
Wiltshire bacon	Bacteria	Development of "sulfide" off-flavor caused by contamination of bacon brines by foreign bacteria.
Minced beef pies	Lactic acid bacteria	Contamination of cooked beef mix by lactic acid bacteria.
Sausages	Lactic acid bacteria	Souring of sausages caused by selective growth of lactic acid bacteria in presence of sulfite.

[a] From Goldenberg and Matheson (1975).

E. Summary

This chapter briefly discusses some of the OF which have been found to occur in foods. Unfortunately, the occurrence of OF is quite common and the causes are very diverse. It is impossible to present a detailed discussion of all of the OF reported in the literature. In some cases, the source of the OF has never been found. Despite the sophisticated instrumentation available to the flavor chemist, identification of OF in foods is literally like "looking for a needle (one or two OF compounds) in a haystack (the normal flavor profile of the food)." When one appreciates the very low sensory thresholds one may encounter in OF flavor compounds [e.g., bis-(2-methyl-3-furyl)-disulfide, 2 parts in 10^{14} parts water; Buttery *et al.* 1984], the task of finding and then identifying the OF becomes very formidable. Hopefully this chapter provides a better appreciation for the diversity and the host of sources of OF in foods.

References

ADAMS, R., FARBER, I. and LERKE, D. 1964. Bacteriology of spoilage of fish muscle: 2. Incidence of spoilers during spoilage. Appl. Microbiol. *12,* 277.

ALLEN, C. and PARKS, O. W. 1975. Evidence for methional in skim milk exposed to sunlight. J. Dairy Sci. *58,* 1609.

ARAI, S. 1980. The bitter flavor due to peptides or protein hydrolysates and its control by bitterness—masking with acidic oligopeptides. *In* The Analysis and Control of Less Desirable Flavors in Foods and Beverages, p. 1, G. Charalambous, ed. Academic Press, New York.

ARAI, S., YAMASHITA, M., KATO, H. and FUJIMAKI, M. 1970. Applying proteolytic enzymes on soybean: Part V. A nondialyzable bitter peptide hydrolyzate of soybean protein and its bitterness in relation to the chemical structure. Agric. Biol. Chem. *34,* 729.

ARTHEY, V.D. 1974. Problems associated with the use of pesticides: Taints. Pestic. Sci. *5,* 113.

AYERS, J. E., FISHWICK, M. J., LAND, D. G. and SWAIN, T. 1964. Off-flavor in dehydrated carrot stored in oxygen. Nature *203,* 81–82.

BEERY, K. E., SINK, J. D., PATTON, S. and ZIEGLER, J. H. 1969. Characterization of the sex odor components in porcine adipose tissue. J. Amer. Oil Chem. Soc. *46,* 439A.

BEMELMANS, J. M. H. and DENBRABER, H. J. A. 1983. Investigation of an iodine-like taste in herring from the Baltic Sea. Wat. Sci. Tech. *15,* 105–113.

BEMELMANS, J. M. H. and TEN NOEVER DEBRAUW, M. C. 1974. Chloroanisoles as off-flavor components in eggs and broilers. J. Agric. Food Chem. *22,* 1137.

BOLLAND, J. and GEE, G. 1946. Kinetic studies in the chemistry of rubber and related materials: III. Thermochemistry and mechanisms of olefin oxidation. Trans. Faraday Soc. *42,* 244.

BRADFIELD, A. and DUTHIE, A. H. 1966. Vt. Agric. Expt. Sta. Bull., p. 645. Cited in Shipe (1980).

BREWER, F. 1972. Catfish experiments in Kansas. Kansas Fish and Game *29,* 5.

BUSER, H. R., ZANIER, C. and TANNER, H. 1982. Identification of 2,4,5-trichloroanisole as a potent compound causing cork taint in wine. J. Agric. Food Chem. *30*, 359–362.

BUTTERY, R. G., HADDON, W. F., SEIFERT, R. M. and TURNBAUGH, J. G. 1984. Thiamin odor and bis-(2-methyl-3-furyl) disulfide. J. Agric. Food Chem. 32, 674.

CHANDRAN, R. C. and SHAHANI, K. M. 1964. Milk Lipases. A Review. J. Dairy Sci. *47*, 471.

CHARPENTIER, N. and MAUJEAN, A. 1981. Sunlight flavours in champagne wines. *In* Flavour '81, p. 609, P. Schreier, ed. de Gruyter, New York.

CLEGG, K. M., LIM, C. L. and MANSON, W. 1974. The structure of a bitter peptide derived from casein by digestion with papain. J. Dairy Res. *41*, 283.

CRAWFORD, L. and KRETSCH, M. J. 1976a. The effects of cooking in air or in nitrogen on the development of fishy flavor in the breast meat of turkeys fed tuna oil with and without alpha-tocopherol supplement or injection. Fishery Bull. *74*, 89.

CRAWFORD, L. and KRETSCH, M. J. 1976b. GC–MS identification of the volatile components extracted from roasted turkeys fed a basal diet supplemented with tuna oil: Some comments on fishy flavor. J. Food Sci. *41*, 1470.

CRAWFORD, L., KRETSCH, M. J., PETERSON, D. W. and LILYBALDE, A. L. 1975. The remedial and preventative effect of dietary alpha-tocopherol on the development of fish flavor in turkey meat. J. Food Sci. *40*, 751.

CURTIS, R. F., DENNIS, C., GEE, J. M., GEE, M. G., GRIFFITHS, N. M., LAND, D. G., PEEL, J. L. and ROBERTSON, D. 1974. Chloroanisoles as a cause of musty taint in chickens and their microbiological formation from chlorophenols in broiler house litter. J. Sci. Food Agric. *25*, 811–828.

DAVIES, W. L. and GILL, E. 1936. Investigations on fishy flavors. Chem. and Ind. (London) *55*, 1415.

DULLEY, J. R. and KITCHEN, B. J., 1972, Phosphopeptides and bitter peptides produced by rennet. Aust. J. Dairy Technol. *3*, 10–11.

DUTTON, H. J., SCHWAB, A. W., MOSER, H. A. and COWAN, J. C. 1948. The flavor problem of soybean oil: IV. Structure of compounds counteracting the effect of prooxidant metals. J. Amer. Oil Chem. Soc. *25*, 385.

EMMONS, D. B., McGUGAN, W. A. and ELLIOT, J. A. 1962. Effect of strain of starter culture and of manufacturing procedure on bitterness and protein breakdown in cheddar cheese. J. Dairy Sci. *45*, 332–342.

ENGEL, C., DeGROOT, A. P. and WEURMAN, C. 1966. Tetrachloroanisole: A source of musty taint in eggs and broilers. Science *154*, 270–271.

FORSS, D. A. 1979. Review of the progress of dairy science: Mechanisms of formation of aroma compounds in milk and milk products. J. Dairy Res. *46*, 691.

FUJIMAKI, M., YAMASHITA, M., OKAZAWA, Y. and ARAI, S. 1968. Diffusible bitter peptides in peptic hydrolyzate of soybean protein. Agric. Biol.Chem. *32*, 794.

GOLDENBERG, N. and MATHESON, H. R. 1975. "Off-flavors" in foods, a summary of experience: 1948–74. Chemistry and Industry *5*, 551.

GRIFFITHS, N. M. and LAND, D. G. 1972. 6-Chloro-*o*-anisole taint in biscuits. Chem. Industry *2*, 904.

GRIFFITHS, N. M. and PATTERSON, R. L. S. 1970. Human olfactory responses to 5 alpha-androst-16-ene-3-one, principal component of boar taint. J. Sci. Food Agric. *21*, 4.

GROSCH, W. 1982. Lipid oxidation products and flavour. *In* Food Flavours, Part A: Introduction, p. 325, I. D. Morton and A. J. MacLeod, eds. Developments in food Science 3A, Elsevier, New York.

HAMILTON, J. S. and HILL, R. D. 1974. A bitter peptide from cheddar cheese. Agric. Biol. Chem. *38*, 375.

HARDWICK, W. A. 1985. Ultimate flavor influence on canned beer caused by lubricants employed in two-piece can manufacture. Brew. Dig. *60*(6), 22.

HARWALKAR, V. R. 1972. Characterization of an astringent flavor fraction from cheddar cheese. J. Dairy Sci. *55*, 735.

HARWALKAR, V. R. and ELLIOTT, J. A. 1971. Isolation of bitter and astringent fractions from cheddar cheese. J. Dairy Sci. *54*, 8.

HASHIMOTO, N. 1976. Rep. Res. Lab. Kirin Brewery *19*,1. Cited by Tressl *et al.* (1980).

HERBERT, R. A. and SHEWAN, J. M. 1976. Roles played by bacterial and autolytic enzymes in the production of volatile sulphides in spoiling North Sea cod (*Gadus morhua*). J. Sci. Food Agric. *27*, 89.

HEYDANEK, M. G. 1978. How to spot a stinker—predicting flavor effects of packaging materials. Food Prod. Devel. (June), p. 17.

HEYDANEK, M. G. 1977. Tracing the origin of off-flavors in a breakfast cereal. Anal. Chem. *49*, 901a.

HEYDANEK, M. G., WOOLFORD, G. and BAUGH, L. C. 1979. Premiums and coupons as a potential source of objectionable flavor in cereal products. J. Food Sci. *44*, 850.

HSIEH, O. A., HUANG, A. S. and CHANG, S. S. 1981. Isolation and identification of objectionable volatile flavor compounds in defatted soybean flour. J. Food Sci. *47*, 16.

HUBER, L. and KLOSTERMEYER, H. 1974. Isolierung und Identifizierung eines Bitterstoffes aus Butterkase. Milchwiss. *29*, 449.

KALBRENER, J. E., WARNER, K. and ELDRIDGE, A. C. 1974. Flavors derived from linoleic and linolenic acid hydroperoxides. Cereal Chem. *51*, 406–416.

KEENEY, M. 1962. Secondary degradation products. *In* Lipids and Their Oxidation, p. 79, H. W. Schultz, E. A. Day and R. O. Sinnhuber, eds. AVI Publishing, Westport, CT.

KUROIWA, Y. and HASHIMOTO, N. 1961. Composition of sunstruck flavor substance and mechanism of its evolution. Amer. Soc. Brew. Chem. *19*, 28.

LABUZA, T. P. 1971. Kinetics of lipid oxidation in foods. Crit. Rev. Food Technol. *2*, 355.

LABUZA, T. P., MALONEY, J. F. and KAREL, M. 1966. Autoxidation of methyl linoleate in freeze-dried model systems: II. Effect of water on cobalt catalyzed oxidation. J. Food Sci. *31*, 885.

LAWRENCE, R. C., HEAP, H. A., LIMSOWTIN, G. and JARVIS, A. W. 1978. Symposium: Research and development trends in natural cheese manufacturing and ripening cheddar cheese starters: Current knowledge and practices of phage characteristics and strain selection. J. Dairy Sci. *61*, 1181.

LERCHE, B. 1936. Geschlechtgeruch bei eberkastraten. Fleisch und Milch Hyg. *46*, 417.

LILLARD, D. A. 1978. Chemical changes involved in the oxidation of lipids in foods. *In* Lipids as a Source of Flavor, p. 68. M. K. Supran, ed. ACS Symp. Ser., Washington, DC.

LOVELL, R. T. and SACKEY, L. A. 1973. Absorption by channel catfish of earthy–musty flavor compounds synthesized by cultures of blue–green algae. Trans. Am. Fisheries Soc. *102*, 774.

MAARSE, H., NIJSSEN, L. M. and JETTEN, J. 1985. Chloroanisoles: A continuing story. *In* Topics in Flavour Research, R. G. Berger, S. Nitz and P. Schreier, eds. Eichorn, Marzling-Hangenham.

MALIGALIG, L. L., CAUL, J. F., BASSETTE, R. and TIEMEIR, O. W. 1975. Flavoring live channel catfish (*Icalurus punctatus*) experimentally: Effects of concentration and exposure time. J. Food Sci. *40*, 1242.

MALIGALIG, L. L., CAUL, J. F. and TIEMEIR, O. W. 1973. Aroma and flavor of farm-

raised channel catfish: effects of pond condition, storage and diet. Food Product Dev. (May), p. 86.
MATOBA, T., NAGAYASU, C., HAYASHI, R. and HATA, T. 1969. Bitter peptides in tryptic hydrolysate of casein. Agric. Biol. Chem. *33*, 1662.
MATOBA, T., HAYASHI, R. and HATA, T. 1970. Isolation of bitter peptides from tryptic hydrolysate of casein and their chemical structure. Agric. Biol. Chem. *34*, 1235.
MENDENHALL, V. T. 1972. Oxidative rancidity in raw fish fillets harvested from the Gulf of Mexico. J. Food Sci. *37*, 547.
MORGAN, M. E. 1976. The chemistry of some microbially induced flavor defects in milk and dairy Foods. Biotechnol. and Bioeng. *18*, 953.
MOTOHIRO, T. 1962. Studies on petroleum odor in canned chum salmon. Memoirs of the Faculty of Fisheries. Hokkaido University *10*, 1.
MURRY, T. K. and BAKER, B. E. 1952. Studies on protein hydrolysis I—Preliminary observations on the taste of enzymic protein hydrolysates. J. Sci. Food Agric. *3*, 470.
NICHOLS, J. D. and LACEWELL, R. D. 1971. A marketing system—the step beyond production. The Amer. Fish Farm. World Aquaculture News. 2, 18.
NOGUCHI, M., ARAI, S., KATO, H., and FUJIMAKI, M. 1970. Applying proteolytic enzymes on soybean: 2. Effect of Aspergillopeptidase A preparation on removal of flavor from soybean products. J. Food Sci. *35*, 211.
PARK, R. J., CORBETT, J. L. and FURNIVAL, E. P. 1972a. Flavor differences in meats from lambs grazed on either lucerne (*Medicago sativa*) or phalaris (*Phalaris tuberosa*) pastures. J. Agric. Sci. (Camb.) *78*, 47.
PARK, R. J., SPURWAY, R. A. and WHEELER, J. L. 1972b. Flavour differences in meat from sheep grazed on pasture or winter forage crops. J. Agric. Sci. (Camb.) *78*, 83.
PARKS, O. W., SCHWARTZ, D. P. and KEENEY, M. 1964. Identification of *O*-aminoacetophenone as a flavour compound in stale dry milk. Nature. *202*, 185–187.
PASSEY, N. 1983. Off-flavors from packaging materials in food products, some case studies. *In* Instrumental Analysis of Foods, Recent Progress, Vol. 1, p. 413, G. Charalambous and G. Inglett, eds. Academic Press, New York.
PATTERSON, R. L. S. 1968a. Catty odours in food. Their production in meat stores from mesityl oxide in paint solvents. Chem. Ind. London. p. 548.
PATTERSON, R. L. S. 1968b. 5-Alpha-androst-16-ene-3-one, compound responsible for taint in boar fat. J. Sci. Food Agric. *19*, 31.
PATTERSON, R. L. S. 1969. Catty odours in food: Confirmation of the identity of 4-mercapto-4-methylpentan-2-one by gas chromatography and mass spectrometry. Chem. Ind. London, p. 48.
PATTON, S. 1954. The mechanism of sunlight flavor formation in milk with special reference to methionine and riboflavin. J. Dairy Sci. *37*, 446.
PEARSON, A. M., LOVE, J. D. and SHORLAND, F. B. 1977. "Warmed-over" flavor in meat, poultry and fish. Adv. Food Res. *23*, 1.
RACKIS, J. J., HONIG, D. H., SESSA, D. J. and MOSER, H. A. 1972. Lipoxygenase and peroxidase activities of soybeans as related to the flavor profile during maturation. Cereal Chem. *49*, 586.
RAMSHAW, E. H. and DUNSTONE, E. A. 1969. Volatile compounds associated with the off-flavor in stored casein. J. Dairy Res. *36*, 215–223.
REINECCIUS, G. A. 1979. Symposium on meat flavor, off-flavors in meat and fish—A review. J. Food Sci. *44*(1), 12.
RICHARDSON, B. C. and CREAMER, L. K. 1973. Casein proteolysis and bitter peptides in cheddar cheese. New Zealand J. Dairy Sci. Tech. *8*, 46.
ROOZON, J. P. 1979. The bitterness of protein hydrolysates. *In* Progress in Flavour Research, p. 321, D. G. Land and H. E. Nursten, eds. Appl. Science Publ., London.

SAMUELSSON, E. G. 1962. Experiments on sunlight flavour in milk—S^{35}-labeled milk. Milchwissenchaft *17,* 401.

SAXBY, M. J. 1982. Taints and off flavours in foods. *In* Food Flavours: Part A. Introduction, p. 439, I. D. Morton and A. J. MacLeod, eds. Developments in Food Science 3A, Elsevier, New York.

SESSA, D. J. and RACKIS, J. J. 1977. Lipid-derived flavors of legume protein products. J. Amer. Oil Chem. Soc. *54,* 468.

SHIPE, W. F. 1980. Analysis and control of milk flavor. *In* The Analysis and control of Less Desirable Flavors in Foods and Beverages, p. 201, G. Charalambous, ed. Academic Press, New York.

SHIPE, W. F., BASSETTE, R., DEANE, D. D., DUNKLEY, W. L., HAMMOND, E. G., HARPER, W. J., KLEYN, D. A., MORGAN, M. E., NELSON, J. H. and SCANLAN, R. A. 1978. Off flavors of milk: Nomenclature, standards, and bibliography. J. Dairy Sci. *61,* 855.

SHIPTON, J., LAST, J. H., MURRAY, K. E. and VALE, G. L. 1970. Studies on a kerosene-like taint in mullet: 2. Chemical nature of the volatile constituents. J. Sci. Food Agric. *21,* 433.

STANSBY, M. E. 1962. Speculations on fishy odors and flavors. Food Technol. *16,* 28.

STARK, W. and FORSS, D. A. 1962. A compound responsible for metallic flavour in dairy products: I. Isolation and identification. J. Dairy Res. *29,* 173–180.

SULLIVAN, J. J., LYNETTE, M., ROOD, J. I. and JAGO, G. R. 1973. The enzymatic degradation of bitter peptides by starter streptococci. Aust. J. Dairy Technol. *3,* 20–26.

THAYSEN, A. C. 1936. The origin of an earthy muddy taint in fish: 1. The nature and isolation of the taint. Ann. Appl. Biol. *23,* 99.

TIMS, M. J. and WATTS, B. M. 1958. Protection of cooked meats with phosphate. Food Technol. *12,* 240.

TRESSL, R., BAHRI, D. and KOSSA, M. 1980. Formation of off-flavor components in beer. *In* The Analysis and Control of Less Desirable Flavors in Foods and Beverages, p. 293. G. Charalambous, ed. Academic Press, New York.

VALE, G. L., SIDHU, G. S., MONTGOMERY, W. A. and JOHNSON, A. R. 1970. Studies on a kerosene-like taint in mullet (*Mugil cephalus*): 1. General nature of the taint. J. Sci. Food Agric. *21,* 429.

WATSON, D. W. 1939. Studies on fish spoilage: 4: The bacterial reduction of trimethylamine oxide. J. Fish Res. Bd. (Canada) *4,* 252.

WHITFIELD, F. B. and SHAW, K. J. 1985. Analysis of off-flavors. *In* Progress in Flavour Research 1984, J. Adda, ed. Elsevier, Amsterdam.

WHITFIELD, F. B. and SHIPTON, J. 1966. Volatile carbonyls in stored unblanched frozen peas. J. Food Sci. *31,* 328.

WIESER and BELITZ. 1975. Z. Lebensm.-Unters. Forsch. *159,* 329. Cited by Arai (1980).

WONG, E. 1975. Mutton flavour. Food Tech. New Zealand. p. 13.

WONG, E., NIXON, L. N. and JOHNSON, C. B. 1975. Volatile medium chain fatty acids and mutton flavor. J. Agric. Food Chem. *23,* 495.

YURKOWSKI, M. and TABACHEK, J. A. L. 1974. Identification, analysis and removal of geosmin from muddy flavored trout. J. Fish Res. Board Can. *31,* 1851.

Part II

Flavor Technology

5

Flavors and Flavoring Materials

A. Flavorings in Foods

Food technology studies, of necessity, place considerable emphasis on plant and processing and technology and on the basic materials used in the making of foods and other comestibles. The purpose of this chapter is to draw attention to the considerable diversity of flavoring materials involved in food formulations and, in particular, to study the nature and range of such ingredients.

The first impression one has of any food or beverage is its appearance, including, of course, its color. This has the greatest impact on choice since visual attractiveness makes nutritionally equal products stand out one from another, whether they are in the pack or prepared ready for eating or drinking. Appearance is a valuable attribute right up to the time of ingestion (Hicks 1979). During eating or drinking, the volatile components are first inhaled through the nose thereby stimulating the olfactory receptors and then this effect is enhanced by volatiles released in the mouth during chewing and swallowing. Simultaneously, the basic taste factors (i.e., salt, sweet, acidic or bitter) are appreciated via the taste buds on the tongue, while the inner linings of the mouth and throat respond to tactile, temperature and/or chemical stimuli (i.e., texture, pungency, astringency, etc.). Only then is the complete flavor recognized. It is at this stage that the judgment passes from external to internal product attributes and the hedonic responses to odor, taste, flavor and other integrated sensations become paramount. It has long been recognized that flavor is an attribute which has no chemical or physiological basis for inclusion in the diet, yet without it one could well starve as the body cannot accept, over the long term, a continuously insipid diet.

In spite of misguided popular notions, flavor is a sensation induced by volatile and nonvolatile chemical compounds, whether these be natural or synthetic, present in what we eat and drink and in equilibrium at the time of ingestion.

The sources of flavor are (a) preexisting in the raw materials such as meats, fish, fruits and vegetables, which comprise the bulk of our diet and are nutritionally necessary; (b) developed from precursors in basic food ingredients by chemicals changes induced by the effects of heat during cooking; and (c) deliberately added in the form of concentrated flavorings or seasonings which may be either natural or artificial (Heath 1980).

Flavor in foods is not a static attribute but is in dynamic equilibrium and capable of significant changes stemming from the raw materials, processing conditions, packaging materials, handling and storage through to the final preparation of the product for consumption. One of the prime concerns of technologists involved in food product formulation and production is the recognition of these changes in order to make use of them wherever possible to improve the acceptability of the product or to eliminate or restrict those which may detract from the quality of the end product. All foodstuffs have a basic flavor profile derived mainly from items of low flavor intensity included for their nutritional or textural contribution. Many of these contain precursors and enzyme systems which may be activated during preparation and subsequent processing of the food product. Not all of these will result in desirable flavor attributes and the elimination of undesirable effects may dictate special processing techniques. Other components may be added to intensify desirable flavor characters, suppress undesirable flavor characters or impose a totally new flavor profile.

B. Added Flavorings

Some of the reasons flavoring materials are added to foods, beverages, confectionery, etc., are (a) to give flavor to an otherwise flavorless base; (b) to impart a different flavor character from that arising from the basic materials; (c) to enhance weak intrinsic flavors; (d) to replace natural flavor lost during processing; (e) to modify an existing flavor profile; (f) to disguise objectionable flavor characters; (g) to overcome seasonal availability or variations in natural flavoring materials; (h) to impart a flavor where the use of natural flavoring materials is technologically impracticable; and (i) to make available, at an economical price, the flavor of natural materials which are of limited commercial availability or are prohibitively expensive.

The nature of the flavoring materials selected and the quantity to be used to achieve a desired profile will depend on consumer expectancy in the end product—a factor which can demonstrate wide regional and ethnic group preferences. It is difficult to generalize, as there are too many variables arising from the raw materials, the composition of the product, manufacturing parameters, packaging materials, marketing requirements, legislative restrictions and consumer acceptability.

Flavorings may be entirely natural in composition, they may be a blend of natural with synthetic aromatic chemicals or they may be entirely compounded from permitted synthetic chemicals. Whatever the precise formulation, all flavorings must satisfy the following criteria: They must be (a) quite harmless to the consumer; (b) technologically and aesthetically compatible with the end product; (c) in strict compliance with all legislative requirements prevailing in the country in which the end product is offered for sale to the public; (d) readily handled in the food processing department; (e) capable of accurate dosage; (f) readily and uniformly dispersible throughout the product mix; (g) stable before, during and after incorporation into a product mix; (h) resistant to adverse handling and storage conditions; and (i) economically viable both to the producer and the user.

All flavorings, whatever their physical character, are composed of highly aromatic materials specifically selected to contribute a particular character or nuance to the odor and flavor profiles. The raw materials available to the creative flavorist include (a) naturally occurring plant materials, e.g., herbs, spices, vanilla, fruits, aromatic vegetables, etc.; (b) products derived directly from such natural sources by physical means (e.g., expression, extraction, distillation)—e.g., essential oils, oleoresins, extracts, tinctures, fruit juices and concentrates, (c) isolates or pure chemicals prepared from natural products—e.g., eugenol from clove leaf oil, citral from lemongrass oil, (d) synthetics chemically prepared from isolates or other natural products—e.g., vanillin from wood lignin; (e) synthetic chemicals identical with those found in natural products; (f) artificial aromatic chemicals which have not yet been identified in nature—e.g., γ-undecalactone, ethyl vanillin, etc; (g) flavor potentiators–e.g., MSG, ribonucleotides; and (h) taste modifiers—e.g., salt, sugars, or sweeteners, food acids or bittering agents.

C. Compounded Flavorings

The blending together of materials to create characteristic and acceptable flavorings, which satisfy the criteria already listed, is a highly skilled process based on long practical experience and a sound knowledge of the physicochemical and sensory properties of many hundreds of individual items recognized as safe for use in foodstuffs. Compounding of such physically dissimilar components as fruit juice concentrates, alcoholic extracts, essential oils, liquid or crystalline organic chemicals and either a solvent or carrier, demands a high level of technical competence in selection and handling.

D. Flavoring Materials

The physical characters of flavoring materials are very diverse, but principally they may be either solids, liquids or viscous extracts or pastes (Table 5.1).

In composition, flavorants may be either simple materials comprising one single ingredient (e.g., herbs and spices, organic chemicals), processed (e.g., essential oils, oleoresins) or compounded from several ingredients (e.g., imitation flavorings).

The physical form of such materials is of considerable importance to the food technologist as it usually determines the suitability or otherwise of a given product for its intended application. For instance, one cannot incorporate cinnamon bark into a product without prior comminution and even then the resulting powder may be incompatible with the end

Table 5.1. Physical Characters of Flavoring Materials

Solids	Semisolids and Pastes	Liquids
Whole and broken botanicals	Soft vegetable extracts	Aromatic distillates
Natural gums and exudates	Oleoresins (natural and processed)	Emulsions
Dehydrated aromatic vegetables	Resinoids	Essential oils and secondary products including: concentrated oils, rectified oils, terpeneless oils, sesquiterpeneless oils, compounded oils and blends, distillation fractions and isolates
Crystalline and noncrystalline chemicals	Concretes and absolutes	Fluid oleoresins
Powdered versions of these materials		Fluid extracts and concentrates
Powdered processed products including: vacuum-oven dried, fluid-bed dried, freeze-dried, spray-dried		Fruit juices and concentrates
Plated or dispersed spice oils/oleoresins		Imitation flavorings
Encapsulated flavorings		Infusions
		Soluble essences
		Spirits

product. Similarly, there are considerable technological difficulties in incorporating a liquid flavorant into a dry powder product mix. A knowledge of the physical nature of flavoring materials is, therefore, necessary in making a choice of product ingredients.

1. Solid Flavoring Materials

Solids may be either crystalline or amorphous. Many organics used as flavorants are crystalline and whereas these are readily handled when dissolved in a solvent in the making of liquid flavorings they must usually be finely powdered prior to incorporation into a dry mix (e.g., vanillin or ethyl vanillin in a pudding mix, MSG in a snack food dusting).

Most natural plant materials, as handled commercially, consist of whole or coarsely broken parts or tissues (e.g., barks, rhizomes, seeds). These are rarely suitable for direct use in food products. The few exceptions include the use of whole peppercorns in certain meat products, of mixed whole spices in various pickles, caraway seeds in cakes and poppy seeds as a surface dressing on baked goods. Most of these applications are traditional and desirable to give a product eye appeal. Prior to use, botanicals must be milled and reduced in particle size. This may range from the coarsely broken herbs right down to microfine powder of about 50 μm. The particle size is determined by the end use. In this form, they are readily handled on the production floor, can be accurately weighed as appropriate and pose no problems of uniform dispersion in a dry product mix although they may not be so easily dispersed in a "wet" product.

Powdered botanicals have the following disadvantages: (a) They are usually hygroscopic and readily pick up atmospheric moisture, and, as a result, after relatively short periods of storage, particularly under poor packaging and storage conditions, they may agglomerate and form intractable lumps. Under extreme conditions they become moldy. (b) The more aromatic materials (e.g., herbs and spices) are subject to flavor loss and degradation. (c) They can be dusty and irritating to handle and may pose problems of plant hygiene such as the adherence of dust to ledges. (d) They frequently carry very high levels of bacteria, mold spores, etc.

Other powdered flavoring materials include the following: (a) dehydrated vegetables (e.g., onion and garlic powders); (b) dried and powdered plant extracts made by various techniques including vacuum oven drying, fluid-bed drying, spray drying and freeze drying (e.g., coffee extract, citrus fruit powders); (c) encapsulated flavorings made by spray drying single or blended concentrated flavorants and encapsulating either in natural hydrocolloids (e.g., gum acacia) or permitted modified food starches; and (d) plated or dispersed flavorings (usually herbs and spices) comprising single or blended essential oils and/or oleoresins spread onto a dry edible carrier (e.g., salt, dextrose).

With the exception of the spray-dried encapsulated products, which have an extended shelf life, all the other powdered products are subject to volatile loss during storage, unless precautions are taken to avoid or limit this loss. Such products should be kept in well-filled, tightly closed, impervious containers stored in a cool dry place. All too often such materials are left open in factory stores and/or the processing department and are exposed to a warm, moist ambient atmosphere. Under such conditions, volatile losses are significant and the product has a very limited effective shelf life.

2. Liquid Flavoring Materials

When it comes to food processing, the liquid flavorants offer very many advantages, particularly in ease of handling, accurate dosage by weight or volume and ready dispersion in a "wet" product mix. Generally they have excellent stability and are designed to have a long shelf life. They are free from problems of hygiene and, with few exceptions, are not dangerous to plant operatives.

Almost all of the essential oils are fluids having various degrees of viscosity. Even those which are solid at room temperatures melt when gently warmed. Some of the oleoresins prepared from herbs and spices are oily liquids owing to their high content of fixed oils. Most liquid flavorings are compounded from various aromatic materials and blended together by the use of one of the permitted food solvents (e.g., ethanol, isopropanol, propylene glycol) to achieve the desired profile and level of flavor.

Ethanol is widely accepted as the solvent of choice but, in most countries, this attracts varying levels of duty which may make the resulting flavoring unacceptably expensive when compared with similar products based on one of the nondutiable solvents. Ethanol is the solvent used in the manufacture of vanilla and other essences as required by U.S. standards.

Of increasing importance are flavoring emulsions. These have been used for very many years in the soft drink industry and their application to other product groups is spreading as the food industry becomes more automated. They comprise blends of flavorants (e.g., essential oils, oleoresins, aromatic chemicals) required to produce the desired flavor profile emulsified by the use of natural gums, dextrins or permitted emulsifying agents and/or stabilizers and homogenized to form a stable, readily pourable, oil-in-water emulsion. Without preservatives such products are unstable and liable to ferment so that it is necessary either to incorporate a permitted preservative, adjust the product to a low pH where it is self-preserving or to pasteurize the product in the final pack. In this latter case the contents must be used within a very short time of opening. Most of these products are best stored under chilled conditions until required. Emulsions offer several marked advantages over other liquid flavoring materials. They are readily handled, capable of accurate dosage by weight or volume and are particularly suitable for automatic metering, are very easily and uniformly dispersed in the aqueous phase of "wet" products and are hygienically excellent. In meat products, the spice emulsions osmotically permeate the meat fiber matrix and ensure a better degree of seasoning than can be achieved with ground spices or plated, dispersed or encapsulated spice extracts.

3. Semifluid or Paste Flavoring Products

The aromatic and other flavoring components of herbs and spices are readily extracted by organic solvents (e.g., acetone or hexane), chlorinated hydrocarbons (e.g., methylene dichloride) or by supercritical gases (e.g., carbon dioxide). In all cases the comminuted spice is exposed to the solvent, the miscella separated from the exhausted solid matter and the solvent completely removed by controlled vacuum distillation. The resulting product has a physical form ranging from a viscous oil (e.g., oleoresin dill) to a viscous butterlike paste (e.g., nutmeg butter) to a hard solid (e.g., deoiled turmeric extract). These forms are not suitable for incorporation directly into a food product mix owing to the difficulty of accurate dosage and the dangers of errors. Various methods are employed to make these valuable flavoring materials more acceptable to the food processor. These include dispersion onto an edible carrier, solution in a permitted food solvent, emulsification or encapsulation. All of these diluted products can be formulated for specific applications and their flavoring quality and strength as well as their application characteristics may be modified to suit the processing parameters involved.

The aromatic vegetables (e.g., onion, garlic) are amenable to extraction with water and consequently a range of thick viscous extracts, based on the evaporation of the expressed juice and having a strongly cooked flavor character, is available commercially. A more recent development involves the liquid/liquid extraction of crushed onion with an immiscible organic solvent; the resulting water-miscible fluid concentrate has the full fresh character of raw onions, is convenient to handle and easy to measure, and mixes readily and uniformly with water.

Tomato paste also finds wide use as a flavoring material. This is a very specialized product which is manufactured in the producing region.

E. The Flavor Industry

Having briefly reviewed the nature of flavoring materials it is now necessary to establish their source and availability to food processors. Although certain materials, notably herbs and spices and imported essential oils, may be obtained from importing agents and brokers, the major needs of the food industry for such items are met by what may be broadly described as the flavor industry. The processing of foods, beverages, sugar confectionery and other comestibles is deep rooted in history whereas the flavor industry has developed only over the past 150 years. Its beginnings were small both in the United States and European countries and comprised numerous small family companies which specialized in the importation, processing and marketing of natural botanicals, principally spices, vanilla beans, vegetable drugs, gums and resins, using these for the distillation of essential oils, extraction and preparation of pharmaceuticals and, to a very limited extent, the isolation of pure aromatic chemicals from natural sources (e.g., eugenol and then vanillin from clove oils). In those days, the main product lines were pharmaceuticals (galenicals) as required by national pharmacopoeias, essential oils used mainly as fragrances and ground spices for seasonings; flavorings as we know them came much later. The early history of the United States flavor industry has been well described by Dorland and Rogers (1977). In the United Kingdom and other European countries the flavor industry developed along similar lines but rapidly became involved in business in America where it became firmly established (Pisano 1973). From Europe, essential oils, compounded flavorings and fragrances were increasingly exported around the world as countries were opened up for trade. With the later setting up or expansion of indigenous food processing operations early in this century, particularly in the United States, many of the larger European flavor houses established local manufacturing facilities in order to serve this growing industry. Although very many of the smaller family companies have gone out of business, several of the original pioneering companies have survived and today have complex international operations offering a wide range of flavoring materials to meet the needs of the equally complex food and related industries as well as fragrances for the toiletries and cosmetic industries. Today the flavor industry is dominated by several very large companies which are mainly prime manufacturers of natural products and/or synthetic aromatic chemicals which are used for the compounding of an almost limitless range of flavoring and fragrance products. There are, of course, many much smaller companies in the same business of compounding flavorings and one has only to scan the list of exhibitors at the annual meeting of the Institute of Food

Technologists (IFT) to realize that these cover the whole of the United States and Canada. Many serve only a limited area or provide flavorings and seasonings to a particular branch of the food industry or maybe handle only a restricted range of specialty products. All ultimately aim to satisfy the demands of the consumer who is still the final judge of the success or failure of any given product.

Some of the very big food processing companies have established or acquired their own flavor compounding departments or companies which not only serve their own immediate needs but also compete on the open market. In today's economic climate the cost effectiveness of such direct involvement perhaps has to be questioned.

The headquarters establishment of the biggest flavor and fragrance houses, many of which are based in Europe, have extensive research and development facilities and their subsidiaries in the United States and other major markets throughout the world are similarly supported either by parallel laboratories or by the dissemination of technical data via computer banks, etc. The aim of this effort is to provide an advanced basis for continuing business and an effective technical service to their end-user industries.

F. Relationships between the Food and Flavor Manufacturers

Flavor plays an essential role in food product acceptance and consumers demand a consistency of flavor in the products of their choice, yet the food manufacturer is expected to provide a sufficient variation in product concepts to allow for changes in eating patterns. This is frequently achieved by providing a range of flavors within a product group. In consequence, flavoring ingredients are in great demand and are certainly the most numerous single group of intentional food additives. In 1971 the market was valued at $227 million, of which approximately 58% were compounded flavorings, with an annual growth rate of about 6%. This represents about 40% of the total ingredients market. Though forecasted figures are unreliable owing to the complexity of the industries concerned and the difficulty of assessing changing consumer demands with any degree of accuracy, Schubert (1978) calculated that in 1985 the estimated sales of flavoring materials would reach $635 million comprising synthetic flavorants, $180 million; natural flavorings, $170 million; vanilla, $70 million; and raw flavoring materials, $215 million.

The manufacture of flavorings demands an expertise based on a sound knowledge, selection and quality control of raw materials originating from all over the world, process technology and quality assurance within the factory, an understanding of related end-user technologies and the processing parameters likely to be met in the production of food, soft drink, bakery, confectionery and several other products which include flavorings. In addition, the flavor manufacturer must be fully aware of current legislation not only in the country of manufacture but particularly in the country in which the goods are offered for sale to the public. He must ensure that flavoring products comply in all respects. To service the food industry efficiently, the flavor manufacturer has to employ a sales force which is capable of presenting essentially technical products in a professional and well-informed manner. Although there is some disagreement, most flavor houses employ technologists who have had a training course in sales rather than nontechnical representatives who may have considerable difficulties in handling technical problems associated with product applications. Not only must the flavor manufacturer offer products which are uniform, able to withstand adverse storage and handling conditions and produce the desired flavor in the end product, but the manufacturer is also frequently called on to offer technological assistance to ensure the products' most effective application.

Over the past 50 years, relationships between the food and flavor industries have become very close. But competition in both is active and, of necessity, transactions are still carried out in the strictest confidence. Fortunately, much of the deep secrecy which once existed has now largely been replaced by a relationship based on mutual cooperation and respect between the supplier and user of flavorings ingredients.

This respect is fostered by close attention to technical collaboration and most flavor manufacturers have established departments staffed and equipped specifically to give expert advice on the choice, suitability and effectiveness of particular flavoring ingredients. They frequently offer direct technical assistance to the end user in the development of a given food product concept. This technical service generally involves a frank and practical dialogue between the technical staffs of the companies concerned, although other commercial interests of both may also be involved. It is the policy of the flavor industry to respect user-company organizational requirements and not to take shortcuts to achieve product establishment.

The provision of technical service calls for a very considerable commitment on the part of the flavor industry and a realization of both its benefits and limits by the user industry. Until quite recently, the food industry tended to expect the flavor industry and other ingredients suppliers to provide a free, unlimited, unconditional technical service involving not only the recommendation of suitable flavorings but also total new product concepts and development. Present day economic reality has caused the industry to rethink the cost effectiveness of this service and most suppliers now provide technical assistance only if it has a real prospect of profitable flavoring business in a successful end product.

G. Selection of Flavoring Materials

In the process of development, the ultimate flavor of the projected food product must be among the first things to be decided. Initially, the precise aroma and flavor profiles may not be known and hence cannot be defined but the flavor type, expressed in broad and clearly understandable terms, is most important as this forms the basis for discussions and ultimate briefing of the flavoring supply houses. Obviously, the profile may be dictated by the other raw materials used, the method of manufacture and the physical character of the end product, but it is part of the development process to evaluate this basic data and to select and apply flavorings in successive evaluations until the desired profile is achieved. Success is necessarily dependent upon the number and quality of flavoring submissions received. If food manufacturers are to make best use of the expertise and facilities of the flavor industry then there must be an efficient system of communication that encourages the maximum interchange of ideas without endangering the competitive nature of new product development. This implies a commitment on the part of both user and supplier.

The flavor industry, on its own initiative, is constantly creating new flavorings based on current research information or updating old flavor formulations to satisfy changing user and/or consumer demands. Clark (1972) examined the reasons why so few new flavorings ever find their way into nationally marketed products and concluded that the major contributors to a low success rate are the following factors:

a. Poor and often complicated lines of communication exist between user and supplier. This is aggravated in many of the bigger companies by the involvement of technical/nontechnical staffs with an inadequate knowledge of each others' area of expertise and only a limited common vocabulary to enable the effective transmission of needs on the one hand and available products on the other.

b. Flavorings submitted may not be suitable for the product being developed. This is usually the result of reticence on the part of the food industry to divulge any new product concept for fear of endangering its commercial success. Such attitudes are understandable but they do leave the flavor supplier guessing which of the many products available will most suit the end product.

c. Flavorings may be incompatible with product base. Again, this is due to a lack of confidence in the supplier who can only advise on an optimum flavoring when all the basic product data has been made available.

d. New flavorings far exceed specific end-product needs so that the supplier may submit several varieties of a flavor type (e.g., fresh strawberry, ripe strawberry, green strawberry, wild strawberry) in order to give the product developer a choice. This hit-or-miss approach to sampling, which results from inadequate briefing, involves a considerable extra load of testing which is time consuming and costly. It can easily be avoided by a precise profile and application brief to the flavor manufacturer. Most food manufacturers are reluctant to evaluate unsolicited flavorings unless these represent a significant technological advance or a unique flavor profile, in which case the supplier should submit them in a ready-to-eat end product.

e. Samples are not evaluated, sometimes because they do not reach the food technologist who is actively concerned in the development project. Overhurried development programs may result in an early but less acceptable flavoring submission being approved, leaving later and perhaps better submissions unevaluated.

f. The food industry and consumers are very conservative and tend to prefer products and flavors with which they are familiar. The introduction of a new product concept or a new flavoring experience usually requires considerable marketing effort to get the product established. Many excellent new products never reach the marketplace because of the high costs of a successful product launch.

g. Users often have a decided loyalty to one supplier who has given good service in the past and tend to approach only that supplier when new flavorings are required.

Relationships between the technical and marketing staffs of food manufacturers and the flavor industry have become significantly closer over the past two decades. Food technologists and flavorists are often members of the same technical institutes and societies and hence meet each other at regular intervals outside the factory environment. There is an increasing understanding of each other's basic technology, and, despite difficulties, a reasonable vocabulary has been adopted for the interchange of ideas, information and data on the sensory properties of both ingredients and end products. The need for the strictest confidence in all these intercompany relationships is now fully recognized and respected and this mutual trust has facilited the exchange of relevant data and close practical collaboration particularly in the early stages of new product development and later in product assessment. This effective interplay of expertise results in a greater confidence in the ultimate commercial success of the new product.

References

CLARK, K. J. 1972. Selection and evaluation of flavourings for foods. Papers Inst. Food Sci. Technol. (UK) Symp., Birmingham, April, 1972.
DORLAND, W. E. and ROGERS, J. A., JR. 1977. The Fragrance and Flavor Industry. Wayne E. Dorland Co., Mendham, NJ.
HEATH, H. B. 1981. Flavours in food products—art or science. Royal Soc. Health J. *101* (Feb.), 6–12.
HICKS, D. 1979. Benefits of added colourings in foods and drink. Int. Flavour Food Additives *10*, 31–32.
LINDSAY, R. C. 1984. Flavor ingredient technology. Food Tech. *38*(1), 76–81.
PISANO, R. C. 1973. The American flavor industry—its growth and development and relations with government. Flavour Ind. *4*, 384, 386, 388.
SCHUBERT, F. W. 1978. Growth of an industry. *In* Flavor: Chemical, Behavioral and Commerical Aspects, pp. 167–174, Westview Press, Boulder, CO.

6

Flavoring Materials of Natural Origin

A. Natural Flavors and Flavorings: Definitions

The Code of Practice of the International Organization of the Flavor Industry (IOFI), of which the Flavor and Extract Manufacturers' Association of the United States (FEMA) is a member body, defines *flavorings* as "concentrated preparations, with or without flavor adjuncts required in their manufacture, used to impart flavor. They are not intended to be consumed as such." It further states that flavorings may be made from the following ingredients (plus others of synthetic origin):

Natural aromatic raw materials. Vegetable and/or animal products used for their flavoring properties, either as such or processed for human consumption.

Natural flavors. Concentrated preparations obtained exclusively by physical means from natural aromatic raw materials.

Natural flavoring substances. Substances isolated from natural aromatic raw materials exclusively by physical means. Those biological processes which occur spontaneously and roasting can be regarded as physical processes within this definition.

Flavoring materials of natural origin described in this chapter embrace all these various forms of products which are used in foods and other products intended for human consumption either as such or after processing or cooking.

B. Sources of Natural Flavoring Materials

Plants and parts of plants which have been recognized for a very long time as of value in foods fall into five categories:

1. Those which are nutritious but have little intrinsic flavor in the raw state but may be made more flavorful by cooking (e.g., cereals, meats).
2. Those which have limited nutritional value and a relatively low level of flavor; these can be eaten raw or cooked and form an essential supplement to the diet (e.g., fruits, nuts).
3. Those which have little or no nutritional value. These may be further subdivided into four classes:

    a. Herbs and spices, which are intrinsically highly aromatic and flavorful and may be used singly or as seasoning blends,
    b. Other aromatic plant materials used mainly as a source of essential oils (e.g., citrus peels).
    c. Vanilla, which only develops its characteristic flavor profile after postharvest fermentation and curing.
    d. Coffee, tea and cocoa all of which require postharvest treatment to achieve an acceptable flavor profile and which are largely used to make beverages.

4. Flavors derived from microbiologically induced reactions intrinsic to the source materials (e.g., wines, cheese),
5. Flavors arising from enzymatic reactions within the source materials and induced by certain conditions (e.g., onion, garlic, horseradish).

Hodge (1975) reviewed the history and botanical classification of flavor-producing plants, the broad extent of which emphasizes the problems of classification of sensory attributes and the difficulty of selecting those of major interest to flavor technologists. Plant materials contain both volatile and nonvolatile constituents which affect their individual odor and/or flavor profiles as well as their overall sensory impact. The volatile components are what give the plant its distinctive odor whereas the nonvolatile fraction is either inert (e.g., cellulose, lignin, starch) or influences some sensory character or reaction (e.g., color, bitterness, sweetness, pungency, astringency) often coupled with some physiological effect (e.g., coffee and tea are stimulants). The intensity and quality of these attributes show wide variations between plant families and species and even between different sources or lots of the same plant species.

The groups of plants just outlined will be studied from the point of view of the characteristics of the raw materials as they occur in commerce, the nature of the products made from them for incorporation into processed foods, beverages, etc., the attributes by which their quality can be assessed and the technological parameters which determine their suitability for use.

Those materials which are used primarily for their nutritional contribution to the diet will not be discussed as these are not generally recognized as flavoring materials, although, of course, they may have a significant impact on the overall flavor profile of the end product.

C. Herbs and Spices

1. Definitions

a. Herbs

Botanically herbs are soft-stemmed plants the main stem of which dies down to ground level and either does not regrow (annuals), grows again in the following year only (biennials) or regrows each year (perennials). Culinary herbs are a restricted group of such plants together with some traditionally used nonherbs such as sage (subshrub) and bay laurel (the leaves of a large tree). All have been used for centuries in the seasoning of foods as well as many other domestic and commercial outlets. Herbs may be used fresh or after dehydration. In the latter case it is usual to separate the leaves, floral parts and seeds from the heavier, harder and less aromatic stems by screening. Such herbs are called "broken" or "rubbed" and it is in this form that they are generally sold for domestic and commercial use. Those of greatest commercial value are listed in Table 6.1.

b. Spices

Spices are all other aromatic plant materials used in the flavoring or seasoning of foods, although this very broad definition admits to several exceptions (e.g., vanilla) which are not generally regarded as spices. Spices are usually only parts of plants. The major spices are listed in Table 6.2.

Condiments are seasonings which are added to food after it has been served. The most widely used of these are pepper, salt, mustard, horseradish and ginger.

2. Historical Associations

The use of herbs and spices in the flavoring of foods is almost as old as man himself. Few other commodities have had such a profound effect upon world history and their role in shaping the destiny of nations is a long and fascinating story. The spice trade is one of the oldest known

Table 6.1. Principal Culinary Herbs

Herb	Botanical Source	Family	Parts Used
Basil	*Ocimum basilicum* L. and other spp.	Labiatae	Leaves and flowering tops
Bay laurel	*Laurus nobilis* L.	Lauraceae	Leaves
Lovage	*Levisticum officinale* Koch	Umbelliferae	Roots, leaves and seeds
Marjoram	*Majorana hortensis* Moench. (= *Origanum majorana* L.)		Leaves and flowering tops
Mints			
Cornmint	*Mentha arvensis* L. var. *piperascens* Holmes and other varieties.	Labiatae	Leaves distilled for essential oil
Peppermint	*Mentha piperita* L.		Leaves distilled for essential oil
Spearmint	*Mentha spicata* (L) Huds. var. *tenuis* (= *M. viridis* L.) and other varieties		Leaves distilled for essential oil
Scotch mint	*Mentha cardiaca* Gerard & Baker		
Oregano		Labiatae	
Mexican sage	*Lippia* spp., usually *Lippia graveolens* HBK.		Leaves
Greek	*Origanum* spp. mostly *Origanum vulgare* L.		
Origanum (wild marjoram)	*Coridothymus capitatus* Richb. (= *Thymus capitatus* Hoffm. & Link.) or *Origanum vulgare* L. or *Thymus masticina* L. and several other spp. dependent on source		Leaves and floral parts
Rosemary	*Rosmarinus officinalis* L.	Labiatae	Leaves
Sage		Labiatae	
Dalmatian	*Salvia officinalis* L.		Leaves
Greek	*Salvia triloba* L.		
Spanish	*Salvia lavandularfolia* (L) Vahl.		
Savory		Labiatae	
Summer	*Satureia hortensis* L.		Leaves and flowering tops
Winter	*Satureia montana* L.		
Tarragon (estragon)	*Artemisia dracunculus* L.	Compositae	Leaves
Thyme	*Thymus vulgaris* L; *T. zygis* L. and other spp. dependent on source	Labiatae	Leaves and floral parts

Table 6.2. Principal Spices

Spice	Botanical Source	Family	Part Used
Allspice (pimento, Jamaican pepper)	*Pimenta dioica* (L) Merr. (= *Pimenta officinalis* Lindly)	Myrtacae	Fruit
Anise	*Pimpinella anisum* L.	Umbelliferae	Fruit
Anise, China star	*Illicium verum* Hook	Magnoliacae	Fruit
Capsicum			
Sweet peppers, capsicum	*Capsicum annuum* L.	Solanaceae	Fruit
Cayenne, tabasco, chili	*Capsicum minimum* Roxb. or *C. frutescens* L.		
Caraway	*Carum carvi* L.	Umbelliferae	Fruit
Cardamom	*Elettaria cardamomum* (L) Maton var. *miniscula* Burkhill	Zingiberaceae	Seed
Cassia	*Cinnamomum cassia* (L) Blume	Lauraceae	Bark
Cinnamon	*Cinnamomum zeylanicum* Nees	Lauraceae	Bark
Saigon	*C. loureirii* Nees		
Korintje	*C. burmanii* Blume		
Clove	*Syzygium aromaticum* (L) Marr. & Perryl. (= *Eugenia caryophyllus* (Spreng) Bullock & Harrison; *E. caryophyllata* Thunberg or *E. aromaticus* (L) Bailley)	Myrtaceae	Flower bud
Coriander	*Coriandrum sativum* L.	Umbelliferae	Fruit
Cumin	*Cuminum cyminum* L.	Umbelliferae	Fruit
Dill	*Anethum graveolens* L.	Umbelliferae	Fruit
Indian	*Anethum sowa* Roxb.		
Fennel	*Foeniculum vulgare* Mill.	Umbelliferae	Fruit
Fenugreek	*Trigonella foenum-graecum* L.	Leguminosae	Seed
Ginger	*Zingiber officinale* Roscoe	Zingiberaceae	Rhizome
Mace	*Myristica fragrans* Houttyn	Myristicaceae	Aril
Mustard		Cruciferae	Seed
Black	*Brassica nigra* (L) Koch		
White	*Sinapis alba* L. (= *Brassica alba* Boissier)		
Indian	*Brassica juncea* (L) Cosson or Hook & Thoms.		
Nutmeg	*Myristica fragrans* Houttyn	Myristicaceae	Seed
Paprika	*Capsicum annuum* L.	Solanaceae	Fruit
Parsley	*Petrosilinum sativum* Hoffm. or *P. crispum* (Mill) Nyman	Umbelliferae	Seed
Pepper, black and white	*Piper nigrum* L.	Piperaceae	Fruit
Poppy	*Papaver somniferum* L.	Papaveraceae	Seed
Saffron	*Crucus sativus* L.	Iridaceae	Stigma
Mexican	*Carmanthus tinctorius* L.	Compositae	Florets
Sesame	*Sesamum indicum* L.	Pedaliaceae	Seed
Turmeric (curmuma)	*Curcuma longa* L.	Zingiberaceae	Rhizome

Table 6.2a. Importation of Spices into the United States in 1984[a]

Spice	Weight (tonnes)	Value ($1,000)
Pepper	38,321	ca 76,000
Capsicums	14,521	18,553
Pimento	849	1,973
Cloves	1,072	6,703
Nutmeg/mace	2,255	4,361
Cinnamon/cassia	13,917	14,984
Ginger	4,208	5,885
Cardamom	80	1,116
Vanilla beans	841	49,934
Turmeric	1,789	2,155
Coriander seed	6,340	2,391
Cumin seed	4,400	5,329
Mixed spices	1,515	3,290

[a] Source: Tropical Products Research Institute, London.

to man and there is mention of spices such as cinnamon, cardamom and ginger in the oldest manuscripts dating back to before 2000 B.C. The spice trade has always been extensive and profitable. The great wealth of King Solomon came from "the traffic of the spice merchants, whose shrewd methods ensured the prosperity of the Middle East for centuries. By the tenth century the great cities of Alexandria and Venice were built and flourished. In the fifteenth century, great wealth was accrued from the commerce in spices, and bitter battles were fought to gain possession of valuable spice caravans and sea cargoes. The restless exploration of the Portuguese and the hazardous voyages of the great navigators, such as Columbus and Magellan, were all closely associated with the supply and commercial exploitation of these commodities. International rivalry reached unprecedented heights in the seeking out and annexing of spice growing areas and trading posts, for it was recognized that the country which controlled the spice trade was not only the richest but the most influential in the world.

The value of these natural flavoring materials, particularly black pepper, in converting otherwise monotonous diets into something appetizing was quickly appreciated and, in consequence, the European demand was almost insatiable, the prices were high and the profits enormous. In the days long before refrigeration and other modern methods for the preservation of fresh food commodities, what one could eat was dictated entirely by what was in season or could be preserved by the simple expedients of of drying, cooking or brining. The use of spices made such food palatable but there were then only a few who could afford the luxury.

Control of the spice trade passed through several hands over the ages. The initial supremacy of the Portuguese was replaced at the end of the sixteenth century by the Dutch who maintained their control of the spice trade until well into the eighteenth century. By the early nineteenth century Great Britain dominated this position but today the United States is the major importer of these materials with New York as its trading center. Although figures vary from year to year, on average the United States imports some 115,000 metric tonnes (t) of herbs and spices worth about $63 million. It also exports about 3000 t of spices worth $4.5 million.

3. Commercial Considerations

The relative importance of spices has declined significantly and today represents only about 0.1% of total world trade. In the Middle Ages the price differential between source and user was in the order of 100:1, now it is under 2:1; an interesting commentary on the relative importance and ready availability of these materials in modern trading terms.

With few exceptions, spices as prime commodities constitute only a small item of the trade and convertible currency earnings of the producing country. They are of major importance in only two regions: Tanzania (Zanzibar), where clove exports account for a large percentage of the country's revenue, producing about two thirds of the world's requirements of approximately 12,500 t per annum; and Granada, where the nutmeg and mace crops produce about half the country's revenue. But here the quantity produced, about 1,500 t of nutmegs and 250 t of mace, is far less than that produced in Indonesia. One can readily appreciate why such countries need aid to establish alternative crops and other sources of income. Many countries are no longer desirous of continuing spice production, which is now barely cost effective since their profitability on the world commodity market has been so eroded.

Universally there are some 70 well-known herbs and spices used in the flavoring or seasoning of foods. Many of these are local and appreciated only in the regional cuisine. The bulk of the world's trade in these commodities is concentrated in only a few of them; black and white pepper being by far the most important in both tonnage and dollar value, closely followed by sesame and mustard seeds. Of the remaining spices, the greatest trade is carried out in clove, nutmeg, capsicums of various types, cardamom, cinnamon or cassia barks, ginger, mace and allspice or pimento berries. Statistics for the trade in herbs is difficult to establish as countries do not report on these separately. Only in the United States are figures given for herb imports. Most countries publish consolidated figures for national import/export statistics and these do not include frequently significant domestic usage in the producing countries. The inclusion of other natural materials which are generally not regarded as "herbs or spices" also tends to distort the figures and make them difficult to interpret. World exports of spices in 1984 are estimated to be between 320,000 and 350,000 t valued at over $1,000 million. In 1984 the United States imported 101,000 t of the main spices representing a value of $198 million. A breakdown of these figures is given in Table 6.2a.

Herbs are predominantly wild or cultivated plants which flourish in temperate regions with few extremes of climate, on the other hand spices still come from semitropical and tropical regions as they have done for centuries. Many are now cultivated in the United States although Central and South American countries, particularly Mexico and Brazil, are becoming increasingly important growing areas. In the United States capital intensive agricultural methods are employed but in most other regions the traditional relatively labor intensive methods persist. Although supplying about 35% of its needs by domestic production, the United States still draws heavily on European countries for herbs and on Africa, India and Indonesia for spices and aromatic seeds. In the United States, California is by far the biggest producer and, if one includes dehydrated vegetables such as onion and garlic, then that region provides more tonnage of products for America's spice shelves than any other area of the world. Latest figures show that California produces over 100 million lb annually of dehydrated onion, over 25 million lb of dehydrated garlic as well as significant quantities of the aromatic seeds such as celery, paprika, chili peppers and red peppers. It has commercial cultivation of basil, tarragon, dillweed, sage, rosemary, marjoram, chervil and savory herbs amounting to about one half to three quarters of a million pounds of dried broken material per year. United States domestic production has been steadily rising and is estimated to be growing at about 10% per annum.

From studies made recently, the American Spice Trade Association concludes that spice consumption in the United States is increasing in excess of population growth. For example, between 1972 and 1976 the annual consumption of spices was 451 million lb as compared with 284 million lb in the 1962 through 1966 period. That represents a gain of 58% which is nearly five times greater than the population increase of 12% between the same two periods. In terms of individual spices, black and white pepper increased by 27%; red pepper increased by 54%; oregano, the herb most widely used in the flavoring of pizzas, increased by 97%; cumin, a major component of curry powders, increased by 106%

and mustard seed by an enormous 201%. Many spices originate in the third world developing countries and trade statistics show that the seven largest spice exporting countries are Indonesia, Brazil, Spain, India, Malaysia, Morocco and Hungary.

4. Spice Importation

Spices and spice seeds are usually imported whole and dried herbs broken, some to be sold in their natural condition others to be further processed by grinding prior to sale in bulk for commercial users or in small containers for the domestic market. Transportation from the producing region is normally in bulk shipments, importation into the United States being either through New York or San Francisco. New York handles about 75% of the total trade. At the receiver ports all imported spices must pass through several checkpoints before acceptance and release onto the domestic market.

The United States has the most stringent specifications of any country, and checking all shipments for compliance imposes a heavy burden on the Food and Drugs Administration (FDA), whose inspectors are required to draw official representative samples for quality assessment. Such samples must be tested for compliance with limits on macrofilth as specified by the ASTA as well as the very strict "defect action levels" operated by the FDA. To ease the burden, the FDA has agreed on the conditional release of herb imports without this formal inspection routine, but importers must guarantee that all shipments will be sampled and assayed by an ASTA-approved laboratory and that bulks will not be released onto the U.S. market until they comply in all respects with the standards laid down. Under the normal procedure, the importer is provided with a duplicate sample to enable an independent assay to be ample in the event of any dispute. Both the ASTA and FEMA recognize the adverse publicity which spices have received over many years due to their acknowledged poor hygenic quality and these organizations are working closely with the FDA inspectorate to produce meaningful guidelines on sampling and test procedures and the establishment of effective "defect action levels." Until all parties are satisfied that the goods do comply with legal requirements then they remain in quarantine in the dock warehouses. Shipments that fail the FDA inspection, or do not meet the standards for cleanliness, must be either upgraded under strict FDA surveillance or be shipped back to origin at the importer's expense. Once cleared, the stocks may be removed from the docks and are usually delivered to one of the major spice processing companies which specialize in these commodities.

Similar arrangements exist in the principal European spice trading centers of London, England; Amsterdam, Holland; and Hamburg, Germany.

5. Standards of Purity

The absence of internationally acceptable standards of purity for spices is the cause of much concern to producers, importers and users. Some progress has been made in this field by the International Standards Organization (ISO) as well as by many national standards agencies. To date standards have been published by the ISO covering cardamom, black and white pepper, chilies, allspice (pimento berries), ginger, mustard seed and curry powder. In addition there are documents dealing with nomenclature, methods of sampling and determination of defined quality attributes. These do not, as yet, have the full international recognition that is necessary for binding commercial transactions but they are being used increasingly as the basis for national standards in many countries. In the United States the ASTA makes a valuable contribution to the quality control of spices by the publication of its "Analytical Methods and Standards for Cleanliness of Unprocessed Spices, Seeds and Herbs."

Spice buying and selling at the industrial level is one of mutual problem solving between buyer and seller. This involves technical and economic performance assessments based on samples rather than purchase by name and price only. The prime quality factor is flavor and, although the quantitative and qualitative determination of the essential oil content is important, the most meaningful assessment must be based on sensory evaluation under carefully defined test conditions preferably related to the ultimate use of the spice under test. Only when all the test results are available can one make a valid judgment concerning the acceptability of the material under consideration.

6. Sensory Assessment

The sensory assessment describes the odor and flavor of aromatic materials.

a. Flavor Profiles

In view of the importance of sensory evaluation as a prerequisite to the commercial acceptance or rejection of most flavoring materials it is necessary to consider here what is implied by this widely used method of analysis. The discussion will be biased towards spices but can be applied equally to other flavoring substances and products. Sensory analysis has developed into a complex topic but here the concern lies only with the value of descriptive analysis or flavor profiling as an understandable aid to commercial transactions.

A method of assessing odors and flavors in terms of their individual characters and intensities was first described by Sjostrom in 1949 and has since received wide acceptance as the basis for evaluating both raw materials and end products. The flavor profile method depends on the independent judgments of a panel of four to six trained assessors and provides both qualitative and quantitative data about the sensations of taste, smell and feeling of the material under examination.

It is extremely difficult to achieve an unambiguous description of even well-known, but purely sensory effects, such as the odor and/or flavor of, say, peppermint, banana, apple or ginger. Certain descriptive terms are readily understood by the majority of those likely to be called upon to evaluate aromatic materials, but, generally speaking, the mere reading of a description, however lucid, gives one very little idea of the reality of the sensory effect. It is fair to say that no odor or flavor can be described in any language in such a way that an uninitiated person would immediately recognize and be able to visualize the material and so identify it when presented with a sample for the first time. Even individual words used in such descriptions do not necessarily convey the correct impression or even the same meaning to different people. For example, one may have a reasonably clear understanding of the word "aromatic" as something which has both an hedonic and a nonhedonic connotation. That is, it is both pleasing and sweet or even descriptive of an organic chemical substance. On the other hand, the term "green," which is so frequently used to describe herbs, is far less precise, ranging from the effect associated with freshly cut grass ("grassy") to that of green, freshly cut leaves ("leafy") or even freshly gathered herbs ("herbaceous"). All of these characters are quite different and in most cases it is necessary to carry out a first hand evaluation in order to appreciate fully the nuances and notes which combine to make an odor or flavor profile.

Another pitfall frequently encountered in specifications is the use of the term "characteristic." It is all too easy to use this expression in describing an odor or flavor, but it is largely without meaning to anyone who has not previously experienced the particular odor or flavor concerned. To say that basil herb has a "characteristic" odor is in effect saying that it smells like "basil" and that is begging the question. Every aromatic plant has its own distinctive odor pattern by which it may be recognized. It is this factor which is of importance in quality assessment as it allows the differentiation of one sample or product from another on a demonstrable norm. The case of recognition is largely the result of one's experience and an ability to memorize odor and flavor effects by association with form and color.

The description of odors and flavors has been the subject of many studies which have been aimed at establishing a reasonable methodology and the development of an effective list of descriptive terms and phrases. Work by Harper *et al.* (1968) was based on the following precepts: (a) abstract terms are generally not acceptable as they are usually meaningless; (b) descriptions are best defined in terms of specific substances or well-known effects which are demonstrable; (c) there is a bias towards first describing undesirable or off-odors or off-flavors in preference to those of more pleasing character, and (d) certain commonly used words are capable of alternative interpretations and may need clarification by reference to the particular context of their use.

The meaning of words has been and is likely to remain the cause of continuing discussion, particularly accentuated when one has to translate terms into another language as so frequently happens in international trade. Using comprehensible terminology is difficult enough for the experts, who do not always agree, but how much more difficult the problem becomes when one tries to convey a sensory impression to someone with perhaps little or no first-hand knowledge and expects them to understand words having special meanings when applied to the sample under examination or discussion. Two specialists rarely come up with the same descriptive profile, hence the need to use panels of a reasonable number of assessors so as to get an acceptable consensus of opinion.

It should be remembered that this is a particularly subjective area of judgment and that a clear understanding of each word or term used must be established by personal appraisal, preferably by reference to some well-defined substance which displays the odor or flavor attribute under review. Some of these are listed in Table 6.3.

In everyday experience, odors and flavors are appreciated as a whole so that one gets an overall impression of both the character and strength of the sensory effect. In creating a profile it is necessary to be more analytical, assessing each note or attribute as it is perceived. But not all aromatic effects are of equal intensity. Some have a powerful impact on the nose giving to the material or product an initial often pleasing freshness but with little real flavor when masticated and ingested; others are dull and heavy, contributing a well-rounded aroma accompanied by a positive full-bodied flavor when chewed and swallowed. It is only by personal experience that such impressions can be established and com-

Table 6.3. Sensory Characteristics of Some Organic Flavorants

Character	Chemical	Character	Chemical
Acrid	*n*-Valeraldehyde	Mint-like	Menthol
Balsamic	Isoamyl cinnamate	Musklike	ω-Pentadecalactone
Burnt	Guaiacol	Musty	*p*-Menth-3-en-1-ol
Buttery	3-Decalactone	Oily/Fatty	Decadienal; isoamyl laurate
Camphoraceous	Camphor; isoborneol	Onionlike	Methyl disulfide
Coconutlike	γ-Nonalactone	Orangelike	Methyl *n*-methylanthranilate
Creamy	2,3-Heptanedione	Rancid	*n*-Butyric acid
Earthy	Hexyl propionate	Roselike	Geraniol
Ethereal	Diethyl ether; ethyl amyl ketone	Smokey	Pyroligneous acid
Fishy	Hexylamine	Soapy	Lauryl alcohol
Floral	Phenyl ethyl acetate	Spicy	Eugenol; cinnamyl propionate
Green (floral)	2-Phenyl propionaldehyde	Sulfurous	Dimethyl sulfide
Green (grassy)	*cis*-Hexyl acetate	Sweaty	Isovaleric acid
Green (leafy)	Heptyl caproate; 2-hexenal; *cis*-3-hexenol	Vanillalike	Vanillin; ethyl vanillin
Herbaceous	Carvacryl methyl ether; octyl butyrate	Winey	Ethyl 10-undecenoate
Lemony	Citral; limonene	Woody	Heptyl isobutryate; β-terpineol
Metallic	Amyl vinyl ketone		

mitted to memory. A practical knowledge of the odor and flavor characters of the herbs and spices is essential for anyone who wishes to use these valuable flavoring materials in the seasoning of food products.

b. Relationships of Components and Profiles

The nature of the constituents responsible for the aromatic character of plant materials, and in particular the essential oils, has been the subject of considerable research. Over the past three decades, since the introduction of gas–liquid chromatography (GC) and more recently coupled mass spectrometry (MS) the number of compounds which have been isolated and identified has increased significantly. Current research is tending to concentrate on the relative importance of such compounds as determinants of the odor and flavor profiles. Techniques such as high performance liquid chromatography (HPLC) are now being applied to studies into precursors and the nonvolatile constituents, many of which contribute significantly to the flavor profile (Maarse 1983). Our present knowledge of the chemistry of natural aromatic plant materials is comprehensive, but much work still remains to correlate chemical composition with relative contribution to the quality and intensity of the odor and flavor profiles.

Combined instrumental techniques such as GC/MS with computerized integrated data processing are increasingly revealing the true complexity of the essential oils and of extracts prepared from aromatic plant materials. The composition of the quantitatively major components has been established for many years by classical analytical techniques but modern methods now refute some of these earlier findings by the more ready recognition of artifacts arising from the relatively unsophisticated methods available. Research findings underline the importance of precursive chemistry activated by enzyme systems and also the disproportionate contribution of trace components, many of which have still to be positively identified. It is necessary to appreciate that each component has a characteristic and often complex profile and threshold of recognition. Every odor and flavor is a combination of the profiles of its components, some acting in unison, others competing to the point of elimination.

In some essential oils one main component may predominate; for example, clove bud oil contains 80–85% of the phenol, eugenol, but the proportion of eugenyl acetate and other eugenyl esters effectively determines the quality of any given sample of clove bud oil. Rogers *et al.* (1971) compared the composition of clove bud, stem and leaf oils and demonstrated the relative proportions of eugenol and eugenyl acetate (see the accompanying tabulation). They concluded that even though the differences were minor the odor characteristics of the three oils were distinctly different and distinguishable.

	Clove Oils		
	Bud	Stem	Leaf
Eugenol	80–85%	87–92%	85–90%
Eugenyl acetate	8–12%	3–5%	0–10%

In other essential oils there is a balance of components so that no one compound characteristically predominates. For example, the essential oil of sweet marjoram contains *d*-α-terpineol (25%), 1-linalool (6%), geraniol (19%), eugenol (8%) and terpinen-4-ol (10%) together with some 40% of terpenoid hydrocarbons. This gives to the oil a well-rounded, sweetly aromatic profile showing little change in character as the oil evaporates from a smelling strip (Heath 1978).

An examination of the odor profile of an essential oil may be directly compared with those of its known constituents and this has been done for oil of rosemary in Table 6.4. The odor pattern may be further studied by an examination of the effluent gases from a GC column (Fig. 6.1), which requires considerable training and a good odor association recall but principally a very quick response to changing conditions (Forss 1981). The relative importance of each component cannot be judged directly from such evaluations but requires a consideration of threshold values and a systematic approach to the determination of the quantitative relationship between composition and sensory effect.

Based on work first published by Guadagni *et al.* (1972) it is suggested that the relative contribution of individual components to the intensity and quality of the odor and flavor profiles could be expressed in terms of *odor units* which are the ratio of concentration to the threshold value (expressed in ppm) of the pure chemical. If the total peak area of the sample is C_m and its determined threshold is T_m then $C_m/T_m = X$ odor units. Individual components can then be summated to give an indication of their contribution.

$$\frac{C_{m1}}{T_{m1}} + \frac{C_{m2}}{T_{m2}} + \frac{C_{mn}}{T_{mn}} = \frac{C_m}{T_m} = X \quad \text{odor units}$$

This concept may be a little too simplistic and may not apply strictly to complex sensory stimulations. The concept of odor units implies that

Table 6.4. Profiles of Oil of Rosemary and Its Main Constituents

Oil of Rosemary	α-Pinene	Cineole (17–30%)	Borneol (16–20%)	Camphor (about 10%)	Bornyl Acetate (2–7%)
Odor					
Fresh	Fresh	Fresh		Ethereal	Fresh
Cooling		Cooling		Cooling	
Slightly warming	Warm			Warm	
Penetrating		Penetrating	Pungent	Penetrating	
Strong		Strong			Strong
Eucalyptuslike (cineolic)		Eucalyptuslike		Slightly minty	
Aromatic					Sweet, balsamic
Camphoraceous, piney	Resinous Piney Turpentinelike	Camphoraceous	Camphoraceous		Piney
Dry-out					
Top notes quickly lost	Very poor tenacity	Poor tenacity		Poor tenacity	
Residuals:					
Herbaceous					Herbaceous
Dry			Dry		
Woody			Woody		
Peppery			Peppery		

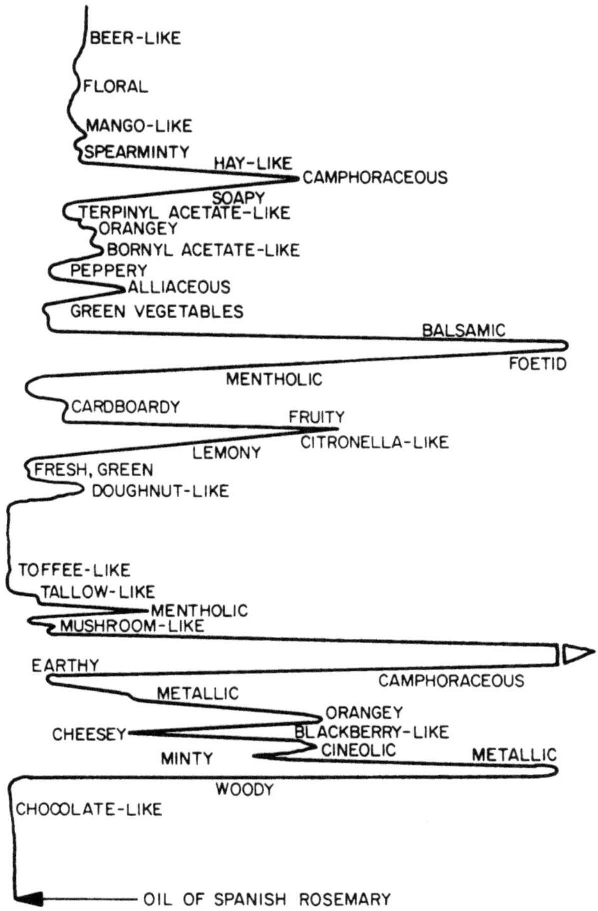

Fig. 6.1. Sensory assessment of the effluent from GLC of Spanish oil of rosemary. *From Heath (1978)*

the total sensation is purely additive but it is well known that in many cases synergistic and antagonistic effects can occur. In addition the odor intensity of chemicals is not strictly linear with concentration. Physical factors also affect perception and the determination of reliable threshold values is far from easy. However, in spite of these objections, the idea has considerable merit as its increasing use in flavor research indicates.

7. Classification of Herbs and Spices

It is possible to classify herbs and spices, and indeed other natural flavoring materials, in four ways depending upon the value of the classification to its application. The four methods used are (a) botanical, (b) agronomic, (c) physiological and (d) organoleptic. One could, of course, add alphabetical which at least facilitates quick reference to available data.

Botanical classification systems may depend either upon plant form or on botanical family associations, although the latter is subjective and depends on which botanist one accepts. The herbs and spices are classified in this way in Tables 6.5 and 6.6.

Agronomic considerations give a classification based on the nature and extent of cultivation. These factors are of little direct concern to the food and flavor technologist.

Physiological classification is based on the traditional and widespread use of natural plant materials many of which have been effective and potent remedies for generations and today are attracting increasing interest in alternative medicine. This very specialized viewpoint is of little further interest in our context.

Organoleptic groupings, based on the sensory attributes which, in turn, are related to the prime aromatic constituents or in some cases the nonvolatile components, give to the food and flavor technologist the most meaningful and directly useful comparative classification. The complexity of the system depends on the specificity of the attributes selected. For example, Arctander (1960) proposed 88 distinguishable groups of sensorily related natural raw materials used in food flavorings and fragrances. Of these 22 groups involved herbs and spices. This is a very effective grouping for the trained flavorist and perfumer but for most general purposes a much simpler grouping is quite adequate as an aid to the use of herbs and spices in blended seasonings. Such classifications are given in Tables 6.7 and 6.8. However, few are in complete

Table 6.5. Classification of Spices Based on Part of Plant Used

Part Used	Spice	Part Used	Spice
Leaf	Bay laurel	Seed	Cardamom
Fruit	Allspice (Pimento)		Fenugreek
	Anise		Mustard
	Capsicum		Poppy
	Caraway		Sesame
	Celery	Bark	Cassia
	Coriander		Cinnamon
	Cumin	Flower bud	Clove
	Dill	Root	Horseradish
	Fennel		Lovage
	Paprika	Rhizome	Ginger
	Pimento (Allspice)		Turmeric
Part fruit (aril)	Mace	Stigma	Saffron
		Flower	Safflower

Table 6.6. Classification of Herbs and Spices by Botanical Associations

Botanical Family	Herb/Spice	Botanical Family	Herb/Spice
Labiatae	Basil, sweet	Myristicaceae	Nutmeg
(Lamiaceae)	Majoram		Mace
	Origanum spp.	Zingiberaceae	Cardamom
	Mints		Ginger
	Rosemary		Turmeric
	Sage	Piperaceae	Pepper
	Savory	Magnoliaceae	Star anise
	Thyme	Cruciferae	Mustard
Lauraceae	Bay laurel	(Brassicaceae)	Horseradish
	Cassia	Leguminosae	Fenugreek
	Cinnamon	(Fabaceae)	
Myrtaceae	Allspice (Pimento)	Papaveraceae	Poppy
	Clove	Pedaliaceae	Sesame
Umbelliferae	Anise	Solanaceae	Capsicum
(Apiaceae)	Caraway	Compositae	Safflower
	Celery		
	Coriander		
	Cumin		
	Dill		
	Fennel		
	Lovage		
	Parsley		

agreement owing to the personal nature of sensory responses and the difficulties associated with the communication of findings. In spite of this the system is of value and can readily be adapted to suit individual requirements.

8. Culinary Herbs

Depending on their purpose most books on herbs describe upwards of 100 different items which may be used in foods, for medicinal pur-

Table 6.7. Classification of Herbs by Main Flavor Component in Essential Oil

Essential Oil Contains					
Cineole	Thymol/ Carvacrol	Alcohols/ Esters	Thujone	Menthol	Carvone
Bay laurel	Thyme	Sweet Basil	Dalmatian Sage	Peppermint	Spearmint
Rosemary	Oregano	Marjoram	Greek Sage	Cornmint	Dillweed
Spanish sage	Origanum	Tarragon			
	Savory				
	Wild Marjoram				

Table 6.8. Classification of Spices by Sensory Associations

Pungent Spices	Aromatic Fruits or Seeds	Aromatic Barks	Phenolic Spices	Colored Spices
Capsicum	Nutmeg	Cassia	Clove bud	Paprika
Ginger	Mace (aril)	Cinnamon	Allspice (pimento)	Saffron
Pepper (black and white)	Cardamom		West Indian Bay	Safflower
Mustard	Fenugreek		Oils of:	Turmeric
Horseradish	Anise		Clove stem and leaf	
	Fennel		Cinnamon leaf	
	Caraway			
	Dill			
	Indian dill			
	Celery			
	Lovage (root)			
	Parsley			
	Cumin			
	Coriander			

poses or in the preparation of toiletries and cosmetics either in their fresh form, as dried broken herbs or as essential oils and/or extracts. The food industry uses only about 20 such plants and products made from them.

Herbs, for the most part, have a light and very distinctive aromatic character although they contain relatively low levels of essential oil. For culinary use the whole herbaceous tops are harvested and may be used fresh or after drying. In the latter case it is usual to remove by mechanical screening the leaves and any floral parts or seeds as may be present from the heavier, harder and less aromatic stems.

The flavor impact of the freshly cut herb is appreciably higher and of a different quality from that of its dried counterpart. This is due to loss or modification of the low boiling fraction of the essential oil made up mostly of terpene hydrocarbons. These do not directly contribute much to the aroma profile but their presence gives the herb a clean freshness which is markedly absent from the dried product, although modern freeze-drying techniques have reduced this difference to a large extent. The characteristic top-notes of the profile of the freshly cut herb are not only lacking in the traditionally dried material but may be overlaid by a dull haylike aroma which is not removed by rehydration but is generally dissipated during cooking.

Within the several groups of herbs given in Table 6.7 there is a close similarity of sensory effect and when blended together these tend to reinforce each other in their main flavoring attributes while supplementing each other in their minor aromatic nuances. Individual profiles are, of course, different and often dictate the use of one herb rather than another in a seasoning blend (Heath 1978).

a. Flavor Characters of Herbs

Herbs containing cineole have a very distinctive note which can readily be recognized as eucalyptuslike although each has its own diagnostic profile. These herbs may be described as "light" or "fresh" and their odor may be "penetrating," initially warming but usually developing a more cooling character. Such herbs are useful in the seasoning of white meats (e.g., veal, pork, chicken, fish) and other protein sources which have a low intrinsic flavor. They give an attractive lift to the overall profile of the cooked product but when used in excess they can overpower the more delicate natural flavors present. All of them can be described as having "green, herbaceous" notes usually with a marked "bitter" after-note which makes them of value in the seasoning of the sweet meats such as pork.

Herbs containing thymol and/or carvacrol are all richly "sweet" and "spicy, aromatic," usually with a distinctive phenolic character. Their odor is penetratingly sharp to the point of being unpleasant and this largely determines the level of their effective use in seasonings. The aromatic character is carried into the flavor profile which may be described as "sweetly phenolic" with a "warm, richly herbaceous" fullness. Seasonings containing herbs such as thyme may leave a distinctly unpleasant "dry, woody" after-note and a certain lingering bitterness.

The so-called "sweet" herbs are characterized by the presence of various alcohols and esters in their contained essential oils. These give an attractive fragrance and generally well-balanced roundness to their odor and flavor profiles. It is this property which makes them of value in blended seasonings. Each has a quite distinctive profile which is considerably more pungently fragrant when freshly cut than after drying. The initial odor impact is not high but is usually penetrating and may be slightly irritating. Many of them show an initial cooling effect backed by a "green, herbaceous" nuance with a hint of "balsamic, floral" notes. The flavor effect is smoother than the odor and can be described as "warmly spicy," but in spite of this they are all pleasantly "fresh" and "herby" probably due to a "camphoraceous" note which has a delayed and somewhat persistent cooling effect. The aftertaste in most cases is bitter but not unpleasant.

The varieties of *Salvia* which are collectively known as *sage* are all highly aromatic herbs having a penetrating odor which is initially cooling but then "warmly spicy." Their essential oils contain 40–60% of thujone modified by about 15% of cineole so that the profile is at first "camphoraceous" with a "musty, dull, sickly sweet" impact. These herbs are not to everybody's taste and regional preferences for them are very marked. The flavor of the sages is overpoweringly "herbaceous" with a "sweet, eucalyptuslike" back-note and a lingering and not very pleasant "bitter-sweet" after effect. Sage is a powerful and effective seasoning and is widely used in stuffings together with onion for such meats as chicken, veal and fish.

The *mints* form another group of herbs distinguished by their wide variety. Several species are of commercial importance as flavoring materials either in the form of the fresh or dehydrated leaves or as essential oil. Their odor and flavor characteristics display wide variation depending on source, growth conditions, harvest and postharvest handling, drying and/or distillation techniques employed. All the mints have a clean cooling odor which is strongly "mentholic" and usually irritating. The attractive flavor is very cooling and refreshing with a marked "mentholic" character which is somewhat pungent but not irritating. Initially the flavor effect is warming but at higher concentrations rapidly becomes cooling and even numbing. Some varieties as well as older samples of essential oil leave the mouth with a heavy "oily-buttery" note, which may even be somewhat "coconut-like," with a slightly "earthy," "unclean" back-note. Spearmint differs completely from peppermint in that the oil contains 50–60% of carvone which gives it a characteristically "balsamic," "creamy" odor and a warm, smooth pleasingly aromatic flavor which is so popular in chewing gum.

b. Preparation of Herbs for Marketing

The main forms in which herbs are handled in commerce are as follows:

a. Whole—cleaned but not otherwise processed.
b. Chopped, cut or sliced—leaves very coarsely broken into pieces about 3–6 mm across. This is a convenient form for bulk transportation prior to further processing.
c. Broken or rubbed—leaves and other soft herbaceous parts are coarsely screened to remove all hard stems. This is the form in which most domestic herbs are marketed.
d. Ground—milled herbs powdered to a specified sieve size.
e. Blended seasonings—broken or ground herbs blended together ready for use in foods. Composition can be varied to suit particular dishes.

Domestically fresh herbs are widely used in the form of *bouquet garni*, *fines herbs*, etc., as these do have a better aroma and flavor than the dried forms. Herbs are, however, seasonal and the difficulties associated with handling fresh produce result in the dried form being extensively used almost exclusively on a commercial scale and very widely on a domestic scale.

The quality of herbs depends on several variables but particularly on the following: plant variety, environmental conditions during growth and harvesting, period of harvesting, drying and postharvest processing methods, storage and transportation, packaging and shelf life. Users of herbs are most concerned with cleanliness, particularly freedom from adhering soil, flavor, aroma and color. Unfortunately there is, as yet, no internationally accepted grading system but obviously herbs derived from cultivated plants are likely to be superior to those gathered from wild plants. Controlled agricultural conditions, mechanical harvesting at the correct time and optimum postharvest handling are most likely to result in a higher level and greater consistency of quality (Heath 1983).

Loss of quality is closely related to loss and/or degradation of volatile oil content which is at most risk during the drying process as the loss of moisture is accompanied by loss of the lighter fractions of the volatile oil. As any residual dampness results in caking and mold infestation the removal of water is very important. From about 60 to 80% in the fresh herb, the water content must be reduced to between 5 and 10% to achieve a stable product. If this is carried out under optimum conditions the color of the herb is retained and the odor and flavor losses are reduced to a minimum. Excessive heat results in an unacceptable degree of browning, the loss of freshness and the imposition of an unacceptable off-odor which detracts from the overall profile. Most herbs are dried naturally under shade but artificial drying at a maximum of 40°C and also freeze-drying are increasingly being used. These are more expensive processes but the resulting products have a much higher quality.

9. Spices

In a manner similar to that used for herbs, it is convenient to classify spices so as to reveal sensory associations which are of value in the formulation of well-balanced seasonings. This is done in Table 6.8.

Spices differ from herbs in being only parts of plants which are very aromatic and, with few exceptions (e.g., capsicum, fenugreek), contain relatively high percentages of volatile oil as well as other powerful, nonvolatile, flavoring and/or coloring components. They are normally derived from plants growing wild in semitropical or tropical regions and, even when cultivated, are often grown, harvested and sun-dried under far from ideal conditions all of which inevitably influences the quality of the marketed product. Of the various factors involved, temperature has been shown to have the greatest significance as a determinant of quality and suitability for use, but insect infestation during storage and mold infection in any damp material may cause serious deterioration in appearance and aromatic profiles to the extent of making the material quite unacceptable for use in food products.

Spice Processing—Milling

Because of their physical characters few spices can be incorporated directly into food products; most require further processing. Of the several processes used, particle size reduction or milling is most widely used. The degree of fineness of the processed spice is determined by its ultimate use (e.g., distillation, extraction, blended seasonings).

It is well recognized that the process of milling spices can directly affect the quality and keeping properties of the resulting material in the following ways: (a) by exposing the material to excessively high temperatures during the actual grinding process; (b) by exposing the volatile oil in the ruptured cellular tissues to loss by evaporation and/or oxidative change; and (c) by altering the physical character of the product (e.g., creating a pasty mass instead of a free-running powder as occurs with nutmeg and mace) thereby influencing its subsequent shelf life and value as a flavoring material.

Milling or comminution of dried plant materials involves subjecting the whole or coarsely broken feedstock to considerable shear forces in specially designed machinery with fixed and high speed moving parts with the inevitable production of heat. Friction between the parts of the milling head and the plant material itself usually results in temperatures of between 40° and 95°C, depending on the fineness of the grinding and the nature of the start material.

Generally, the finer the grind the higher the temperature created in the mill head but this is offset by the more ready availability of the flavor and increased ease of incorporating the milled spice into a food mix. Finely milled spices generally have a very limited shelf life and in manufacturing practice they should not be retained in store for longer than about 3 months before use.

The critical deterioration temperature at which undesirable changes occur in the product (i.e., volatile losses, oxidation, discoloration or physical transformation) is only about 36°C and in some spices, such as nutmeg and mace, may be as low as 32°C.

Many milling operations involve high speed impact machines such as disintegrators, pin-mills, etc., which induce a rapid airflow through and over the material being milled. The occlusion of such air inevitably increases the risks of loss and/or damage to the sensitive essential oil. To minimize this problem, spice milling is normally carried out with as little grinding as possible or, where this cannot be effected, under conditions which minimize both air flow and elevated temperature.

Various techniques have been developed to limit the formation of heat during grinding; two of particular interest in commercial use are (a) prechilling of the feed material and (b) water-cooling or refrigeration of the milling chamber (i.e., cryogenic grinding).

Both methods are aimed at maintaining the spice at a temperature below 25°C throughout the process. Problems associated with these cold-milling techniques include undesirable atmospheric condensation on the milling head, absorption of moisture into the milled product leading to "balling" and the possibility of soft-iron rust contamination. Cohodas (1969) patented a process for cryogenic grinding of spices in which the feed material and liquid nitrogen at −125°C are simultaneously fed at a controlled rate into a specially designed milling head. The resulting ground spice and vapor discharge are regulated to 25°C throughout the milling operation. Under such conditions even the most sensitive herbs retain their full aroma and display little discoloration.

Certain spices require special treatment, notably the following:

a. Those having high volatile and/or fixed oil content (e.g., caraway, cumin, nutmeg, mace.).
b. Ground white pepper if prepared from whole black pepper by a process known as *decortication.*
c. Cardamoms, the seeds of which must be first removed from the tough flavorless capsule in which they normally remain until required.
d. The blending of curry powders which calls for a multistage milling to achieve the desired color and texture.
e. Mustard seeds which must be partially deoiled and desolventized prior to grinding for use in prepared mustard powders.

Loss of volatiles and autoxidation of fixed oils are not the only problems arising from milling. Most mills are fabricated from iron and only the fast moving parts from various grades of toughened steel. Recent studies have shown that many ground spices contain relatively high levels of particulate iron presumably originating from this source.

10. Microbiology of Spices

Spices are particularly bad microbiologically and are a rich source of viable bacteria, thermophilic aerobic bacteria and spores, yeasts and molds, a few of which may be pathogenic or toxigenic (Krishnaswamy *et al.* 1973). Typical figures for microbiological counts are given in Table 6.9. The food processing industry has long demanded more hygenic raw materials and, as spices are extensively used in processed foods and in meat products which are sold fresh (e.g., sausages), it is obviously essential that they should be subjected to close scrutiny. It has already been noted that spices originate in tropical regions where hygiene and cleanliness are not much practiced. Little or no attempt is made to limit contamination. In spite of increased attention to gross contamination of imported spices and the imposition of limits on "filth," the pattern of microbiological contamination remains largely unchanged. The International Standards Organization (ISO) has so far only specified microbiological requirements for onion and garlic powders and has not issued any guidelines on other spices. There are so many sources of infestation that it is not economically feasible to prepare in the growing areas commercial batches of spices having better bacterial/mold counts than at present.

For food processing, particularly in chilled foods having a limited uncooked shelf-life, the microbiological spoilage due to spices can be significantly improved or indeed eliminated by the following processes: (a) exposure of the spices to sterilant gases, (b) exposure of the spices to gamma-irradiation, (c) sterilization by heat processing, or (d) distillation

Table 6.9. Microbiological Contamination of Spices

Untreated Ground Spice	Total Bacterial Count (per gram)
Cayenne	2.1×10^6
Celery seed	1.1×10^6
Cinnamon	1.1×10^5
Clove	1.90×10^2
Coriander	4.1×10^5
Ginger, African	5.5×10^6
Nutmeg	2.0×10^4
Mace	2.3×10^5
Onion powder, Californian	6×10^3
Onion powder, Egyptian	2×10^6
Paprika	1.6×10^6
Pepper, black	1.04×10^7
Pepper, white	4.2×10^4
Pimento (allspice)	3.7×10^6
Sage (rubbed herb)	5.5×10^5
Thyme (rubbed herb)	3.5×10^4

and/or extraction techniques to isolate the flavoring components of the spice.

a. Gas Sterilization of Spices

Of the possible methods of controlling and reducing the bacterial load of spices, fumigation with sterilant gases offers an effective technique usually carried out by the major spice processing companies in the user countries.

The process most widely used involves precise operating conditions particularly with regard to temperature and humidity control. The technique involves the following stages: (a) loading the spices, in suitable bags, into a purpose-designed retort; (b) evacuation of the chamber; (c) raising the temperature of the material to about 35°C (maximum 60°C); (d) introduction of a predetermined quantity of the sterilant gas; (e) maintenance of conditions to affect sterilization and (f) removal of the gas and sparging of the retort with air several times to remove all residuals.

The gases used are either ethylene oxide or propylene oxide which, for practical and safety purposes, are normally diluted with either carbon dioxide (10%), sold under the name ETOX, or methyl formate (50%). Mixtures containing other sterilants such as dichlorofluormethane are also available.

The application of gas sterilization is increasingly the subject of restrictive legislation due to suspected toxic hazards arising from the formation of chlorohydrins in the treated material. Under the Food Additives Amendment, the treatment time when using propylene oxide is limited to 4 h at a temperature not exceeding 125° F (50°C); the use of ethylene oxide is not so precisely defined, but U.S. regulations limit the residual propylene oxide to 300 ppm and ethylene oxide to 50 ppm.

From a bacteriological point of view, the concentration of gases used rarely penetrates sufficiently to kill all the bacteria and spores present. Under optimum conditions there is no change in the profile of the treated spice although some loss of volatile oil may occur from finely ground spices.

b. Gamma Irradiation

Ionizing irradiations are produced either by mechanical acceleration of electrons to very high velocities and energies, by X-ray machines or by emission of gamma radiation from radioactive isotopes such as cobalt 60 or cesium 137. Of these, cobalt 60 is most favored for commercial processing and several plants are now in operation using this source. The units of energy absorbed by material exposed to gamma rays is measured and expressed in terms of Megarads (Mrads) or as Grays (Gy) and 1 Mrad = 10 kGy.

This technique is now receiving increasing approval and in many countries its application is permitted for the irradiation of specified food materials or products at prescribed irradiation levels. Currently experts recommend that legislative approval be given to all major foods which have been exposed to 1 Mrad (or 10 kGy) or less and there is ample evidence of the safety of foods treated in this manner.

In the United States the FDA proposes to allow (a) any food to be irradiated up to 100 krad (1 kGy) without submission of additional safety data; (b) any food class whose composition is less than 0.01% of total diet to be processed up to 5 mrad (50 kGy) without submission of additional safety data and (c) food clearance for other applications to be based on successful results of two short-term (90-day feeding tests) and several short-term mutagenicity tests (chemi-clearance) (Fed. Reg. 27, March, 1984). These have been further modified by later FDA proposals to allow the use of irradiation (a) at doses not to exceed 100 krad (1 kGy) for inhibition of the growth and maturation of fresh fruits and vegetables; (b) similarly for the microbial disinfestation of foods and (c) at doses not to exceed 3 Mrad (30 kGy) for microbial disinfestation of spices (Fed. Reg. 14, February, 1984).

A scientific status summary by IFT's Expert Panel on Food Safety and Nutrition has been produced to ensure wide publicity of this possibly controversial technique (Anon. 1983).

c. Heat Treatment

The volatile components of herbs and spices are generally thermolabile and any heat treatment is likely to affect stability and quality. Normal thermal processing such as retorting effectively sterilizes any spices present but techniques aimed at sterilizing spices prior to incorporation into food mixes have not proved acceptable.

d. Distillation or Extraction

The products resulting from distillation (i.e., essential oils) or extraction (i.e., oleoresins) are essentially free of bacterial contamination and form the basis of a whole range of products for direct incorporation into food mixes.

The techniques in the preceding sections a–d give to the food processor a wide choice of materials. There is still some adherence to spices in their natural form but, increasingly, manufacturers are recognizing the advantages of seasonings based on spice extractives. The advantages

Table 6.10. Advantages and Disadvantages of Traditional Ground Spices

Advantages
Slow flavor release in high temperature processing
Easy to handle and weigh accurately
No labeling declaration problems
Presence of natural antioxidants in many herbs
Disadvantages
Variable flavor strength
Variable flavor profile
Unhygienic
Often contaminated by filth in one form or another
Ready adulteration with less valuable materials
Presence of lipase enzymes
Flavor loss and degradation on storage
Undesirable appearance characteristics in end products
Flavor distribution poor, particularly in thin liquid products (sauces)
Discoloration due to tannins
Dried herbs usually have unacceptable haylike aroma
Dusty and unpleasant to handle in bulk

and disadvantages of natural spices in various forms are given in Table 6.10.

11. Distillation of Volatile Oils

The bulk of all spices is cellulose which contributes nothing to the aroma and flavor of the spice. As we have seen the aromatic profile is largely determined by the essential oil content, quantitatively as a measure of aromatic strength and qualitatively as a determinant of aromatic character. In most instances, the volatile oil preexists in the plant and is usually contained in special secretory tissues the nature of which is often associated with particular plant families (e.g., Zingiberaceae—oil cells; Umbelliferae—vittae or oil canals, Labiatae—secretory hairs). For this reason one cannot overgeneralize on distillation techniques as optimum conditions will vary between different spices.

The distillation of herbs and spices for the recovery of their contained essential oil is the subject of several standard texts, notably Guenther (1948–1952) and Gildermeister and Hoffman (1965–1967). The process has been well reviewed by Pruthi (1980) and regular articles on essential oils are published by Lawrence in the journal *Perfumer and Flavorist*.

a. Spice Essential Oils

The recovery of aromatic constituents from plant materials, in the form of essential oils, dates back to the ninth century. Although the actual process has not changed much over the centuries, the plant used has been constantly refined and modified to give optimum yields under conditions which can be precisely controlled. Actual yields and quality depend on the nature of the feed material, the methods used in its handling, pretreatment and speed of processing after comminution. Yield figures and aromatic quality are most likely to vary as a result of differences in distillation time, rate of distillation, efficiency of vapor condensation and method of separation from the distillation water.

Modern distillation techniques fall into the following three categories:

a. Water distillation—in which the plant material is loaded into a still fitted with a slow-speed paddle stirrer; covered with water which is brought to the boil by submerged steam coils. Distillation is continued until all the essential oil has been recovered, the condensed water being continuously returned to the still (a process called *cohobation*) (see Fig. 6.2.).

b. Water and steam distillation—in which the plant material is loaded onto a frame within the still body, fixed above a layer of water which can be brought to the boil by submerged steam coils. This technique is widely used for the distillation of green herbaceous materials (e.g., peppermint) (see Fig. 6.3).

c. Steam distillation—in which the material is loaded into a suitable still through which steam may be injected from the base. A reboiler unit is often fitted to enable condensate to be recycled. The shape of the still body depends on the nature of the feed material. For example, the dis-

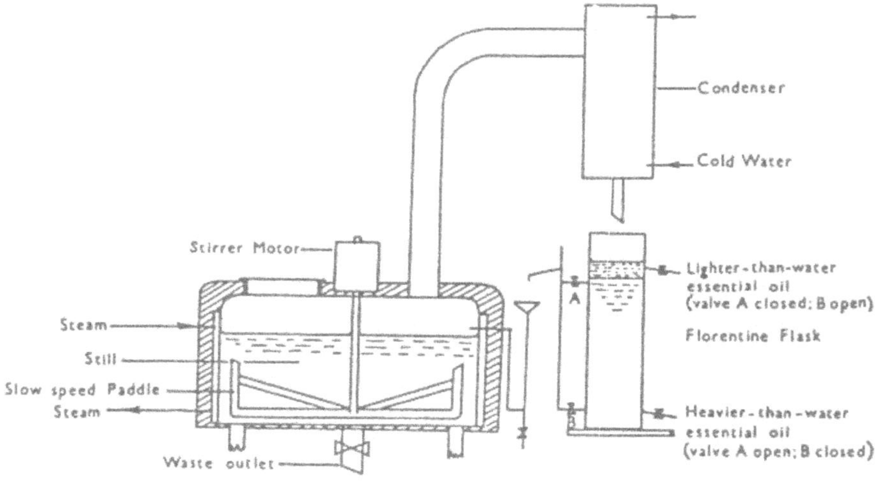

Fig. 6.2. Plant used for water distillation of spices.

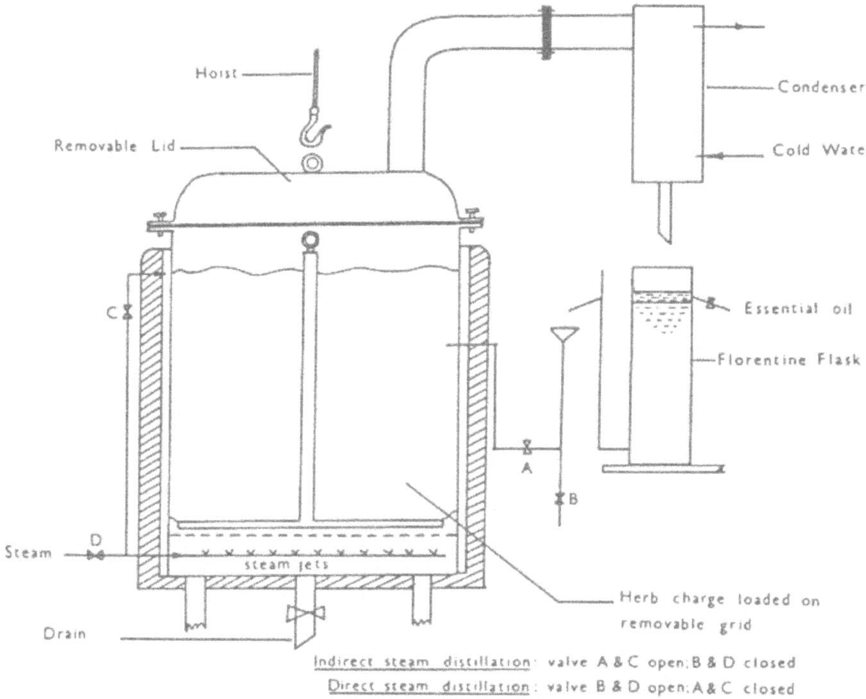

Fig. 6.3. Plant used for water and steam distillation of herbaceous materials.

tillation of peppermint herb in the far west states is carried out in stills formed from the back of container trucks which can be loaded directly in the fields; these trucks being driven to a center where a still head is bolted on and steam injected. Static steam stills tend to be tall and are often divided internally by grids to ensure uniform loading.

Most herbs and spices require some preparation prior to distillation. The unit operations involved include (a) cleaning; (b) comminution; (c) soaking, particularly of hard material such as cinnamon bark; and (d) loading and charging—which requires close attention to avoid channeling or uneven exposure to the steam as this can have a very significant effect upon the sensory characters of the oil.

The method of distillation employed depends on the nature of the feed material. Over the years all the prime aromatic materials have been investigated to establish optimum conditions. Generally water distillation gives the finest quality oil as in this technique there is less chance of damage to sensitive components or the imposition to the essential oil of a burnt character which can result from internal condensation and flowback of an aqueous extract onto the exposed steam injection coils.

Although apparently simple, the separation of essential oil from the distillate can result in a reduction in quality depending on the following factors:

a. The chemical composition of the essential oil, particularly the ratio of low to high boiling constituents.
b. The temperature gradient of the condensate within the condenser, the outflow temperature of which determines the relative density of the oil and water phases. Optimally this should be between 45° and 50°C.
c. The flow rate which determines the time available to achieve effective phase separation.
d. The separator design, preferably allowing continuous cohobation of the separated water which usually contains some residual unseparated oil globules.

Some essential oils are more effectively isolated by solvent extraction within the separator. For this purpose very pure grades of hexane and/or pentane are used. The essential oil is recovered by removing the solvent under high vacuum distillation (Provatoroff 1972).

The literature covering the distillation of essential oils is extensive if now somewhat out of date. But techniques have not changed fundamentally. The texts quoted in the bibliography should be used as a source of further basic information on techniques and specific essential oils.

b. Application of Spice Essential Oils

The use of essential oils goes a long way of overcoming one of the major objections to the use of ground spices, i.e., flavor variability both in strength and character. Most essential oils obtained from materials from a designated source are reasonably constant in their flavoring effects and if used at a fixed rate will give an end product with a consistent profile. However, in many instances, essential oils only represent a part of the total available flavor in the spice, and this can limit their usefulness. For example, in black pepper and ginger, the volatile oil only gives the characteristic odor of the spice; the pungent principles, being nonvolatile, are not present in the distilled oil.

The advantages and disadvantages of using spice essential oils in food processing are listed in Table 6.11.

Essential oils are widely used in the making of plated spices most of which are formulated to replace the ground spice on a weight-for-weight basis. When making dispersions of essential oils onto an edible carrier

Table 6.11. Advantages and Disadvantages of Spice Essential Oils

Advantages
Hygienic, free from all microorganisms
Flavoring strength within acceptable limits
Flavor quality consistent with source of raw material
No color imparted to the end product
Free from enzymes
Free from tannins
Stable in storage under good conditions
Disadvantages
Flavor good but often incomplete
Flavor often unbalanced
Some readily oxidize
No natural antioxidants
Readily adulterated
Very concentrated so difficult to handle and weight accurately
Not readily dispersable, particularly in dry products

it is necessary to reduce the dispersion rate to account for the proportion of oil which is naturally retained within the unbroken cells of ground spices. The precise equivalence must be determined experimentally but may range within a 30–50% reduction. For example, if the yield of essential oil from 100 kg of allspice (pimento berries) is 4 kg then a dispersion containing 2.5% oil of pimento, instead of 4%, will give an equivalent flavor when used weight for weight to replace the ground spice in a seasoning formulation.

c. Essential Oil Content of Spices

The essential oil content of the most widely used culinary spices are listed in Table 6.12.

12. Extraction and Oleoresins

Although they are valuable flavoring agents, the spice essential oils frequently lack the full, rounded profiles associated with the use of ground herbs and spices. In certain instances they are totally devoid of characteristic flavoring attributes contributed by nonvolatile components in the natural material. These can be recovered by extraction of selected, dried and ground herbs and spices. In practice, organic solvents are chosen which optimally dissolve both the essential oil and the desirable nonvolatiles present; the solvent is removed and the resulting concentrated extract is known as an oleoresin.

Table 6.12. Volatile Oil Content of Spices

Spice	Volatile Oil Content (%v/v)
Allspice (pimento)	3.0–4.5
Anise	1.5–3.5
Anise, China star	About 30
Capsicum	Nil
Caraway	2.7–7.5
Cardamom seed	4.0–10.0
Cassia (cinnamon)	0.5–4.0
Celery	2.5–3.0
Cinnamon	0.5–1.0
Clove	15–20
Coriander	0.4–1.0
Cumin	2.5–4.5
Dill	2.0–4.0
Fennel	4.0–6.0
Fenugreek	Trace
Ginger, Jamaican	1.0–3.0
Ginger, African	2.0–3.0
Mace	12–15
Nutmeg	6.5–15
Paprika	Nil
Pepper, black	2.0–4.5
Pepper, white	1.5–2.5
Turmeric	4.0–5.0

a. Solvents

Two main categories of solvents are widely used and these are listed in Table 6.13.

a. Polar solvents contain hydroxy or carboxy groups and are relatively reactive chemicals which are miscible with water. They are generally powerful solvents of a wide range of ingredients. Some pose fire hazards.

b. Nonpolar solvents are generally hydrocarbons or chlorinated hy-

Table 6.13. Solvents Used in Flavor Extraction

Solvent	Boiling Point at 760 mm (°C)	Solvent	Boiling Point at 760 mm (°C)
n-Pentene	36.1	Ethyl acetate	77.1
Methylene chloride	40.0	Ethanol	78.4
Acetone	56.5	Benzene	80.1
Methanol	64.7	Cyclohexane	80.7
Trichloroethane	74.1	Isopropanol	82.5
Carbon tetrachloride	76.8	Ethylene chloride	83.7

drocarbons which are relatively unreactive and immiscible with water. They tend to have limited solvent properties being better for oils and fats. The hydrocarbons may pose fire hazards while the chlorinated solvents are nonflammable but may be relatively toxic.

Modern techniques using liquified gases such as carbon dioxide at low temperature (0°–10°C) and high pressure (8–80 atm) as a selective solvent are currently being developed and adopted industrially. Under these conditions carbon dioxide liquifies to become a nonpolar selective solvent which is odorless, tasteless, colorless, easily removed without residues and nonflammable. Capital investment in the plant is high but the extraction is claimed to be more effective with the added advantage that there are no solvent residues, no off-odors, low monoterpene hydrocarbon levels, more top-notes and more back-notes giving a fuller but different profile than with traditional extraction methods (Calame and Steiner 1982; Brogle 1982; Peyron 1984; Moyler 1984).

b. The Extraction Process

The most widely used extraction process involves the following unit operations:

a. *Preparation of the raw material.* Dried herbs and spices must first be ground to ensure solvent penetration. The smaller the particle size the greater the number of ruptured cells, the greater the interfacial area and the greater the speed of achieving equilibrium in the system. Too fine grinding can lead to slow solvent penetration, slow percolation of the solvent through the mass and difficulties in recovering solvent from the spent material.

b. *Exposure of the material to solvent.* This involves three phases: (i) addition of solvent to ensure even wetting of the dry mass, (ii) period of maceration to allow absorption of the solvent and the attainment of extraction equilibrium between solutes and solvent and (iii) continuous replacement of the miscella with fresh solvent.

These conditions are influenced by the nature of the start material, the design of the extractor and whether the system is operated under gravity or forced-pump circulation. The system of recycling the solvent is economically important and is dictated by the solvent selected, the degree of heat treatment feasible without damage to labile constituents, the established flow rate and the optimum drug/solvent ratio. When the bulk quantity warrants it, a system of continuous countercurrent extraction offers many advantages.

c. *Separation of miscella from extracted material.*

d. *Removal of solvent.* The stripping of solvent from the desired extractives is critically important to the character and flavoring quality of the resulting oleoresin. Many spices contain highly volatile and hence low boiling constituents so that the removal of solvent must be carried out under low temperature conditions so as to avoid loss or damage of these. The use of high vacuum is indicated but this can also lead to volatile losses. For this reason, the lower the boiling point of the solvent the better, hence the attraction of supercritical gases which do not have these problems.

Desolventization is a two stage process involving (i) removal of about 98% of the solvent in a standard falling-film, rising-film or calandria-type evaporator operated under carefully controlled vacuum conditions and (ii) scrubbing, sparging with nitrogen and high vacuum treatment of the concentrate to remove final traces of solvent down to the prescribed levels demanded by U.S. legislation (Table 6.14).

c. Quality of Oleoresins

The physical charcter of oleoresins ranges from viscous oils to thick tacky pastes making their direct incorporation into food mixes difficult. To assist in handling, many oleoresins may be admixed with a permitted diluent such as propylene glycol or a selected vegetable oil, such additions being clearly indicated in the product specification.

It is the balance of volatile to nonvolatile resinous matter that is the best indication of quality and this should be closely similar to that in the original spice. In many commercially available products this is so, but not in all.

Many major oleoresin producers offer products under a generic name of "standardized oleoresins" having specified characteristics different from those found in the equivalent natural spice. These products are designed for particular end uses and their production involves the blending of basic oleoresins with the appropriate essential oil obtained by distillation, usually, but not necessarily, from the same source ma-

Table 6.14. Limits for Residual Solvents in Spice Oleoresins[a]

Solvent	Maximum Permitted (ppm)
Acetone	30
Methanol	50
Isopropanol	50
Hexane	25
All chlorinated solvents	30

[a] *Source:* United States Federal Register (1980), Food Additive Regulations 173.210–173.290.

terial. The purchasing division of the United States Quartermaster Corps has published specifications for several such standardized products [Federal Specification EE-S-00645 (ARMY-GL), 23 October, 1963] designed for dispersion on edible carriers at a fixed percentage to give a range of so-called dry soluble spices for use in food processing. Certain oleoresins are also included in the Essential Oil Association (EOA) Specifications and Standards (1975) and various other official international standards (British Pharmaceutical Codex 1973). Stahl (1972) expressed concern over the lack of agreement on quality attributes for spice oleoresins. In spite of critical reviews by Salzer (1975) there is still considerable diversity of data making quality assessment difficult and misleading.

d. Application of Oleoresins

The advantages and disadvantages of using spice oleoresins in food processing are listed in Table 6.15.

As oleoresins pose so many problems to the food manufacturer various secondary products, suitable for direct incorporation into food products, have been commercially developed. These include the following:

Essences are prepared by the maceration of the ground spice materials with 70% ethanol. The resulting product may be fortified by the addition of *topping notes* compounded from essential oil and/or oleoresin. In the United States the term *extract* is often applied to a flavoring essence but has a defined meaning under FDA regulations.

Emulsions are liquid seasonings prepared by the emulsification of blended essential oils and/or oleoresins with gum arabic (acacia) or other

Table 6.15. Advantages and Disadvantages of Spice Oleoresins

Advantages
Hygienic, free from all microorganisms
Can be standardized for flavoring strength
Many contain natural antioxidants
Free from enzymes
Long shelf life under good storage conditions
Disadvantages
Flavor quality good but as variable as the raw material
Flavor profile dependent on the solvent used
Very concentrated so difficult to handle and weigh accurately
Range from liquids to viscous fluids and pastes making difficult to incorporate into food mixes without "hot spots"
Tannins present unless specially treated

permitted emulsifying agents. A stabilizer may be added to prevent creaming.

Solubilized spices are blended essential, oils and/or oleoresins mixed with one of the polysorbate esters (e.g., Polysorbate 80) or other dispersing or solubilizing agents in such a concentration as to give a clear solution when mixed with water.

Dispersed, plated or "dry-soluble" spices are prepared by dispersing the standardized oleoresin or a blend of essential oil and oleoresin onto an edible carrier such as salt, dextrose, crackermeal, or rusk to give a product having a flavor strength, but not necessarily the same flavor character as that of a good quality freshly ground spice.

Encapsulated spices are prepared by spray drying premade emulsions using gum arabic (acacia) or one of the permitted modified starches. On drying these become the encapsulant. Such products have a long shelf life and the flavor is only released when they are mixed with water.

Heat-resistant spices are specialty encapsulated products in which the outer capsule is rendered water insoluble. Such products have a long shelf life and are designed to release their flavor only at high temperatures such as are achieved during baking.

Fat-based spices are blends of essential oils and/or oleoresins in a liquid edible oil or hydrogenated fat. These latter products may be offered in bulk or as spray-cooled powders. They are designed for ready incorporation into high fat products such as mayonnaise or cream soups.

Most flavor manufacturers offer one or more of these groups of products hence reference to the supplier's literature is necessary to establish their equivalent flavoring strength, suggested applications and use rates. Sensory evaluation of such products in specific end applications is strongly recommended as processing conditions may seriously affect their claimed performance.

One should not assume that handling herbs and spices and the several products manufactured from them is elementary. In fact, it is full of pitfalls for the unwary. In handling spices, which have always been highly speculative items of commerce, it is not surprising to find ambiguities in names, descriptions and grades offered commercially. The lack of "official" or internationally agreed upon specifications does little to help the potential purchaser. There are many examples of substitution of materials of a different geographical source or even of different botanical identity. All of this emphasizes the importance of a good first-hand practical knowledge of genuine herbs and spices—what they look like, smell like and taste like, coupled with an understanding of the characteristics of the several products made from them. The advantages and disadvantages of using these secondary products in food processing are listed in Table 6.16.

Table 6.16. Advantages and Disadvantages of Processed Spices

Dispersed Spice Products
Advantages
Standardized flavor profile by specifying source of raw material
Standardized flavoring strength
Hygienically acceptable
Free from enzymes
Readily handled and weighed with accuracy
Readily dispersed in food mixes
Usually free from dust during handling
Many contain natural antioxidants
Unwanted specks or color not contributed to the end product
Water activity low
Disadvantages
Allowances necessary for the base used
Volatiles lost on storage, particularly if ambient temperature is high
No resistance to high temperature processing under open conditions
Encapsulated Spices
Advantages
Aromatics fully protected from loss and degradation
Where appropriate, natural color retained over long periods (e.g., paprika)
Long shelf life under all conditions
Readily incorporated into food mixes
Free from objectionable odors (applies particularly to garlic)
Hygienically acceptable
Free from enzymes
Low water activity
Do no contribute unwanted specks or color to the end product
Disadvantages
Flavor profile may differ from that of equivalent natural spice
Concentration usually tenfold so that accurate weighing is difficult
Relatively expensive

13. Seasonings

Herbs and spices are rarely used singly and often complement each other when used in well-balanced mixtures or seasonings. Depending on the intrinsic flavor character and strength of the main food components of a dish (i.e., beef, pork, poultry, fish), the seasoning used should aim to enhance any natural flavors present but not to overpower them. However, there are certain well-known dishes in which high levels of spicing are traditional, though sometimes the exported versions of these dishes differ very considerably from their true native concept. Chili con carne and curry are excellent examples of this. Mexican dishes are generally hot but more subtly spiced than their fiery Texan equivalents. In India, curries are generally only moderately spiced though they may demonstrate a wide range of piquancy. It is the judicious use of capsicums which can give an effective level of pungency without unduly affecting the more delicate flavors. In Mexican cooking, full flavored chilies are extensively used whereas in Texas the very pungent and less flavorful varieties of chili are preferred.

A seasoning may be considered to have four main flavoring contributions: (a) light, sweetly herbaceous or aromatic notes which give instant impact when the food is served and when being conveyed to the mouth; (b) medium aromatic, herby and spicy notes selected to bring out the finer flavors in the main food ingredients (e.g., the sweetness of pork with coriander, the delicate flavor of chicken with sage and thyme); (c) heavy, full-bodied spicy notes which add depth, richness and full flavor character to the dish; and (d) pungency or piquancy which, in the case of ginger and pepper, is accompanied by a characteristic flavor, but with chilies the additional flavor is minimal.

Flavor Index and Formulation

Not all spices have the same flavoring power and this is at once obvious when one considers the flavor impact of say an equal weight of chilies and sage. It is possible to rate or classify herbs and spices allocating to each an approximate *flavor index* and to list these in increasing order of flavoring strength. The system cannot be precise as the judgments are entirely subjective and may vary in different hands under different test conditions. But, using such a classification, one can convert any seasoning formulation from a weight relationship to a *flavoring profile*. To achieve this, the percentage weight of each item is multiplied by the respective flavor index (Table 6.17) to give the flavor contribution, which in turn can be converted into a percentage figure. The reverse calculation can be made to formulate a seasoning having specified sensory attributes. For instance, a well-balanced seasoning for a full-flavored pork sausage might have profiles within the following ranges:

a.	Light, sweet, top-notes	5–10%
b.	Medium aromatic notes	5–20%
c.	Full-bodied, spicy notes	20–25%
d.	Pungency	45–65%

These notes would be selected from within the herbs and spices listed in Table 6.18. The formulation in Table 6.19 would lie within this profile character.

Table 6.17. Flavor Impact of Herbs and Spices

Type	Flavor Index	Type	Flavor Index
Freshly cut herbs		Spices	
Chervil	45	Paprika	50
Chives	50	Cardamom seeds	125
Coriander leaves	80	Dill	160
Fennel leaves	80	Fenugreek	200
Sage	95	Coriander	230
Sweet marjoram	100	Allspice (pimento)	260
Thyme	125	Cumin	290
Rosemary	130	Celery	300
Sweet bay leaves	140	Anise	320
		Caraway	330
Dry, broken herbs		Fennel	330
Garden mint	50	Mace	340
Savory	60	Nutmeg	350
Tarragon	60	Turmeric	400
Dalmatian sage	80	Black pepper	450
Marjoram	80	White pepper	460
Oregano	85	Cinnamon/cassia	460
Thyme	85	Ginger (dried)	475
Rosemary	90	Clove	560
Sweet bay	100	Mustard	800
		Cayenne (red pepper)	900
		Chilies	1000

Table 6.18. Flavor Characteristics

Light, sweetly herbaceous, fresh	*Heavy, full-bodied, spicy*
Coriander	Allspice (pimento)
Marjoram	Anise
Rosemary	Cumin
Sweet bay	Clove
Spanish sage	Fenugreek
Medium aromatic	Mace
Caraway	Nutmeg
Cardamom	Origanum (wild marjoram)
Celery	Thyme
Cinnamon/cassia	Turmeric
Dalmatian sage	*Spicy, pungent*
Lovage	Ginger
Mint	Horseradish
Oregano	Mustard
Paprika	Black and white pepper
Sweetly piquant	Capsicum
Basil	Cayenne (red pepper)
Savory	Chilies

Table 6.19. Formulation for a Pork Sausage Seasoning

Spice	Type[a]	% by Weight	Flavor Index	Flavor Contribution	Flavor Contribution (%)
Coriander	a	10	230	2,300	7.0
Celery	b	2	300	600	1.8
Dalmatian sage	b	10	80	800	2.4
Allspice	c	10	260	2,600	7.9
Mace	c	5	340	1,700	5.1
Nutmeg	c	10	350	3,500	10.6
Thyme	c	5	85	425	1.3
Black pepper	d	45	450	20,250	61.2
Cayenne pepper	d	1	900	900	2.7
		100		33,075	100

[a] *Profile:* a. Light, sweet top notes 7.0%
b. Medium aromatic notes 4.2%
c. Full-bodied spicy notes 24.9%
d. Pungent spicy notes 63.9%

One must stress that too high a degree of quantification cannot be placed on this system but the idea, coupled with an understanding of flavor balance, is capable of giving a much better impression of what a seasoning can achieve than can be obtained by merely examining the percentage formulation. There is still a real need for common sense in spice usage as peoples' palates vary considerably, particularly their sensitivity to flavor. Undoubtedly some "like it hot" but the majority prefer a seasoning which follows the old rule—the sweeter the meat the lighter the seasoning.

D. Plants as Sources of Essential Oils

The terms volatile oil and essential oil are synonymous and are used interchangeably.

As a practical definition, an essential oil is a complex mixture of volatile components recovered from odiferous plant materials by some physical process. The oil is designated to indicate the plant, or part of a plant, from which it was derived (e.g., oil of peppermint, oil of cinnamon leaf).

Most essential oils are obtained by distillation using one of the processes already described in Section C.11.a dealing with herbs and spices, but other methods are also applied commercially. These include the following: (i) enfleurage—extraction of floral oils using fats, (ii) solvent extraction and (iii) mechanical pressing, particularly applied to the recovery of citrus peel oils.

Generally, younger plants produce more oil than older ones but the quality of such oils may lack the fuller aromatic attributes normally associated with oils from mature plants.

In plants, essential oils are still a functional enigma; some may act as insect attractants as an aid to pollination, others may protect the plant tissues against parasites and animal depredation. Current expert opinion suggests that they are merely the waste products of plant metabolism, any functional value being coincidental. Chemotaxonomy of spice plants using the composition of the essential oil as a characterizing factor is of some interest (Malingre 1982).

Although several thousand plants are known to contain volatile constituents which could be recovered as essential oils, only about 200 are processed commercially and of these only about 60 are produced in really significant quantities.

In this section two prime sources of essential oils, citrus fruits and the mints, will be studied.

1. Citrus Fruits

The worldwide popularity of citrus fruits is based largely on their nutritional value, flavor, aroma and other intrinsic attributes, such as texture and color. Citrus fruits are an acknowledged source of vitamin C as well as of other dietary requirements. The characteristic color of the fruits is derived from carotenoids and flavonoids. The diagnostic odor is due to essential oil in the peel and the flavor of the juice is determined by the ratio of sugars to organic acids, mainly citric acid, overlaid by the presence of low levels of aromatics. In this context we are only concerned with citrus fruits as a source of essential oils which have a significant value as flavoring agents in several branches of food processing and soft drink manufacture.

a. Classification

The citrus fruits belong to the family *Rutaceae*, those of commercial value are shown in the following tabulation.

Grapefruit	*Citrus paradisi*, Macf.	Mandarin or Tangerine	C. *reticulata*, Blanco
Lemon	C. *limon* (L) Burm.	Orange, bitter	C. *aurantium*, L. subsp. *amara*.
Lime, acid	C. *aurantifolia*, Swingle	Orange, sweet	C. *aurantium*, L. var. *dulcis*
Lime sweet	C. *limetta*, L.		C. *sinensis*, Osbeck.

b. Citrus Essential Oils

In all the citrus fruits, the essential oil is contained in numerous oval, balloon-shaped, oil sacs or glands situated irregularly just below the surface of the colored portion of the peel (the flavedo). The white inner mesocarp (the albedo) does not contain any oil sacs but does carry the bitter glycosides such as hesperidin in lemon, orange and tangerine or naringin in grapefruit.

With the exception of lime oil, which is recovered by distillation, the citrus peel oils are obtained by mechanical extraction using a purpose-designed plant which differs in detail depending on the nature of the fruit to be handled. Other parts of the citrus plant also yield essential oils when distilled and those from the flowers and leaves are of importance in fragrance compounding but of less value in flavorings. The oil from citrus leaves is called oil of petitgrain and that from citrus flowers, oil of neroli. The source is designated in the name, e.g., oil of neroli, Bigarade, is derived from the flowers of the bitter orange, C. *aurantium* L.

Commercially, the citrus oils are recovered directly from the peels of fresh fruits using machinery which has been developed over very many years. The precise design of the plant and the processing techniques employed differ considerably depending on the nature of the fruit and indeed on the main growing regions. The latest techniques have been described by Nagy *et al*. (1977), but it is likely that some of the machinery may have been updated since then. The methods described include the following:

a. *Sponge pressing*. The separated half peels are handpressed, the extruded oil being collected in sponges. This very laborious method is now hardly used but is still claimed to give the highest quality oil.

b. *Machine pressing*. Two main types of machines are used. (i) Sfumatrici machines do automatic pressing of the separated peels after removal of juice and pulp. Such machines can handle about 1 tonne/h. (ii) Pellatrici machines are designed to handle whole fruit with the oil being released by rotary rasping of the fruit surface. Larger versions of these machines can handle 2 tonne/h. The liberated oil is continuously removed by a jet of water and the oil separated from the coarse emulsion by standing; further oil is separated from the aqueous phase by centrifugation and any intractable emulsion is distilled to recover an additional quantity of low grade oil.

c. *Combined juice/oil presses*. This machinery was developed in the United States and designed for the automatic extraction of juice and oil as a single operation without contact between the two products. Two principal systems are (i) the fruit is halved, the juice reemed out and the cleaned peel is subjected to pressure between fluted rollers and (ii) the surface of the whole fruit is first grated and the oil removed in a jet of water, at the same time a hollow pipe is inserted into the base of the fruit and the juice extracted under pressure (Warner *et al.* 1981).

Lime oil is traditionally obtained as a by-product of lime juice manufacture and is recovered by distillation of the acid liquors obtained by crushing the whole fruit. A cold-pressed lime oil is produced commercially; this has a much finer flavor quality than the more familiar distilled oil.

Citrus oils are characterized by having a high percentage of terpenoid hydrocarbons and a relatively low content of terpenoid oxygenated compounds which are mainly responsible for their aromatic profiles. As with the herbs and spices it is necessary to become acquainted first hand with these products. Their characteristics are listed in Table 6.20.

The yield of oil obtained commercially depends on numerous factors other than the type of machinery used, but rarely does it equate to the percentage of oil determined analytically. The thickness of the fruit skin is a major determinant as thick-skinned fruits yield much less oil than thin-skinned varieties owing to the absorption of oil by the spongy albedo. The delay between harvesting and processing is also important as flacid peel yields less oil.

On average, yields vary as follows:

Lemon	2–7 lb/t (tonne)
Orange	1–8 lb/t
Grapefruit	1–2 lb/t
Lime	0.1–0.3 lb/t
Tangerine	1–2 lb/t

c. Composition of Citrus Oils

Citrus oils are characterized by the presence of large percentages of terpenes (C_{10}/H_{16}) and smaller amounts of sesquiterpenes (C_{15}/H_{24}).

Table 6.20. Composition of Citrus Oils

Oil	Terpenes (%)	Principal Oxygenated Compounds (%)	
Lemon oil	About 90	Citral	2.5–5.5
Sweet orange oil	About 90	*n*-Decanal	0.9–3
Bitter orange oil	90–92	*n*-Decanal	0.8–1
		Linalyl acetate	2–2.5
Lime oil	75–80	*n*-Decanal	1–5
Tangerine oil	95–96	Methyl-*N*-methyl anthranilate	About 1
Grapefruit oil	About 90	*n*-Decanal	1–2

Both are the carriers of the oxygenated compounds comprising alcohols, aldehydes, ketones, acids and esters, which are responsible for the characteristic odor and flavor profiles. The terpenoid composition of the various citrus oils is closely similar, the principal component being d-limonene. The terpenes possess little intrinsic odor or flavor value but it would be incorrect to say that they have no flavoring effect. A citrus oil from which the terpenes have been removed is significantly flatter and lacks the characteristic freshness associated with a complete peel oil.

The composition of the principal citrus peel oils are given in Table 6.21.

d. Processed Citrus Oils

Terpene hydrocarbons are characterized by their relative insolubility in diluted alcohol and a ready tendency to oxidize and resinify with an associated deterioration in the odor and flavor of the oil in which they are present. Cold-pressed citrus oils may be further processed to remove all or part of the terpenes and sesquiterpenes, the resulting products being known as *terpenefree* or more usually *terpeneless* or *sesquiterpeneless* oils. As the content of the sesquiterpenes is relatively low the term terpeneless has come to be used in a general sense to describe these concentrated products. Citrus oils from which terpenes have been removed are also called *folded oils* as the remaining flavorful oxygenated compounds are more concentrated. The degree of concentration is often

Table 6.21. Principal Constituents of Citrus Peel Oils[a]

Monoterpenes:	*Oxygenated compounds:*	
α-Pinene	*Acids:*	*Aldehydes:*
β-Pinene	Acetic	Acetaldehyde
α-Terpinene	Capric	Citral
γ-Terpinene	Caprylic	*n*-Decanal
d-Limonene	Decylic	Geranial
Myrcene	Formic	Octanal
p-Cymene	Octylic	
Terpinolene	*Alcohols:*	*Esters:*
Sabinene	Citronellol	Geranyl acetate
Camphene	Geraniol	Linalyl acetate
	Linalool	Methyl-*N*-methyl anthranilate
Sesquiterpenes:	*l*-Nonanol	
Bisabolene	Octanol	Octyl acetate
Cadinene	α-Terpineol	
Caryophyllene	Terpinen-4-ol	

[a] Source: Lawrence (1978).

calculated from the ratio of the principal constituent in the concentrated oil to that in the prime oil from which it was made.

The citral content of a natural lemon oil varies between 4 and 4.5% and that of a concentrated or terpeneless oil may be quoted as 40–45%. On these figures one would assume a tenfold concentration of flavoring effect but this is not borne out in practice. Such oils are considerably more concentrated than "tenfold." The long-established practice of selecting just one component fails to take into account the several other oxygenated compounds which contribute to the total aromatic profile and which may significantly affect its flavor strength and quality.

The quantitative yield of concentrated or terpeneless oil obtained from a known weight of whole oil is the best basis for stating the degree of concentration or *fold* achieved. In practice, the comparative evaluation of the oil in an end product is the only really satisfactory way of establishing its flavoring concentration and hence its relative commercial value.

Specifications and precise usage data for terpeneless or concentrated citrus oils should be obtained from the manufacturer. They are used mainly in beverages because of their better storage properties and improved flavoring effects.

e. Other Citrus Peel Oils

In order to achieve maximum recovery of oil, the citrus processing industry applies several techniques which increase the yield. These include the following:

a. The distillation of the aqueous phase from which no more oil may be recovered by centrifugation.
b. The steam distillation of the pressed residual peels.
c. The superheated steam treatment of "citrus molasses" obtained from cured citrus wastes.
d. The deoiling of canned single-strength citrus juices.

All of these products are much inferior to cold-pressed citrus oils in both flavor and keeping properties.

f. Methods of Deterpenation

Three methods widely used for the production of deterpenated oils are the following:

a. Fractional distillation under vacuum.

b. Selective extraction of the soluble oxygenated compounds in diluted alcohol.
c. Chromatography using silica-gel.

i. Fractional distillation The bulk of terpeneless oils are produced by fractional distillation under a high vacuum (1–2 mm Hg) so that the oil boils at a sufficiently low temperature to ensure the minimum of damage to the heat sensitive constituents during the relatively long distillation period required to remove the large percentage of terpenes present. At these low temperatures the boiling points of the constituents are lowered making efficient separation difficult, but as the terpenes are rarely individually separated, this is a problem of still design and reflux balance which can readily be solved should it be necessary.

An alternative method to dry distillation is one in which steam injection is used. Unfortunately the distillation time is increased leading to possible damage to the flavoring components. In addition, the presence of water may lead to hydrolysis of certain constituents with consequent damage to the aromatic profile of the end product.

A combination of these techniques is applied in Sicily, the bulk of the terpenes being removed by normal vacuum distillation, the residual oil then being subjected to vacuum steam distillation to give the terpeneless product. It is claimed that this method gives an oil having a much cleaner and fresher top note.

ii. Selective solvent extraction Two methods used to produce terpeneless citrus oils having excellent solubility in diluted ethanol (30–35%) are the following:

a. The whole oil is distilled with three to four times its volume of diluted alcohol in a still fitted with a separator and a return trap to permit recycling of the distillate. During the process, the terpenes distill over with the alcohol vapor and ultimately separate on the surface of the distillate; the required oxygenated fraction is progressively dissolved in the weak alcohol in the still. When no further separation of terpenes takes place the recycling process is discontinued and the oxygenated constituents are recovered by fractional distillation.

b. The whole oil is mixed with alcohol (95% ethanol by volume) and then with a calculated volume of water to reduce the alcohol strength to about 35%. The mixture is vigorously agitated and then allowed to stand to effect separation. The terpenes being insoluble in this strength of alcohol separate as an upper layer. The lower level may be run off and processed directly as a soluble essence for use in the soft drink industry, or it may be further diluted with water, saturated with sodium chloride to break any emulsion, and the terpeneless oil separated and dried. Such oils have a good bouquet but lack the freshness of the original oil.

iii. Chromatographic separation Braverman and Solomiansky (1957) developed a chromatographic method of deterpenation for the production of high quality terpeneless orange oil using silicic acid as an adsorbent for the oxygenated fraction. The technique involves passing the natural citrus oil through a column of silica gel followed by elution of the column with a nonpolar solvent (e.g., hexane). This removes any residual terpenes and allows the oxygenated fraction to be recovered from the silica gel by extraction with ethyl acetate or some other low-boiling polar solvent which can then be removed by vacuum distillation. The process is exothermic so that the column must be cooled and precautions taken to ensure minimum heat damage to the flavoring constituents.

Terpeneless and concentrated citrus oils from which only a part of the terpenes have been removed are widely used as flavoring materials as they have improved stability and a longer shelf life, a lower usage rate and improved solubility making them of particular value in the flavoring of soft drinks and liqueurs.

g. Citrus Leaf and Flower Oils

Although the essential oils obtained from citrus peels play a major role in food and beverage flavorings the quite different oils obtained by steam distillation of citrus leaves, twigs and flowers have an equal importance in fragrances. The oils obtained from the leaves and twigs are collectively known as oil of petitgrain; those obtained from the flowers are called oil of neroli. In each case the source is indicated in the precise designation.

The most widely used of these oils are derived from cultivated bitter orange trees (*C. aurantium* var. *amara*) and are designated bigarade from the French word for bitter orange tree la bigaradier. The oils concerned are (i) oil of petitgrain bigarade—a pale yellow oil having a pleasantly sweet, fresh, floral odor with a woody herbaceous background character and (ii) oil of neroli bigarade–a pale yellow oil, which darkens on keeping, has a very strong, fresh floral odor when newly distilled but loses its top-notes on storage.

These oils have only limited value in flavorings and are used mainly to contribute interesting flavor nuances in imitation fruit flavorings.

2. The Mints—Peppermint, Cornmint, Spearmint

a. Classification

The mints are a very complex and diverse group of perennial plants belonging to the family *Labiatae* which have hybridized to such an extent that their botanical parentage and classification is confusing, even to the experts. For practical purposes the following are recognized as of major commercial importance as flavoring materials either in the form of freshly cut leaves, dried and rubbed flakes or as essential oils.

a. Peppermint [*Mentha x piperita* (L)] which is considered to be a natural hybrid of *M. aquatica*, L and *M. x viridis* (L). It exists in two varieties: *vulgaris* (black or Mitcham mint) and *officinalis* (white mint).
b. Cornmint (*Mentha arvensis*, L), characterized by having an essential oil containing a high percentage of menthol. Two varieties are of commercial importance: *piperascens*, Holmes (Japan and Brazil) and *glabrata*, Holmes (China).
c. Spearmint (*Mentha spicata*, L or *M. cardiaca*, Ger.)—the essential oil of which is rich in 1-carvone.
d. Garden or culinary mint which represents a wide spectrum of species, varieties and cultivars including: *M. viridis* (garden mint), *M. rotundifolia* (applemint), *M. gentilis* (American applemint) and *M. spicata* var. *crispata* (curley mint).

b. Peppermint

Peppermint is cultivated almost exclusively for distillation of its essential oil. Although the so-called Mitcham mint is still regarded as having the finest flavor, this variety is now rarely grown in the United Kingdom where it originated. Plants of it were introduced into the United States in the early nineteenth century and prospered so well that this has become the world's biggest peppermint growing enterprise. Initially cultivation was carried out in the northwestern states but soon spread to the midwestern states of Michigan, Indiana, Ohio and Wisconsin. Unfortunately the "white mint" variety developed various diseases resulting in poor oil yields. This lead to the development of hardier "black mint" cultivars which were introduced into the far west states of Oregon and Washington which now form the prime center of peppermint oil production.

Significant crops of peppermint are also cultivated in Argentina, Australia, Bulgaria, France, Germany, Hungary, India, Italy, Morocco, Netherlands, Poland, Spain, USSR and Yugoslavia. Not all of these are of international importance. There are now so many varieties and strains that the commercially available oils display a wide spectrum of profiles and flavoring quality. Such wide differences are brought about by factors intrinsic to the plant and its environment and also to the methods of harvesting and distillation employed in the various growing areas.

Peppermint is probably one of the world's most popular flavors being used in a wide range of sugar confectionery, chewing gum, chocolate centers, liqueurs, oral toiletries and medicines. The fresh herb is popular in certain regions (notably West Germany and North Africa) for making a refreshing mint tea.

For most purposes the essential oil is the preferred flavoring agent and is commercially available in the following forms:

a. *Natural*—the prime distillation of the leaves and flowering tops of the freshly harvested herb.
b. *Rectified* (*or double distilled*)—a redistillation of the prime oil in which the highest boiling fractions are rejected.
c. *Double rectified* (*or triple distilled*)—a rectification in which 5–10% of the lowest boiling fractions and 2–5% of the high boiling still residues are rejected.

For most purposes the single rectified oil is used and is the one described in the Food Chemicals Codex (1972) and is official in many pharmacopeias. In products where solubility is important, or where a clean peppermint flavor is essential (e.g., crême de menthe, chocolate fillings), the triple distilled oil is preferred.

i. Cultivation and distillation Peppermint is propagated by root cuttings and, as the plants continue to thrive for about 5 years, it is necessary to select the cultivar very carefully. Under optimum soil and climatic conditions, the best yield and quality of oil is produced in the second and third years. The plants are normally replaced after the fifth harvest.

Harvesting takes place when the plants are in full flower, the precise time being critical to ensure maximum yield and a well-balanced oil. The yield decreases rapidly after full bloom due largely to leaf fall. If cutting is delayed, the oil loss can be significant but too early cutting results in a poor quality oil having an indifferent flavor due to an incorrect balance of menthol/menthone. This relationship will be discussed later.

Distillation is carried out on the wilted or partially dried herbaceous tops. In most countries this is carried out in field pot stills, but in the main producing regions of the United States use is made of sophisticated mobile stills the body of which is the truck used to carry the cut herb to the distillation center. In the far west states of Oregon and Washington the whole peppermint operation is fully mechanized and the process has been well described by Quarré (1975) and Green (1975).

In the far west states the average yield of prime oil is upwards of 100 lb/acre (50–60 kg/ha). Percentage yields are rarely quoted as these depend on the state of dryness of the herb when distilled, hence it is more usual to quote yields as weight of oil per area of ground cut.

ii. Rectification Bulks of single distilled peppermint oils as received from the prime producers often have marked and objectionable sulfurous, "weedy" off-notes due to the presence of dimethyl and other polysulfides. These undesirable taints, which detract from the clean, fresh, minty profile, are best removed by a process of rectification. Rectification of any essential oil is a process of redistillation in which high and/or low boiling fractions are separated to an extent necessary to provide an end product having defined physical and sensory properties. The process, although generally considerably improving the sensory attributes of an oil, cannot convert a poor quality oil into a good one. As redistillation involves some degree of concentration the process may accentuate certain abnormalities or imbalance in the flavoring constituents rather than elimination of them.

Rectification is carried out in one of two ways:

a. Dry distillation—the oil is redistilled under high vacuum (5–10 mm Hg) under which conditions the oil boils at a sufficiently low temperature to ensure minimum damage to any thermolabile constituents. The aromatic vapors rise through a suitable fractionating column which enables fractions of a defined boiling range to be separated and collected. Depending on the nature of the start material and the desired character of the end product, the various fractions may be reblended as necessary and any unwanted fractions rejected.

b. Steam injection—this is the older method and is carried out in purpose-designed, jacketed stills fitted with a suitable separator to permit either continuous recycling of the distillate where water is used in the still, or its rejection where high-pressure live steam is injected. Contact of oils with boiling water and/or steam can lead to decomposition of certain constituents so that conditions must be very carefully controlled.

Dry distillation *in vacuo* is now the more widely used technique.

As applied to peppermint oil, rectification results in two principal products:

a. *Oil of peppermint, single rectified* (also called *double distilled*)—a colorless oil prepared by rectification in which the last 2–5% of the oil remaining in the still is rejected. This eliminates the harsh, "weedy," resinous off-notes of the natural oil.

b. *Oil of peppermint, double-rectified* (also called *triple distilled*)—a water-white oil obtained by rectification in which the first fractions of distillate, up to whatever percentage is required, but usually within the range 5–10%, are rejected along with the last 2–5% of still residues. This process removes both the low boiling terpenes and the high boiling odorous compounds. The resulting product has a much improved aromatic profile, better solubility and increased shelf life.

Heath (1978) quotes the following comparative profiles for oils obtained from Oregon:

Single Distilled	Double Distilled	Triple Distilled
Minty	Sharply minty	Strongly minty
Toffeelike	Clean, cooling	Initially cooling then warm
Full-bodied	Smooth	Smooth
Green, herbaceous	Creamy	Sweetly rounded

The rectification of dementholized oil derived from cornmint (*Mentha arvensis*) also significantly improves its odor and flavor profiles. The pungent, harsh flavor character, which resides mainly in the low boiling components, and the unpleasantly dull, oily notes which arise from high boiling constituents are both removed by double rectification. Such oils have profiles much nearer those of *piperita* oils with which they blend well. However, *M. arvensis* oil does have certain characteristic notes which a trained panel can readily detect and as little as 10% of rectified *arvensis* oil can be recognized in a blend with *piperita* oil.

c. Cornmint

Cornmint (*Mentha arvensis*, L) grows wild in China and in parts of Japan but is widely cultivated in both these countries as well as in Taiwan, Brazil, Argentina, India and South Africa. The oil distilled from the harvested herb contains 60–80% of menthol and is usually solid at room temperature. This prime oil is not commercially available and is a rich source of natural menthol. The commercial oil is a dementholized product but still retains between 45 and 55% of total alcohols (as menthol). In the United States such oils are not recognized as peppermint oil and must be clearly labeled as mint oil, cornmint oil or fieldmint oil. When used in any food product, such oils must be similarly declared. In international transactions, however, these oils are frequently called peppermint oil and usually are distinguished by their source (e.g., Brazilian peppermint oil, Chinese peppermint oil).

i. Cultivation and distillation Cultivation of these mints differ considerably with the growing regions and one of the major texts on essential oils (Guenther 1948–1952) should be referred to for more specific information. The following notes merely indicate the differences.

Japan. Plants are raised from root cuttings and give their maximum yield of oil during the second and third years but must be renewed after the fourth year's cropping. In some areas annual replanting is practiced. Harvesting, which varies between one crop per year in the north and three crops per year in the south, is carried out on bright, sunny days when the menthol content of the leaves is at its highest. The cut plant is bundled and hung up, under shade, to air dry for 10–30 days during which it loses about one third of its wet weight. Distillation facilities are somewhat primitive and the numerous small grower/distillers sell their oils to well-equipped central processing companies. Oil yields vary considerably with location and season but average 50 lb/acre (50–60 kg/ha).

China. Mint is usually grown as an annual crop which is harvested twice per year, in June and September. Cutting takes place when the plants are about 3 ft high, the cut material being bundled and sun cured prior to distillation. Field distillation is carried out in small but relatively modern stills, the product being sold to major central processors.

Brazil. Mint was introduced into the São Paulo region by Japanese settlers early in this century. Propagation is by root cuttings planted in cleared ground in late August/early September. The plantations have to be renewed every 3 years owing to limited soil fertility and the onset of rust diseases which result in uneconomic yields. Two crops are harvested per year, cutting taking place when the plants are in full bloom. The cut material is bundled and air dried prior to distillation. Prime oil produced in local field stills is transported to central processors for dementholization.

ii. Dementholization The oils derived from the distillation of the partially dried cut herb contain between 60 and 80% of total menthols. The dementholization process employed in the major processing centers consists of (a) drying the oil and filtering to remove particulate matter,

(b) filling into metal containers, (c) immersing the containers into brine baths prior to reducing the temperature slowly to −5°C—this slow cooling encourages the growth of large menthol crystals—and (d) separating the dementholized oil by draining or centrifugation.

The separated crystals are further purified by recrystallization.

d. Spearmint

The designation spearmint is applied commercially to several species and varieties of *Mentha* possessing a characteristic odor profile which is quite different from either peppermint or cornmint, although the plant form, its cultivation and harvesting are very similar. The main species involved are the following:

Mentha spicata (L) Huds. var. *tenuis* Mich., which was formerly classified as *Mentha viridis* L. (USA).
M. spicata var. *trichoura* Briq. (UK).
M. spicata var. *crispata* Benth. (Germany and Japan).
M. verticillata L. var. *strabola* Briq. (USSR).
M. cardiaca, Gerard ex Baker, called 'Scotch mint' (USA).

The cultivation of spearmint is a relatively small operation in the United States mainly centered in the far west states of Oregon and Washington, although the mint flavor, as provided by the essential oil, is extremely popular particularly in chewing gum, toothpaste and some sugar confectionery. Most of the crop is distilled in the growing regions where yields of up to 40 lb/acre are obtained. Only about 250 tonnes per annum are dried and rubbed for domestic use in the United States. About an equal quantity of the dried broken herb is imported mainly from Egypt, West Germany, Yugoslavia and Bulgaria and much smaller quantities from Spain and Argentina.

There is still some demand for the so-called 'Scotch' spearmint which is cultivated in the midwestern states, as many consider this oil to have a sweeter and more rounded profile.

e. Blended Peppermint Oils

The commercial designation of essential oils is not always an unequivocal indication of either source or quality. This is particularly the case with peppermint oils where there is not only a wide range of materials available but also a broad spectrum of application parameters. Peppermint oil may be single, double or triple distilled, and the quality and flavoring properties of the resulting products depend not only on the start material but also on the processing techniques employed. Alternatives between *piperita* and *arvensis* oils are also available for use where price may be a significant factor. These oils may be offered in their genuine form or as blends formulated to satisfy particular manufacturing needs. In this specialized area of trading it is not unusual to find essential oils which have been sophisticated, even to the extent of the addition of synthetic versions of major components.

Fortunately, current labeling legislation has significantly reduced this practice and today most manufacturers and dealers willingly supply specifications which list the components of blended oils. It is, of course, an offense in the United States to mislabel as natural an essential oil, or other flavoring material, which has been modified by the addition of synthetic components. Such products must be clearly described as imitation.

The quality and suitability for use of blended peppermint oils should be judged on three criteria: sensory evaluation in their intended end use product, physicochemical characteristics and price. When dealing with blends of any essential oils, it is a matter of taste and the effect required which determines their acceptability. There is no need to use a very expensive oil if a cheaper grade will do the job just as well, perhaps, even better. In many cases it is a matter of what one expects to get for one's money. Whatever the nature of the blend, it is almost certain that any pretensions to quality will depend to a large extent on the percentage of the genuine oil which is present. Careful quality assessment is necessary, particularly the evaluation of the odor profile of the oil on a smelling strip over a period of 24 h, supported by an examination of the oil by gas chromatography. Such an examination will draw attention to any abnormalities, which can then be questioned. The physical characters of the principal commercially available essential oils are published in EOA Specifications and Standards (1976).

Where *arvensis* oil is used in a blend the title should indicate this, often by the use of the abbreviation WONF (with other natural flavors).

f. Composition of Mint Oils

There are considerable differences in the composition of the various mint oils, determined by species and environmental conditions during growth, harvesting and postharvest handling. It is convenient to consider these oils in two groups: (a) peppermint and (b) spearmint. Peppermint (*M. piperita*) and cornmint (*M. arvensis*) oils have much in common, the principal constituents of each being menthol, menthone, menthyl acetate and other esters. It is the ratio of these components which determines their distinctive profiles. The relative composition of oils from various sources is given in the following tabulation:

Source	Menthol (%)	Menthone (%)	Menthyl Esters (%)
M. piperita oils			
American, midwest	45–61	14–30	4.6–9.6
American, far west	51–65	17–21	3.1–9.5
English 'Mitcham'	46–68	9–12	3–21
M. arvensis oils (dementholized)			
Brazilian	33–50	21–30	5–20
Chinese	36–39	25–28	2–10
Japanese	36–39	22–25	10–12

Menthofuran, which is present in *piperita* oils is absent from *arvensis* oils. Based on this observation, several tests have been proposed to distinguish between the more expensive *piperita* oils and the considerably cheaper *arvensis* oils. The best known test is that described in the USP XX(1980), but unfortunately the distinguishing color, which is due to the presence of menthofuran, is produced in blends containing as little as 15% *piperita* oil. Gas chromatography and critical sensory appraisal are more reliable methods of detecting undisclosed blending of these oils.

Research into the biosynthesis of mint oil constituents has established that menthone is formed from piperitone as the plant matures and that this is further converted into menthol, the percentage of which continues to increase in the leaves until the plant is in full bloom. It is the final balance of menthol/menthone which is the major determinant of the flavoring quality of the distilled essential oil. Smith and Levi (1961) have proposed certain compositional ratios as measures of quality. Those of particular value include the following terpenes/other constituents; menthone/isomenthone; limonene/cineole; menthofuran/menthone-related constituents; neomenthol/menthyl acetate; menthol-related compounds/neomenthol and finally menthone-related compounds/menthol-related compounds (i.e., menthofuran + menthone + isomenthone/neomenthol + menthol + menthyl acetate). Reference should be made to the original literature for the significance of these ratios in the determination of the quality and genuiness of any given oil sample.

Spearmint (*M. spicata*) oil is characterized by having a high content of *l*-carvone which is present at about 56%. Its sensory and physicochemical characteristics are entirely different from those of peppermint.

The compositional data of the various mint oils is given in Tables 6.22 and 6.23.

Table 6.22. Compositional Data for Mint Oils[a,b,c]

	M. piperita (%)	*M. arvensis* (%)
Terpenes:		
α-Pinene	0.5–0.8	0.5–1.2
Camphene	0–0.2	0
β-Pinene	0.9–1.7	0.7–2.7
Limonene	3.2–4.3	4.9–11.2
1,8-Cineole and γ-terpinene	6.4–13.5	0.2–0.1
Oxygenated compounds:		
3-Octanol and menthofuran	3.2–9.4	0.3–3.6
Linalool	0–0.4	0.1–0.8
Menthone	8.9–23.8	21.5–29.7
Isomenthone	2.2–4.5	2.9–8.9
Neomenthol	2.8–4.2	2.8–5.0
Menthol	40.2–48.7	33.2–39.3
Pulegone	0.9–2.6	0.5–4.9
Menthyl acetate	4.4–11.6	0.9–6.2
Piperitone	0.7–3.1	1.8–11.6

[a] Source: Smith and Levi (1961).
[b] Minor components amount to about 1.5%.
[c] Precise composition is determined by variety and geographical source as well as by imposed environmental factors.

Table 6.23. Compositional Data (%) for Japanese Spearmint Oil[a,b]

Terpenes:		*Oxygenated compounds:*	
α-Pinene	0.7	3-Octyl acetate	0.2
β-Pinene	2.3	3-Octanol	0.4
Limonene	20.2	*cis*-3-Hexenyl isovalerate	0.2
1,8-Cineole	0.8	Menthone	0.4
		Linalool	1.1
Sesquiterpenes:		Menthol	0.5
β-Bourbonene	0.2	Dihydrocarvone	1.8
Caryophyllene	0.3	Dihydrocarvyl acetate	1.6
δ- and γ-Cadinene	0.3	Borneol	0.1
		Neohydrocarveol	0.1
		Carvone	1.3
		trans-Carvyl acetate	63.3
		cis-Carvyl acetate	0.3
		trans-Carveol	1.8[c]
		cis-Carveol	0.2
		Perillyl alcohol	0.4
		cis-Jasmone	0.1

[a] Source: Nagaswara *et al.* (1974).
[b] Trace quantitities have also been reported of the following: camphene, iosmenthone, β-elemene, farnescene, β-humulene, farnescol, elemol, cadinol, carvacrol and eugenol.
[c] The characterizing odor component.

Table 6.24. Principle Essential Oils Used in Flavorings[a]

Common Name	Botanical Source	Principal Producing Countries	Main Crop Period	Average Yield of Essential Oil (%)	Chief Aromatic Constituents	Regulatory Status
Allspice (pimento)	Fruits of *Pimenta dioica*, L.	West Indies (Jamaica)	July-Aug.	3–4.5	Eugenol	GRAS
Ambrette	Seeds of *Hibiscus abelmoschus*, L.	Java Martinique		0.2–0.6	Farnesol Ambrettolide	GRAS
Angelica	Root and seeds of *Angelica archangelica*, L.	Europe (Germany, France)	July	Root: 0.3–1 seeds: 0.6–1.5	β-Phellandrene	GRAS
Angostura	Bark of *Galipea officinalis*, Hancock. (*G. cuspania*, DC)	South America (Venezuela)		1–2		GRAS
Anise	Fruits of *Pimpinella anisum*, L.	Europe U.S.S.R.	Aug.-Sept.	1.5–4	Anethol	GRAS
Artemisia	Herbaceous tops of *Artemisia absinthium*, L.	Central Europe Asia North America	Aug.	0.3–0.4	Thujone	GRAS, thujone free
Asafetida	Oleo-gum-resin from *Ferula asafoetida*, L. and related species	Iran Afganistan	July	7–9	Disulfides	GRAS
Balm (lemon balm)	Herbaceous tops of *Melissa officinalis*, L.	Southern and Central Europe North Africa	Aug.	0.1	Geraniol Citronellol Citral	GRAS
Balsam Peru	Oleoresin from *Myroxylon pereirae*, Klotzsch.	El Salvador	Nov.–June	45–55 (from the extracted balsam)	Benzyl cinnamate Benzyl benzoate	GRAS
Basil, sweet	Herbaceous tops of *Ocimum basilicum*, L. Several varieties are distilled:	Mediterranean U.S.A.	Sept.	0.1	Methyl chavicol Linalool Cineol	GRAS
	Réunion-type	Malagasy Republic				
	Methyl cinnamate-type	Bulgaria Haiti				
	Eugenol-type	Java U.S.S.R.				
Bay laurel (sweet bay)	Leaves of *Laurus nobilis*, L.	Mediterranean	Sept.–Mar.	0.5–1	Cineol	GRAS
Bay, West Indian	Leaves of *Pimenta racemosa* (Mill) J. W. Moore (formerly *Myrcia acris*, Wight	West Indies (Dominica) Venezuela	Jan.–June	1–2	Eugenol	GRAS
Bergamot (bergamot orange)	Peels of *Citrus bergamia*, Risso or *C. aurantium*, L. sub-species *bergamia*, Wright and Arn.	Calabria (Italy)	Dec.–Mar.	0.5 (approx)	Linalyl esters	GRAS
Birch bark (sweet birch)	Bark of *Betula lenta*, L.	Canada U.S.A. Eastern Europe, U.S.S.R.	Oct.–Apr.	0.5	Methyl salicylate	
Bitter almond	Partially de-oiled press-cake from the kernels of *Prunus amygdalus*, Batsch, var. *amara*.	U.S.A. Middle East Southern Europe North Africa		0.5–0.7	Benzaldehyde (oil is treated to remove HCN)	GRAS, free from HCN
Bois de Rose	Wood of *Aniba rosaeodora*, Ducke.	South America (Brazil, Peru, Guiana)	Jan.–June	0.7–1.2	Linalool	GRAS
Buchu	Leaves of *Barosma betulina*, Bartl and Wendl, *B. crenulata* (L) Hook., or *B. serratifolia*, Willd.	South Africa		0.8–3.5 (depending on species)	Diosphenol	Good mfg practice
Cacao	Fermented and roasted seeds of *Theobroma cacao*, L.	Central and South America Africa		0.001 (approx)	Linalool	GRAS

[a] From Heath (1978).

Continued.

Table 6.24. Continued.

Common Name	Botanical Source	Principal Producing Countries	Main Crop Period	Average Yield of Essential Oil (%)	Chief Aromatic Constituents	Regulatory Status
Guaiac	Wood of *Guaicum officinale*, L. *G. sanctum*, L. or *Bulnesia sarmienti*, Lor.	South America Jamaica Cuba		approx 5 (much lower in local field stills)	Guaiol	Good mfg practice
Hemlock (spruce)	Needles and twigs of *Tsuga canadensis* (L) Carr. *T. heterophylla*, (Raf) Sarg. *Picea glauca* (Moench) Voss. or *P. mariana* (Mill) B.S.P.	North America China Japan			α- and β-pinenes (terpenes approx 45%) Bornyl acetate	Good mfg practice
Hops	Flower catkins of *Humulus lupulus*, L.	Europe North America	Sept.	0.3–0.5	Myrcene Humulene	GRAS
Horsemint (Monarda)	Herbaceous tops of *Monarda punctata*, L.	U.S.A.		1–3	Thymol (approx 60%)	GRAS
Hyssop	Herbaceous tops of *Hyssopus officinalis*, L.	Southern Europe	Aug.	0.1–0.3	α-Pinene Sesquiterpene alcohols	GRAS
Immortelle	Flowering tops of *Helichrysum angustifolium*, DC.	Mediterranean countries (Yugoslavia)		less than 0.1	Nerol Neryl acetate	GRAS
Jasmine	Flowers of *Jasminum officinale*, L. or *J. grandiflorum*, L.	Mediterranean countries Asia	June–Sept.	3 of concrete containing 40–50% essential oil	Nerol Terpineol Jasmone	GRAS
Juniper berry	Berries of *Juniperus communis*, L.	Europe (Italy) Asia North America	Aug.–Sept.	0.5–0.6 (up to 2.5 from dried berries)	α-Pinene	GRAS
Laurel (sweet bay)	Leaves of *Laurus nobilis*, L.	Mediterranean countries	Sept.–Mar.	0.5–1	Cineole	GRAS
Labdanum	Gum resin from the leaves and twigs of *Cistus ladaniferus*, L. and other *Cistus* sps.	Spain Cyprus	July–Aug.	1–2 (from the gum resin)		Good mfg practice
Lavandin	Flowering tops of *Lavandula hybrida*, Rev. being a hybrid of *L. latifolia*, Vill. and *L. officinalis*, Chaix.	Southern France	Aug.	0.5–1	Linalyl acetate (20–22%)	GRAS
Lavandin (Abrialis type)	Flowering tops of a selected strain of *L. hybrida*, Rev.	Southern France	Aug.	0.5–1	Linalyl acetate (30–32%)	
Lavender	Flowering tops of *Lavandula, officinalis*, Chaix.	Mediterranean countries U.K. North Africa	July–Aug.	0.5–1	Linalool Linalyl acetate	GRAS
Lavender, spike	Flowering tops of *Lavandula latifolia*, Vill. (*L. spica*, DC)	Spain France	Aug.	0.75–1	Linalool Linalyl acetate	GRAS
Lemon	Peel of *Citrus limon*, (L) Burm. fruit by cold expression.	Italy California Spain Israel	Dec.–Apr. through to Oct.	4 (approx)	Limonene (90%)	GRAS
Lemongrass	Herbaceous tops of *Cymbopogon citratus*, DC or *Andropogon nardus* var. *ceriferus*	West Indies (Malagasy Guatemala)	Aug.	0.3–0.4	Citral (approx 70%)	GRAS
	C. flexuosus, (Nees.) Stapf. or *Andropogon nardus* var. *flexuosus*	East Indies (India Indonesia Sri Lanka)	May–June	0.2–0.3	Citral (80–85%)	
Lemon verbena	Leaves and twigs of *Lippia citriodora*, H.B.K.	Europe	June–Oct.	0.1–0.2	Citral	Good mfg practice in alcoholic beverages only
Lime	Peels of *Citrus aurantifolia*, Swingle fruit by cold expression or by distillation as a by-product of lime juice production	West Indies Mexico Florida	Throughout the year	0.1 (approx)	Citral	GRAS

Table 6.24. Continued.

Common Name	Botanical Source	Principal Producing Countries	Main Crop Period	Average Yield of Essential Oil (%)	Chief Aromatic Constituents	Regulatory Status
Cumin (cummin)	Ripe fruits of *Cuminum cyminum*, L.	Mediterranean countries India U.S.S.R.	Sept.	2.5 (approx)	Cuminic aldehyde	GRAS
Davana	Flowers of *Artemisia pallens*, Wall.	Southern India	Apr.–May	0.1–0.6 (approx)		Good mfg practice
Dill (seed)	Fruits of *Anethum graveolens*, L.	Europe Pakistan U.S.A.	Sept.	2.5–4	Carvone	GRAS
Dill (weed)	Green fruiting tops of *Anethum graveolens*, L.	Eastern Europe U.S.A.	July	0.3–1.5	Carvone	GRAS
Dill, Indian	Fruits of *Anethum sowa*, Roxb. or *Peucedanum graveolens*, Benth & Hook.	India		1.2–3.5	Carvone Limonene Dillapiol	Good mfg practice
Elder flowers	Flowers of *Sambucus canadensis*, L. and *S. nigra*, L.	Hungary Germany France	June			GRAS
Elemi	Resin from *Canarium commune*, L. or *C. luzonicum*, Mig.	Phillipines	June–Oct.	20–30 (from the crude resin)		GRAS
Erigeron	Herbaceous tops of *Erigeron canadensis*, L.	Canada U.S.A. Southern Europe	July–Nov.	0.3–0.6	Terpinyl acetate	
Estragon (tarragon)	Herbaceous tops of *Artemisia dracunculus*, L.	Central and eastern Europe	July–Aug.	0.3–1.5	Methyl chavicol (estragole) Anethol	GRAS
Eucalyptus	Leaves of *Eucalyptus globulus*, Labill. and many other *Eucalyptus* sps. (Guenther 1949)	Spain Portugal Brazil Zaria		0.75–1.25	Cineole	GRAS
Fennel, sweet	Fruits of *Foeniculum vulgare*, Mill. var. dulce	Germany Romania U.S.S.R. North Africa India	Sept.	4–6	Anethol	GRAS
Fir (pine)	Needles, twigs and cones of: Siberian Fir (*Abies sibirica*, Lebed); Silver Fir (*A. alba*, Mill.); Japanese Fir (*A. Sachilenensis*, Masters or *A. mayriana*, Miyabe & Kudo)	U.S.S.R. North America Europe Japan		0.3 (approx)		Good mfg practice
Galangal	Rhizomes of *Alpinia officinarum*, Hance.	China		0.5–1		GRAS
Galbanum	Dried gum-resin from *Ferula galbaniflua*, Boiss & Buhse and other *Ferula* sps.	Iran Turkestan		10–22 (from the crude resin)		Good mfg practice
Garlic	Bulbs of *Allium sativum*, L.	U.K. Italy Egypt U.S.A.		0.1–0.25	Allyl sulfides	GRAS
Geranium	Leaves and stems of *Pelargonium graveolens*, L'Her. and other *Pelargonium* sps.	Réunion North Africa Central and western Europe	Mar.–Oct. (depending on region)	0.15–0.2 (from fresh material)	Geraniol Citronellol	GRAS
Ginger	Rhizomes of *Zingiber officinale*, Roscoe	Jamaica West Africa India China Australia	Jamaica: Febr.–Mar.	0.25–1.25	Zingiberene Citral	GRAS
Grapefruit	Peel of *Citrus paradisi*, Macf. fruit, by cold expression.	U.S.A. East Africa West Indies Brazil				GRAS

Continued.

Table 6.24. Continued.

Common Name	Botanical Source	Principal Producing Countries	Main Crop Period	Average Yield of Essential Oil (%)	Chief Aromatic Constituents	Regulatory Status
Camomile (German chamomile)	Flowers of *Matricaria chamomilla*, L.	Europe Asia U.S.A. Australia	May–July	0.3–1	Chamazulene	GRAS
Camomile (Roman or English chamomile)	Flowers of *Anthemis nobilis*, L.	Europe	July–Oct.	0.3–1		
Camphor, tree	Wood of *Cinnamomum camphora* (L) Eberm.	Far East (China, Taiwan, Japan)			Camphor and linalool depending on variety	
Cananga	Flowers of *Cananga odorata*, Hook and Thoms.	Islands in Indian Ocean Java	Oct.–Dec.	0.5–1	Sesquiterpene alcohols	GRAS
Caraway	Fruits of *Carum carvi*, L.	Europe (Holland, Hungary) North America	Aug.–Sept.	3–6	Carvone	GRAS
Cardamom	Seeds of *Elettaria cardamomum*, Maton.	India Guatamala	India: Aug.–Oct. Guatamala: Jan.	5 (approx) 3–6	Cineole d-α-Terpineol Terpinyl acetate	GRAS
Carrot	Seeds of *Daucus carota*, L.	France	Sept.	0.4–0.8	Carotol	GRAS
Cascarilla	Bark of *Croton eluteria*, Bennett	Bahamas		1.5–3		GRAS
Cassia, Chinese	Bark of *Cinnamomum cassia*, Blume.	Southeast China	June–July	0.3 (approx)	Cinnamic aldehyde	GRAS
Cedar, white (thuja)	Leaves and twigs of *Thuja occidentalis*, L.	U.S.A. Canada Japan	Apr.–Nov.	0.6–1	α-Pinene Thujone	GRAS (thujone free)
Celery seed	Fruits of *Apium graveolens*, L.	France Holland Hungary India	Sept.	1.5–2.5	d-Limonene Sedanolide	GRAS
Cherry, wild bark	Bark of *Prunus serotina*, Ehrh. or *P. canadensis*, Poiret.	Canada U.S.A. Europe				GRAS (mainly as extract)
Chervil	Herbaceous tops of *Anthiscus cerefolium* (L) Hoffm.	Southern Europe U.S.A.				GRAS
Cinnamon bark, Ceylon	Bark of *Cinnamomum zeylanicum*, Nees.	Sri Lanka	July–Aug. and Dec.–Jan.	0.5–0.8	Cinnamic aldehyde	GRAS
Cinnamon bark, (cassia)	Bark of *C. cassia*, Blume *C. loureirii*, Nees. *C. burmanni*, Blume	China Saigon Batavia		0.5–3 (depending on species)	Cinnamic aldehyde	GRAS
Cinnamon leaf, Ceylon	Leaves of *Cinnamomum zeylanicum*, Nees.	Sri Lanka	July–Sept.	0.5–0.7	Eugenol	GRAS
Citronella, Ceylon	Fresh or dried herb *Cymbopogon nardus*, Rendle	Sri Lanka	Three crops per year	0.5 (approx)	Geraniol Citronellal	GRAS
Citronella, Java	Fresh herb *Cymbopogon winterianus*, Jowitt.	Java	3–4 crops per year	0.7 (approx)	Geraniol Citronellal	GRAS
Clove bud	Unopened buds of *Eugenia caryophyllata* Thunb.	Tanzania Malagasy Republic East Indies	Throughout the year	14–21	Eugenol	GRAS
Clove leaf	Leaves of *Eugenia caryophyllata*, Thunb.	East Indies		1.5–2	Eugenol	GRAS
Clove stem	Stems of *Eugenia caryophyllata*, Thunb.	East Indies		5 (approx)	Eugenol	GRAS
Clover	Flowers of *Trifolium* species	Europe Central Asia		0.005–0.02		GRAS
Cognac oil	Wine lees and dregs or pressed grape skins	France		0.04–0.1 (depending on source)	So-called ethyl aenanthate	GRAS
Coriander	Fruits of *Coriandrum sativum*, L.	North Africa Russia India	Sept.	0.1–1	Linalool	GRAS
Cubeb	Unripe berries of *Piper cubeba*, L.	Indonesia Africa	Apr.–May	13 (approx)	Cineole	Good mfg practice

Table 6.24. Continued.

Common Name	Botanical Source	Principal Producing Countries	Main Crop Period	Average Yield of Essential Oil (%)	Chief Aromatic Constituents	Regulatory Status
Linaloe	Wood of *Bursera delpechiana*, Poiss. and other *Bursera* sps.	Mexico	Jan.–May	8 (approx)	Linalool (60–75%)	Good mfg practice
Lovage	Rhizomes of *Levisticum officinale*, Koch.	France Germany Czechoslovakia	Sept.	fresh: 0.1–0.2 dried: 0.5–1	α-Terpineol	Good mfg practice
Mace	Arilus of the fruits of *Myristicum fragrans*, Houtt.	West Indies East Indies	Throughout the year	4–15		GRAS
Mandarin (tangerine)	Peels of *Citrus reticulata*, Blanco. fruit by cold expression or by steam-distillation.	Italy U.S.A.		0.5 (approx)	Decylaldehyde	GRAS
Marjoram, sweet	Herbaceous tops of *Majorana hortensis*, Moench. (*Origanum majorana*, L.)	Europe	Aug.–Sept.	0.2–0.3	Terpen-4-ol α-Terpineol	GRAS
Mentha Arvensis (corn mint)	Leaves of *Mentha arvensis*, L. var. *piperascens*, Holmes or *Mentha arvensis*, L. var. *glabrata*, Holmes.	Brazil Japan China	Brazil: Jan.; Mar.–Apr. Japan: Aug. and Oct.	0.5–1 (depending on source)	Commercial oil is partially de-mentholated to a residual menthol level depending on source	Not permitted in U.S.A. but widely used elsewhere
Mustard	De-oiled press-cake of the seeds of *Brassica nigra* (L) Koch.	U.K. Europe	Sept.–Oct.	0.6–1.2	Allyl isothiocyanate	GRAS
Myrrh	Oleo-gum-resin from *Commiphora molmol*, Engl., *C. abyssinica*, Engl. and other *Commiphora* sps.	Somalia Ethiopia		3–8 (from the gum)		Good mfg practice
Myrtle	Leaves of *Myrtus communis*, L.	Mediterranean countries	July–Aug.	0.2–0.3	α-Pinene Camphene Cineole	Good mfg practice
Nutmeg	Kernels of the fruits of *Myristica fragrans*, Houtt.	East Indies West Indies Sri Lanka		6.5–13 (depending on source)	α- and β-Pinenes Limonene 4-Terpineol Safrole Myristicin	GRAS
Ocimum	Herbaceous tops of *Ocimum* sps. Thymol-type: *O. gratissimum*, Eugenol-type: *O. gratissimum*,	 West Africa Seychelles Indonesia Brazil			 Thymol Eugenol	
Onion	Bulbs of *Allium cepa*, L.	U.K. Europe U.S.A.		0.005–0.01	n-Propyl disulfide	GRAS
Orange, bitter	Leaves and twigs (petitgrain); flowers (neroli bigarde) or peel of *Citrus aurantium*, L. subspecies *amara*, L.	Sicily Spain Guinea Brazil		Petitgrain: approx 0.2 Neroli approx 0.1 Peel: 0.4–0.5	Linalyl acetate	GRAS
Orange, sweet	Leaves and twigs (petitgrain); flowers (neroli) or peel of *Citrus sinensis*, L. Osbeck. or *C. aurantium*, L. var. *dulcis*, L.	Italy Spain North Africa California Florida		Peel: 0.4–0.5	Limonene (90%) Decylaldehyde	GRAS
Origanum (wild marjoram)	Herbaceous tops of *Origanum vulgare*, L. and other *Origanum* sps. (Arctander 1960)	U.S.S.R. Central Europe Italy	Aug.	1 (approx)	Thymol or carvacrol depending on sps. or variety	GRAS
Orris	Peeled rhizomes of *Iris germanica*, L. or *I. pallida*, Lam.	Italy	Peeled rhizomes are aged for 2 yr	0.2–0.3	Irone, fatty acid esters	Good mfg practice

Continued.

Table 6.24. Continued.

Common Name	Botanical Source	Principal Producing Countries	Main Crop Period	Average Yield of Essential Oil (%)	Chief Aromatic Constituents	Regulatory Status
Palmarosa	Herbaceous tops of *Cymbopogon martini*, Stapf. var *motia*.	Indonesia India Seychelles	Java: every 4 months India: Oct.–Sept.		Geraniol (approx 90%)	GRAS
Parsley Seed	Ripe seeds of *Petroselinum sativum*, Hoffm	Europe U.S.A.		1.5–3.5	Apiol	GRAS
Parsley herb	Young green tops of *Petroselinum sativum*, Hoffm.	U.K. Hungary Germany U.S.A.		0.05 (approx)		GRAS
Patchouly	Leaves of *Pogostemon cablin*, Benth. or *P. heyneanus*, Benth.	Indonesia Malaysia Brazil	Dec.–Jan.	1.5–3		Good mfg practice
Pennyroyal	Herbaceous tops of *Hedeoma pulegioides* (L) Pers. (American) or *Mentha pulegium*. L. (European)	U.S.A. Mediterranean countries	May–Sept.	0.5–1	Pulegone	Good mfg practice
Pepper, black	Dried immature berries of *Piper nigrum*, L.	India Indonesia Brazil	Aug. and Nov. (depending on region)	1–2.7	Terpenes	GRAS
Peppermint	Flowering tops of *Mentha piperita*, L.	U.S.A. Italy Japan	Sept.	0.3–0.7 (depending on source and variety)	Menthol Menthone Menthyl acetate	GRAS
Petitgrain	Leaves and twigs of Citrus sps. (*q.v.*)					
Pimento (allspice)	Fruits of *Pimenta dioica*, L. (*Pimenta officinalis*, Lindl.) An oil is also distilled from the leaves	West Indies	July–Aug.	3–4.5	Eugenol (70–80%)	GRAS
Pine, dwarf	Needles and twigs of *Pinus mugo*, Turra. var. *pumileo* (Haenke) Zenari	Central Europe		0.3–0.4	α-Pinene	Good mfg practice
Pine, Scotch	Needles and twigs of *Pinus sylvestris*, L.	Europe Asia North America			Borneol Bornyl acetate	Good mfg practice
Pine, white	Needles and wood of *Pinus strobus*, L.	North America (but of no commercial value)			α- and β-pinene	Good mfg practice in alcoholic beverages only
Rose	Flowers, buds and leaves of *Rosa damascena*, Mill., *R. alba*, L., *R. centifolia*, L., *R. gallica*, L. and vars. of these sps.	Bulgaria Turkey Morocco France Italy	June	Depends on distillation method (Guenther 1949)	Rhodinol (30–60%)	GRAS
Rosemary	Whole plant or leaves of *Rosmarinus officinalis*, L.	Spain and other Mediterranean countries	May–Sept.	0.5–1.2	Cineole Borneol	GRAS
Sage, Dalmatian	Leaves of *Salvia officinalis*, L.	Yugoslavia	Sept.	0.5–1	Cineole	GRAS
Sage, Greek	Leaves of *Salvia triloba*, L.	Greece Turkey	Sept.		Cineole	GRAS
Sage, Spanish	Leaves of *Salvia lavandulaefolia*, Vahl.	Spain	Sept.		Cineole	GRAS
Sandalwood, white (East Indian)	Wood from *Santalum album*, L. or *S. spicatum* (R. Br) and *Eucarya spicata*, (R. Br) Sprag & Summ.	India (Mysore) Australia		Mysore: 4.5–6 Australia: 1.4–2.6	Santalol (90%)	Good mfg practice
Sandalwood, West Indian (Amyris)	Wood from *Amyris balsifera*, L.	South America Haiti Jamaica		2.5–3	β-Caryophyllene	Good mfg practice
Sassafras	Root bark and leaves of *Sassafras albidum*, (Nutt) Nees.	U.S.A. Canada	Aug.–Sept.	2–3	Safrole	Not permitted in food in many countries

Table 6.24. Continued.

Common Name	Botanical Source	Principal Producing Countries	Main Crop Period	Average Yield of Essential Oil (%)	Chief Aromatic Constituents	Regulatory Status
Sassafras, Brazilian	Wood from *Ocotea cyrubarum*, H.B.K. (*O. pretiosa*)	Brazil	July–Oct.	1.2–1.5	Safrole (90%)	
Savory, summer	Herbaceous tops of *Satureia hortensis*, L.	Southern Europe U.S.A.	July–Aug.	0.1 (approx)	Carvacrol	GRAS
Savory, winter	Herbaceous tops of *Satureia montana*, L. freed from woody portions	Spain Morocco Yugoslavia	Aug.	0.2 (approx)	Carvacrol	GRAS
Schinus Molle (Californian or Peruvian pepper)	Dried berries of *Schinus molle*, L.	Mediterranean countries	Aug.	5–7	β-Phellandrene	GRAS
Spearmint	Leaves of *Mentha spicata*, Houds or L.	North America Europe	Sept.	0.6 (approx)	1-Carvone	GRAS
Spike lavender	Flowering tops of *Lavandula latifolia*, Vill (*L. Spica*, DC)	Spain France	Aug.	0.75–1	Linalool Linalyl acetate	GRAS
Star anise	Fruits (seeds) of *Illicium verum*, Hook.	China Vietnam	Throughout the year; max. Aug.–Sept.	8–9	Anethole (85–90%)	GRAS
Storax (styrax)	Balsamic exudate from *Liquidambar orientalis*, Mill. (Levant or Asian) or *L. styraciflua*, L. (American)	Asia Minor Southern U.S.A. Mexico Honduras Guatemala		0.5–1 (from the crude balsam)		Good mfg practice
Tagetes (marigold)	Flowering tops of *Tagetes patula*, L., *T. erecta*, L. or *T. minuta*, L (*T. gladulifera*, Schrank)	India Southern Mexico South Africa Australia		0.3–0.4	Ocimene Tagetone (50–60%)	Good mfg practice
Tarragon (estragon)	Herbaceous tops of *Artemisia dracunculus*, L.	Central and eastern Europe	July–Aug.	0.3–1.5	Methyl chavicol (estragole) Anethol	GRAS
Thyme	Herbaceous tops of *Thymus vulgaris*, L., *T. zygis*, L. var. *gracilis*, Boiss., (white thyme) or *T. serpylllum*, L (wild thyme) or *T. capitatus*, Hoff & Link. (Spanish 'origanum') (Arctander 1960)	Mediterranean countries Central Europe North America	May–Sept.	0.5–1.5 (depending on source)	Thymol Carvacrol	GRAS except Spanish "origanum" which is good mfg practice
Tolu balsam	Balsamic exudate of *Myroxylon balsamum* (L) Harms.	Venezuela		2–7 (from the crude balsam)	Benzyl benzoate Benzyl cinnamate	
Tuberose	Flowers of *Polyanthes tuberosa*, L.	France Morocco	Aug.–Sept.	3–6 (from the extracted concrete)		GRAS
Turmeric (curcuma)	Dried rhizomes of *Curcuma longa*, L.	India Malaysia Sri Lanka Japan		1.5–5	Turmerone	GRAS
Wintergreen	Leaves of *Gaultheria procumbens*, L.	Canada U.S.A.	June–Sept.	0.5–0.7	Methyl salicylate	
Ylang-ylang	Flowers of *Cananga odorata*, Hook. and Thoms.	Phillipines Malagasy Réunion		1 (approx)	Alcohols and esters (50–60%) Sesquiterpenes (30–35%)	GRAS
Zedoary	Rhizomes of *Curcuma zedoaria*, Rosc.	India		1 (approx)	Cineole	GRAS

Source: In addition to Arctander (1960) and Guenther (1949), data was procured from numerous articles and papers which have appeared in *Flavor Industry* and *International Food Additives and Flavours*.

3. Other Commercially Important Sources

Many other aromatic plant materials are collected or cultivated as a source of essential oils, many being of considerable commercial importance. Such essential oils are used as flavoring agents in a wide range of food, beverage and confectionery products and also in numerous fragrance applications. These sources and oils have been fully described in several standard texts, particularly in Guenther (1949–1952); Gildermeister and Hoffmann (1965–1967); Arctander (1960); and Fenaroli (1975). Reference should be made to these works as necessary.

For convenience, Table 6.24 shows the principal essential oils used as flavoring materials [from Heath (1978)].

E. Fruit, Fruit Juices and Concentrates

Fruits have formed an essential dietary supplement since the earliest times, providing nutritionally valuable components coupled with a wide range of very acceptable flavors; they may be regarded as nature's convenience foods.

Botanically, fruits are the matured ovary with or without seeds and sometimes with the remains of floral parts. The fruit wall, which may be either dry or fleshy is known as the pericarp and it is this which forms the edible part of most fruits. In the case of strawberries it is the swollen receptacle which forms the edible fruit. Most common fruits are used for dessert purposes but what are not sold on the domestic market may be used as the start material for several products of value in food processing. These include (a) fruit pieces, pulp and puree; (b) dehydrated fruit; (c) fruit juices and concentrates; (d) fruit powders; (e) preserved fruits (canned or bottled) and (f) jams and jellies.

Nuts are also fruits but it is usual to treat these separately as their use and applications are so very different.

1. Classification of Fruits

Fruits are classified into six categories, as follows:

a. *Berries*—fruits in which the edible part is pulpy and more or less juicy, containing numerous seeds: black currant, blackberry (bramble), blueberry, boysenberry, cranberry, gooseberry, huckleberry, kiwi fruit, loganberry, raspberry, red currant, and strawberry.

b. *Citrus*—fleshy edible portion is segregated into sections surrounded by a thick skin which itself contains a strongly aromatic essential oil: bergamot, citron, grapefruit, kumquat, lemon, lime, tangerine (mandarin), orange, sweet, and orange, bitter.
c. *Stoned fruit*—fruit contains a single seed (the stone or pit) surrounded by a fleshy pericarp which is normally the edible portion: apricot, cherry, damson, peach, and plum.
d. *Grapes*—small multiseed or seedless fruits which grow on vines in clusters.
e. *Melons*—large fruit containing either a mass of seeds in the central cavity within a soft pulpy flesh or embedded throughout a watery cellular flesh: cantaloupe, honeydew, musk, and water.
f. *Pomes*—fruits in which the body is formed by an elongation and swelling of the receptacle surrounding the seed capsule: apple, pear, and quince.
g. *Tropical fruits*—classified together for convenience but embracing many fruit forms: avocado, banana, date, fig, mango, pineapple, and pomegranate.

Most edible fruits consist of a skin or peel which encloses the soft pulpy tissues and it is this fleshy pericarp which is of most importance as a source of flavor. This tissue contains the juice comprising sugars (e.g., dextrose, fructose and sucrose), fruit acids (e.g., citric, tartaric, malic acids), coloring matter and, in some fruits, phenolic bodies including tannins. There is a correlation between the sugar content, color, acidity and flavor and the balance changes as the fruit ripens. In general, the flavor is at a maximum when the sugar content is highest and the skin color at its brightest. The flavor changes and deteriorates significantly as the fruit becomes overripe.

2. Fruit Juice and Flavor

A fruit juice is defined as the clear or uniformly cloudy, unfermented liquid recovered from sound fruits by pressing or other mechanical means. The precise technique employed in its manufacture depends on the nature of the fruit and not all fruits are of equal value as a commercial source of juice. The selection of fruit for juice processing calls for considerable expertise and an understanding of the balance between variety, ripeness and physical quality of the fruit used and the character and stability of the separated juice. The expressed juice does not contain the flavor of the whole fruit as the characteristic volatile components, which are the source of the aromatic profile, are present in the freshly prepared juice in less than 0.07 part per thousand. The flavor normally associated with fresh fruit can only be achieved by blending juice with aromatics recovered from the rest of the fruit.

a. Fruit Juice Extraction

The processing of fruit juice lies outside the scope of this book and reference should be made to standard works including Nelson and Tressler (1980) and Nagy *et al.* (1977).

Basically, the extraction process involves loading the soft fruit into a stainless steel vertical basket hydraulic press. As pressure is slowly applied, the juice is gradually expressed, passes through a coarse mesh screen and is collected. Such simple batch operations are necessarily slow and continuous presses are increasingly employed. In these, the fruit is fed into a screw conveyor within a tapering perforated cage. The juice escapes as the pulp is increasingly compressed as it passes along the screw bed. After expression, the dry press-cake is usually discarded except in some instances when it is economically viable to remix the cake with water and repress, but such additional products are of very low quality and only fit for blending.

Fruits differ considerably in the ease with which they can be processed. The presence of pectins and proteinaceous matter may give rise to a viscous juice which is difficult to handle, slow to express and poses problems when screening. Pectinase and other enzyme systems to break down the pectin result in better yields and a more fluid juice but any such enzymes must be completely removed from the final product.

Extraction of citrus juices The processing of citrus fruit calls for a special plant designed to minimize the content of peel oil in the resulting expressed juice. Two main types of plant are widely used throughout the industry:

a. *Brown juice extractor.* This simulates hand reaming of halved fruits. The fruit is cut in half by a sharp blade and each half is caught between the halves of a rubber cap and is then held against a rotating cone which removes the juice and a proportion of the cellular matter.

b. *FMC "in-line" extractor.* The whole fruit is received in a row of serrated cups and squeezed by a similar cup which descends and meshes with the stationary one. The juice and cellular matter pass through a 1-in. hole cut in the bottom of the fruit by the sharpened end of a stainless steel tube.

In each case the initial pressings are passed through a fine screen to remove all pips and unwanted solids.

Limes, being very small fruits, are processed by crushing the whole fruit between rollers or in screw presses, the extracted juice is screened to remove solid matter and is then centrifuged to separate the peel oil.

Freshly expressed citrus juices always contain extraneous matter comprising coarse cellular tissues and a fine suspension of pectins, gums and proteinaceous substances much of which may be colloidal in character. The nature of the finished juice depends on demand—orange and grapefruit juices desirably contain a uniform cloud of particulate solids; lemon and lime juices are usually presented as clear products and the juice may have to be centrifuged to ensure this.

b. Preservation of Fruit Juices

Fruit juices are very susceptible to deterioration and spoilage due to yeast growth hence products not intended for immediate use must be preserved to prevent fermentation. Throughout the processing and subsequent storage of fruit juices, rigid standards of cleanliness and hygiene are essential. The juice must be treated in a manner that will keep it from fermenting or undergoing any other form of deterioration. This is achieved in one of three ways:

a. *Pasteurization.* The juice is heated, either in batches or as a continuous process, at 60°–88°C for periods of time ranging from 5 to 30 min. This inevitably leads to some changes in the sensory quality of the product and the degree of heat treatment is limited to that necessary to destroy microorganisms in a relatively low pH medium and to inactivate any enzyme system which could lead to flavor deterioration and cloud instability during ultimate storage.

b. *Freezing.* The product is kept under refrigeration at below 4°C, under which conditions the flavor is retained almost indefinitely.

c. *Chemical preservation.* Where permitted by legislation, fruit juices may be preserved by the addition of sodium benzoate (0.1–0.3%) or sulfur dioxide (0.03–0.08%). In the latter case, there is a gradual loss of SO_2 during storage and in bulk containers it is often necessary to add further quantities of sodium metabisulfite to maintain an adequate level of preservative.

Sulfur dioxide as a preservative has certain advantages, particularly in lemon, grapefruit and other pale colored juices, as it inhibits browning on storage. A disadvantage of its use is that it binds with carbonyls so that compliance with legal limits may prove difficult.

Benzoic acid does not inhibit browning but otherwise is the preservative of choice. Care is necessary in its use as it may precipitate during mixing and is then difficult to redissolve.

c. Concentrated Fruit Juices

Most juices consist of about 85–90% water and have a flavor level which makes them acceptable to drink per se. As a source of flavor in beverages and other processed foods, most natural juices are much too weak and it is necessary to concentrate them taking care that the delicate aromatic profiles and sugar/acid balance are not unduly changed.

Ideally, any concentration process should remove water without altering the characteristics of the natural juice, either by loss of essential volatiles or by the development of off-notes in either the aroma or flavor. Several methods have been developed to achieve this, including:

a. *Vacuum distillation.* This is carried out in heat-exchange units under carefully controlled conditions. This results in a concentration of soluble solids and extractive matter but with little aroma. The aromatic substances in the original juice are stripped out during the distillation, part being recovered by condensation and low-boiling fractions being lost. This technique leads to concentrates lacking much of the fresh character of the start material. This can be partially replaced by recovery and incorporation of the volatiles from the aqueous distillate, but the end product will still be deficient of top-notes. Besides deterioration of aroma and flavor, the color of the juice may caramelize and darken with increasing concentration if the temperature is permitted to rise above 60°C.

b. *Freeze concentration.* When dilute juices are frozen the ice crystals formed are pure water until the eutectic point is reached when the whole mass solidifies. As no heat is applied, using this technique results in a superior product having a well-balanced profile closely similar to that of the start material. The freezing temperatures employed range from −12° to −18°C. The concentrate is recovered either by allowing the slush-frozen mass to warm gently and running off the concentrated liquor as it thaws or, preferably, by centrifugation or vacuum filtration.

c. *Continuous freeze concentration.* This uses a highly efficient, low volume, scraped surface heat exchanger known as a votator and involves a continuous feed of prechilled natural juice which is slush frozen and then centrifuged to separate the ice phase. The resulting concentrate is collected and may be recycled to effect further concentration until the processing cycle is steady. In general this sytem will handle juices containing 6–10% solids and will concentrate these to 40% solids in a single-stage operation.

d. Recovery of Aromatics

The technique employed for the recovery of volatiles was developed by the Eastern Regional Research Laboratory of the Department of Agriculture in 1944. The process has since been widely applied to improve the flavor of most fruit concentrates. The operation is carried out at atmospheric pressure and heat damage is minimized by high velocity, turbulent flow in a preheater unit. Essence recovery is effected by condensing the fractionated vapors combined with scrubbing of the noncondensable gases before venting to atmosphere.

It is usual to retain the recovered fruit essences separate from the main bulk of juice concentrate and to mix this in the correct proportion shortly before use.

e. Brix Value

Specifications for fruit juice products usually include a value for soluble solids expressed as *degrees Brix.* This figure is based on the relationship between the specific gravity of the product and the percentage of dissolved solids it contains. The Brix tables quoted in the literature are based on an accurate determination of these relationships using aqueous solutions of pure sucrose. It is assumed that all the dissolved solids in a juice will have a similar effect upon the specific gravity, but this is not strictly so, although it is sufficiently close for all practical purposes. In quality control, the specific gravity of the product is determined by using a hydrometer calibrated into degrees Brix or a refractometer, the determined refractive index then being converted into degrees Brix from the appropriate tables. The two methods are not strictly comparable but, again, are sufficiently accurate for control purposes. The Brix value is really a measure of sucrose concentration but since fruit juices contain more sugars than other soluble solids the results act as a good ready guide.

f. Blending of Fruit Juices—WONF

Many natural and concentrated juices are either too low in available flavor or are too expensive to use in low-priced products. To cater to the needs of food processors and beverage manufacturers the following two types of product have been developed and are widely used.

a. *WONF products* (with other natural flavors). These comprise blends of juices or juice concentrates so that at least 50% of the flavor is derived from the named fruit and not more than 50% from other natural flavors.

b. *Fortified concentrates.* These are products in which the flavor strength is fortified or the aromatic profile intensified by the addition of either other natural flavor constituents or of flavorings compounded from synthetics which are generally, but not necessarily, identical to those found in the natural fruit.

g. Depectinized Juices

All fruit juices contain pectic substances. These are hydrocolloids which not only affect the viscosity of the juices but also stabilize the suspension of any insoluble solids or cloud. As many juices are required to be clear, these pectic substances must be removed during processing. This is achieved by enzymatic action of naturally occurring pectinases or by the addition of such enzymes. Depectinization has the added advantage that the juice or concentrate is miscible with diluted alcohol of the strength normally used in the manufacture of liqueurs whereas the natural juice would result in a heavy precipitate.

In the case of orange and grapefruit juices, which are required to have a stable cloud, the pectinases must be inactivated otherwise the juice will separate into a sediment and a clear supernatant. The cloud of orange juice is generally stabilized by heating, the time and temperature being determined by the pH of the start material (e.g., pH 3.8—2 min at 89°C or 15 sec at 94° C; pH 3.3—2 min at 85°C or 15 sec at 90°C).

Depectinized juices are usually preserved by the addition of 0.05–0.06% benzoic acid.

h. Dehydrated Fruit Juices

There is an increasing demand for flavorings in powder form for use in a range of convenient foods and many attempts have been made to present natural fruit juices in this form, particularly for use in beverages and instant drinks. Dehydration as a unit operation is not difficult but the sugar/acids present in fruit juices tend to dehydrate as a plastic mass rather than as discrete granules and the resulting products are invariably hygroscopic and must be packed in hermetic containers. It is usual to mix substances such as lactose with fruit juices in order to produce a free-flowing powder, and to include a sachet of silica gel in the pack as a drying agent. The following processes have been developed with varying degrees of success:

a. *Roller drying*. This is the standard method of drying a film of the juice on heated rollers. It has a limited application as the resulting product develops an unacceptable cooked note and is abnormally dark in color.

b. *Vacuum drying*. This technique is similar to roller drying but the rollers are operated under vacuum. This enables lower evaporation temperatures to be achieved and results in a product which is friable, easily reconstituted and of good flavor quality.

c. *Spray drying*. This is carried out in a purpose-designed plant and involves drying the juice, with or without the addition of a carrier such as dextrose, at a relatively high temperature for a short time. Several types of spray driers are available (Young 1984), some designed specifically for this purpose.

d. *Foam-mat drying.* This involves the formation of a foam stabilized with proteinaceous matter or glyceryl monostearate. The foam is dried on a conveyor belt either in a stream of hot air or by blowing air through the foam. The product has a good fresh quality when first made but tends to oxidize on storage owing to its increased surface area.

e. *Freeze drying.* This operation involves the sublimation of ice directly from the frozen juice which is maintained at a very low temperature under high vacuum with just sufficient heat input to ensure the removal of the water. This results in products having an exceptional high flavor quality but these are expensive as the process costs are relatively high.

Dried fruit juice products are difficult to handle owing to their hygroscopicity and the advice of the manufacturer should be sought to ensure optimum storage and usage conditions.

3. Fruit Pastes and Comminutes

For use in ice cream, frozen confectionery and certain types of beverages, citrus fruits may be passed through a coloid mill to form a homogenous paste. These finely ground products are made from the whole fruit and may be further blended with juice or natural peel oil emulsion in order to achieve a desired flavor profile and some degree of standardization. Such products are usually flash pasteurized to inactivate any enzyme systems and are then put aseptically into sterile cans which are sealed and stored under refrigeration until required. As the product does not contain any added preservative it is essential that the contents of the cans be fully used immediately after opening.

Similar products made from soft fruits are generally known as nectars.

F. Vanilla

1. Historical Introduction

Vanilla is an orchid which is indigenous to southeast Mexico and Central America where it grows wild in the moist forests. When the Spanish conquistadors under Hernando Cortes invaded Mexico in 1520 they found the emperor Montezuma drinking *chocolatl*, a beverage based on ground cocoa beans and corn flavored with *tlilxochitl*, a blend of ground black vanilla pods and honey. Cortes was so impressed by its flavor that he sent the beans back to Spain. By the latter part of the sixteenth century factories had been established in Spain for the manufacture of chocolate flavored with vanilla. The products soon gained wide popularity and by the beginning of the seventeenth century vanilla was being enjoyed in England and many other European countries not only for its original use but also a flavoring agent in tobacco and medicines.

Attempts to grow the plant in the botanical gardens in Paris and Antwerp were successful but the plants did not fruit owing to the absence of natural pollinators. It was not until 1841 that a practical method of artificial pollination, using a small bamboo stick to transfer the sticky pollen, was successfully developed, opening the way to commercial production in other suitable growing regions.

During the early nineteenth century vanilla plantings were established in Java, Reunion, Mauritius, the Malagasy Republic (Madagascar), the Seychelles, Tanzania (Zanzibar), Jamaica and several other tropical regions. Today, the Malagasy Republic provides some 80% of the world's crop of vanilla beans which amounts to about 1400 tonnes with a market value of about $20 million. The number of suppliers remains relatively small and although over 95% of the demand for vanilla flavor is met by the use of synthetic vanillin, there is still a very healthy demand for the natural product. The principal commercial sources are the Malagasy Republic, Reunion, the Seychelles, Mexico, Tahiti, Java and Indonesia.

2. The Vanilla Plant

Vanilla beans are the fully grown fruits of the orchid, *Vanilla fragrans* (Salisb.) Ames (= *V. planifolia*, Andrews), harvested before they are fully ripe and then fermented and cured. Two other species are of some commercial value:

a. *Vanilla pompona* Schiede is a wild vanilla found in southeast Mexico, Central America, the West Indies and northern parts of South America. It is cultivated in Guadaloupe, Martinique and Dominica, the products being sold under the name *vanillons*. These are thicker and more fleshy than genuine vanilla beans and are generally considered to be of poor quality for use in flavorings.

b. *Vanilla tahitensis* Moore is indigenous to Tahiti where it is cultivated. This species has a quite distinctive profile which differs from that of *V. fragrans*.

Vanilla is a fleshy, herbaceous, perennial vine which climbs to a height of 12–15 m. It thrives in a hot, humid climate up to about 2,500 ft. The plant needs protection from wind. Under cultivation the supports are normally growing trees or shrubs planted at 8-ft intervals. The plant is propagated by cuttings which are planted close to the base of each support tree and regularly pruned to enable easy collection of the fruit. The plant flowers after 2 years, starts fruiting after 3 years and reaches a maximum yield in 10–12 years, after which time it is usually replaced by fresh stock. Vanilla flowers once a year over a period of about 2 months. The fully developed vine bears up to 1000 greenish yellow, waxy, fragrant flowers which bloom for 1 day only. The shape of the flower is tubular with an intricate structure so formed that self-pollination is impossible. Of these, only about 50 on each plant are selected for hand pollination, this being carried out using a needle or sharp pointed stick, early in the day on which the flower opens. The flowers occur in clusters and pollination is limited to about 8 flowers in each to ensure adequate growth room as the pods develop. After successful fertilization, the pendulous, cylindrical pods take 6–9 months to mature and are harvested, while still unripe, as their color changes from green to yellow. At this stage, the beans are 12–24 cm long and about 2.5 cm in circumference and lack the characteristic odor and flavor which is only developed by postharvest processing. Yields are very variable and Purseglove *et al.* (1981) report yields of 500–800 kg of cured beans per hectare per annum (450–710 lb per acre per annum) during a crop life of 7 years, but lower yields are not uncommon in some regions.

3. The Curing Process

Although procedures differ between the various growing regions, the postharvest treatment of vanilla beans to produce a commercial product having an acceptable color and flavor is basically carried in four stages:

a. Killing or wilting, to stop further vegetative growth and to initiate enzymatic reactions.
b. Sweating, to promote these enzymatic reactions, to stimulate the development of the dark brown color and to make the beans supple.
c. Drying, to reduce the moisture content necessary for stable storage.
d. Conditioning, to enable the flavor to develop fully.

The postharvest fermentation and curing process is tedious and time consuming and the methods used in the various producing regions significantly affect the quality and profiles of the resulting beans. The methods are described in detail by Muralidharana and Balagopol (1973), Rossenbaum (1974) and further reviewed by Purseglove *et al.* (1981). Briefly, the two main methods used are as follows:

a. *Bourbon method.* The sorted beans are first immersed in water at 65°C (150°F) for 2–3 min to rupture the inner cell walls and probably to stimulate the existing enzyme systems present. After draining, the wet beans are packed in cloth-lined boxes and allowed to sweat for 2 days. Thereafter, the beans are spread out on cloth-covered trays supported on a platform and sun cured for about 6 hr/day for about 6–8 days. The beans are covered over with a blanket each night. This sun-curing process is uncertain and may take up to 3 months to complete, depending on prevailing weather conditions. Finally, the beans are slowly air dried in sheds over a period of 2–3 months or until they are judged to be in a correct condition for sorting and packing.

b. *Mexican method.* The harvested beans are first stored in sheds until they start to shrink after which they are sorted and spread out on mats in the sun for about 5 hr. As the beans begin to cool they are rolled in a blanket and the rolls taken indoors and packed into air-tight boxes where they sweat. After 12–24 h the beans are removed and inspected and those which have changed color are removed for the next stage of processing. An alternative *oven-wilting* technique is also employed using a specially designed hot-room called a calorifico. This technique is gaining in favor as it gives greater control of temperature and humidity throughout the process. In both cases the next stage consists of alternate exposure to the sun and sweating spread over a period of about 6 days. Some 20–30 days after the initial treatment, the beans will be sufficiently supple and have the characteristic odor and appearance. Very slow drying is then carried out indoors during which the beans are regularly inspected and sorted.

In Tahiti the method is quite different as the beans are collected when fully ripe as they do not split and dehisce like the Mexican species. Here the beans are stacked in piles until their color has changed to brown after which they are spread out on blankets resting on wooden platforms and exposed to the sun for 3–4 h each day. The beans are well wrapped in the blanket during the nights. This process is continued for about 20 days and is followed by slow air drying prior to sorting and packing.

The traditional methods of vanilla preparation and improvements in the techniques employed have been studied by Theodose (1973).

When freshly gathered, vanilla beans are largely odorless and flavorless having none of the characteristics which one associates with the commercial product. Vanilla beans of commerce are blackish brown in color and vary in length from 15 to 25 cm (6 to 10 in.) and from 6 to 12 mm ($\frac{1}{4}$ to $\frac{1}{2}$ in.) in width. On storage good quality vanilla beans may become frosted with a mass of glistening white needlelike crystals of vanillin (the *givre* as it is sometimes called). This is not necessarily a criterion of quality as is so often assumed. The crystals do not appear in the producing areas where the climate is very hot and moist but develop only after shipment to cooler drier countries. The moisture content of most samples lies between 30 and 50% but batches dried to a much lower moisture content are available and are preferred for the production of vanilla extracts.

4. Classification and Grading of Vanilla Beans

Beans are generally graded into three categories: unsplit beans, split beans and *cuts* (resulting from the chopping of beans prior to fermentation or the removal of damaged portions). Each producing region sets its own standards for grading, the details of which can be obtained from importers. Grades depend on bean length, overall appearance particularly color, surface blemishes, moisture content and aroma quality. They are usually tied into bundles ranging from 70 to 130 beans depending on size, accordingly the weight of each bundle varies between 150 and 500 g. Bundles are packed into tin boxes lined with waxed paper, these usually contain 20–40 bundles in rows of ten. The tins are normally packed into wooden cases for shipment.

5. The Flavor of Vanilla

Good quality vanilla beans have a quite distinctive aroma which is diagnostic for their country of origin. The differences in aromatic profiles are set out in Table 6.25. Common to all is an intrinsic profile which can be described as sharply acidic with a slightly bitter back-note and a pronounced pungency. This is not normally associated with vanilla preparations owing to the modifying effects of added sugar but it can be observed in vanilla oleoresins. The presence of nonvanillin volatile and nonvolatile flavor components is the reason for a marked preference for natural vanilla preparations over synthetic vanilla or compounded vanilla flavorings.

6. The Chemistry of Vanilla Flavor

Vanillin is the most abundant volatile aromatic constituent of cured vanilla beans, being present from 0.5 to 2.5% within the following ranges:

Bourbon vanilla	1.9–2.5%
Mexican vanilla	1.3–2.0%
Tahiti vanilla	1.0–2.0%

Table 6.25. Comparative Profiles of Vanilla from Different Sources[a]

Country of Origin	Profile
Bourbon vanilla:	
Malagasy Republic	Richly smooth, full bodied, sweet, spicy; less fine than Mexican vanilla.
La Reunion	Closely similar to Madagascar beans.
Comoro Islands	Smooth, balsamic with dry back-notes.
Seychelles	Creamy, full bodied, slightly woody.
Java	Deep, full bodied, very sweet (almost sickly).
Mexican vanilla:	
Mexico	Sharp, slightly pungent, sweetly spicy lacks body associated with Bourbon beans.
Tahiti vanilla:	
Tahiti	Strongly perfumed, characteristically fragrant, heliotropelike.
Gaudaloupe vanillon:	
Guadaloupe, Martinique, Dominica	Strongly perfumed floral, heliotropelike, lacks body.

[a] Source: Heath (1978).

The quality of vanilla beans is not entirely defined by their vanillin content. Klines and Lamparsky (1976) have identified some 170 volatile components in Bourbon (Madagascar) beans but of these only about 26 were present at greater than 1 ppm. Quantitatively, the principal aromatic constituents are as follows: vanillin (3-methoxy-4-hydroxy benzaldehyde), about 2%; *p*-hydroxy benzaldehyde, about 0.2%; *p*-hydroxy benzyl methyl ether, about 0.02%; and acetic acid, about 0.02%.

Recent work by Schulte-Elte *et al.* (1978) has indicated the importance of two diasteroisomeric vitispiranes, which are present at less than 1 ppm, as contributing significantly to the aroma.

Tahiti vanilla beans have a quite different aromatic composition and contain a higher percentage of anisyl alcohol, anisaldehyde anisic acid and piperonal (heliotropin). Comparison of the chemical composition of the three main types of vanilla beans have been made by several authors and these are reviewed by Purseglove *et al.* (1981). The indications are that the same components exist in all of the types, any differences affecting the aroma and flavor profiles being quantitative rather than qualitative.

Other constituents include reducing sugars (about 10%), fats and waxes (about 11%), gums and some resinous matter.

7. Precursors and the Development of Flavor during Curing

When harvested, vanilla beans are totally lacking in what we recognize as vanilla flavor. This only develops during postharvest fermentation and curing. The chemistry involved during this treatment has been known for many years but modern instrumentation has enabled many of the theories to be confirmed. Methods for the study of aromatic components have been reviewed by Oliver (1973).

The presence of three glucosides, which yield vanillin and related phenols on hydrolytic cleavage, was confirmed in green vanilla beans as far back as 1947. Of these, the most abundant is glucovanillin with lesser quantities of glucovanillyl alcohol and the glucoside of protocatecuic acid (3,4-dihydroxy benzoic acid). The initial enzymatic reactions which take place during the curing process result in the formation of the respective phenol plus glucose.

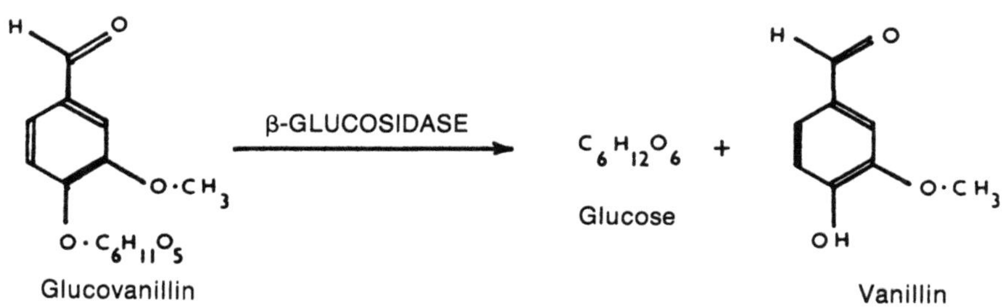

Further peroxidase and polyphenoloxidase reactions in the various substrates, as well as nonenzymatic transformations, result in a complex mixture of aromatic derivatives and the formation of brown pigments. Wild-Altamirano (1969) observed that these enzyme systems reach a maximum at or near the ripening stage of the beans and that the time of harvesting is a major determinant in their ultimate flavoring quality.

Conditions imposed during fermentation and curing ensure that these reactions are encouraged resulting in a high vanillin content but this is not necessarily a prerequisite of aroma and flavor quality. The production of vanilla extracts.

Before use, vanilla beans must be further processed as their form, color, texture and fatty nature make direct incorporation into a food product of very limited applicability. An homogenized puree of vanilla beans has been used in the flavoring of ice cream but the presence of unsaturated fatty matter poses problems of stability and shelf life. It is also very difficult to reduce the seeds so that they do not leave black specks. It is customary to use vanilla in the form of an alcoholic essence or extract and many such products, differing in flavor quality and strength, are available commercially.

In the United States, vanilla extract is defined in the legislation. This requires that the finished product shall contain the extractable matter from 13.35 oz (377.8 g) of vanilla beans (either *V. fragrans (planifolia)* or *V. tahitensis*), having a maximum moisture content of 25%, per U.S. gallon (3.79 L). The term *unit weight of vanilla beans* means that weight containing not more than 25% moisture. In metric terms this is equivalent to not less than 10 g of vanilla beans at 25% moisture, or 7.5 g on a moisture-free basis, per 100 ml. The ethanol content of such extracts should be not less than 35% by volume. The only other permitted ingredients are water and glycerine, propylene glycol, sugar, dextrose or corn syrup as appropriate to improve the sensory qualities and shelf life of the product. Such extracts are known as *singlefold*. Extracts prepared from double the quantity of beans are known as *doublefold*. The maximum concentration which can be prepared directly from the beans is fourfold. Much stronger tenfold extracts are usually made by dissolving a calculated weight of oleoresin vanilla in the appropriate strength alcohol.

Nonalcoholic extracts are also available; these are based on an acceptable solvent such as propylene glycol, according to the demands of the user industries.

The quality of vanilla extracts is determined by several factors, including the following: correct storage of the beans prior to extraction, blending of different quality beans as appropriate, degree of comminution of beans, method of extraction, correct storage of the extract and aging for at least 90 days to ensure full flavor development.

Typically, vanilla beans are extracted as follows (Cowley 1973): Vanilla beans are diced or minced and carefully blended with half their weight of sugar which helps with the extraction and ultimate maturation of the product. The mass is percolated in a suitable stainless steel extractor, using a mixture of ethanol and water of approximately 57% by volume, over a period of 3–4 weeks. The percolate is usually collected in fractions and later blended to produce the final product which is stored in stainless steel containers until sufficiently mature. After this period the extract is filtered or centrifuged and the alcohol content adjusted depending on analytical results.

Other techniques have been described by Merory (Heath 1981).

a. Oleoresin Vanillin

Oleoresin or resinoid vanilla is a dark brown semifluid extract obtained by alcoholic or other solvent extraction of the finely chopped beans ("cuts" are often used). The process involves the complete removal of the solvent under vacuum. Yields depend on the strength of alcohol used and Cowley quotes a 40% yield with 50% ethanol and only 30% with 65% ethanol. Purseglove *et al.* (1981) tabulates yields of oleoresin from beans of different origins. Inevitably the finer top-notes of the aroma are lost or modified by this heat treatment and diluted extracts made from oleoresin do not have the full flavor character usually associated with extracts made by direct percolation of the bean.

b. Vanilla Absolute

In perfumery work it is desirable to have a vanilla extract which is completely soluble in ethanol and/or essential oils. Such products are made by selective solvent extraction and inevitably do not contain all the aromatic components present in the bean. The solvents used vary but it is not uncommon to carry out a first extraction using a nonpolar solvent (e.g., benzene) followed by a polar solvent (e.g., ethanol). A high quality absolute is now being made commercially using liquid CO_2 extraction.

c. Vanilla Sugar

This food ingredient is prepared by mixing ground vanilla beans with sugar or by blending concentrated vanilla extracts with sugar. Imitation products are available, these comprise mixtures of vanillin and ethyl vanillin with dextrose and are often artificially colored.

A minimum of 30% sugar is necessary to obtain a satisfactory product and up to 2% of a permitted anticaking agent may be incorporated.

d. Authenticity of Vanilla Extracts

Vanilla extracts are very expensive flavorants the composition of which is defined in the U.S. legislation. However, to satisfy a user demand, numerous products are offered which contain both natural extract and synthetic vanillin or other flavoring chemicals having a flavor resembling or compatible with that of vanilla. Such products are very much cheaper than the authentic material and the problem of establishing the authenticity of vanilla extracts has been the subject of much research over the past 10 years. Traditionally, tests concentrated on vanillin content, color, lead number and organic acid content, but these could quite easily be circumvented. The advent of chromatography enabled several new tests to be devised and adopted by the Association of Official Analytical Chemists (AOAC), whose methods include the determination of (a) foreign plant matter, (b) flavoring additives, (c) organic acids and (d) nonvanillin vanilla volatiles. Based on work by Bricout *et al.* (1974), the isotopic differences of carbon in natural and synthetic vanillin have been made the basis for a definitive stable isotopic ratio test procedure. This has been described by Martin *et al.* (1977) and by Hoffman and Salb (1979). The quantity of synthetic vanillin in a product can be determined (Martin *et al.* 1981).

e. Imitation Vanilla Flavorings

Imitation vanilla flavorings that replace the natural extracts have now reached a very high level of sensory authenticity and such products can be used with complete confidence to provide a totally acceptable vanilla flavor profile in a wide range of food products. Their use can result in a considerable cost saving and an assurance of continuity of supply. It is usual for such imitation flavors to be offered in strength equivalent to a natural tenfold extract although other strengths can readily be obtained to suit manufacturers' needs.

f. Vanillin and Ethyl Vanillin

Vanillin has been known since about 1816 and by 1858 the pure chemical had been obtained from alcoholic extracts of vanilla beans. It was not until 1872 that Carles established its correct formulation and in 1874, Tiemann and Haarmann reported it as 3-methoxy-4-hydroxy-benzaldehyde. Finally, Reimer synthesized vanillin from guaiacol and thus proved its chemical structure.

For very many years the most important source of vanillin was eugenol, present at about 80–90% in oil of clove, from which it was obtained by oxidation. Later processes involved first converting the eugenol to isoeugenol and this remained the standard method until the source material was changed to the sulfite waste liquors, some 2 million tons of which result from paper manufacture each year. Actual production of vanillin from this new source was first carried out in both the United States and in Canada in 1937 although the patented methods now used differ somewhat from those originally adopted. Today this is the prime source of vanillin.

Ethyl vanillin is a closely related compound, 3-ethoxy-4-hydroxy-benzaldehyde, which is not found in nature but is prepared synthetically from safrole. It has an intense vanillalike odor and is about three to four times more powerful than vanillin as a flavoring agent. Like vanillin, it is widely used in the preparation of imitation vanilla flavorings but can give a somewhat harsh "chemical" character in higher dosage levels. In practice, a maximum of 10% of vanillin may be replaced by ethyl vanillin without this objectionable note being obvious.

G. Beverage Flavors

The term *beverage* embraces many types of liquid products including milk, the many varieties of soft drinks, hot drinks such as chocolate, coffee and tea as well as many alcoholic brews or distillations. All of these are universally enjoyed and form a pleasant and necessary part of our daily intake of liquids. From a flavor point of view only certain of these are of value either as a source of flavor, as a target for imitation or as a carrier for other flavoring effects in food products. Comments will, therefore, be limited to those whose flavor profile has a wider interest to food technologists (i.e., cocoa, coffee, tea).

1. Cacao (Cocoa)

The roasted seeds of the cacao tree have been used as the basis for a pleasant beverage since long before the Spanish conquest of Mexico in 1519. It was Cortez and the conquistadors who realized the potential of the drink called *chocolatl* and introduced the beans to Spain and ultimately Europe together with vanilla, which was used in the same preparation. In due course the source of cacao was studied and the tree classified as *Theobroma cacao* L.

To avoid confusion, it is necessary to note the following popular usage of well-known descriptions: (i) *Cacao* refers to the plant *Theobroma cacao* L. (ii) *Cocoa* refers to the beans and to the product made from them which is ready to mix into a beverage; and to its use as a powdered flavoring material or in the making of a chocolate beverage. (iii) *Chocolate* refers to the solid product based on ground roasted cocoa beans.

The cacao tree grows in most countries within the tropical belt where temperatures range between 25° and 30°C associated with a high humidity of about 90% and adequate ground water. It is a tall perennial evergreen bearing shiny leathery leaves and podlike fruits on the main branches and along the trunk. Each fruit has a 2–4-in. diameter and is 7–12-in. long, with a dark green leathery skin which changes from yellowish orange to purple–red as the fruit ripens. Each contains up to 50 seeds embedded in a pinkish pulp. These seeds, after processing, become the cocoa beans of commerce. Cocoa beans are designated according to their country of origin and/or the port of export. Major sources are listed in Table 6.26.

The ripe pods are harvested, split open and the pulp and seeds removed. At this stage the seeds are very bitter due to the presence of tannins and are only made palatable by a process of fermentation which not only removes the bitterness but also changes the color to a brownish

Table 6.26. Principal Sources and Types of Cacao Beans

Source	Commercial Designation
Forestero-type beans	
Brazil	Bahia
Cameroon	Douala
Congo	Vandeloo
Dominican Republic	Sanchez
Ecuador	Arriba, Guayaguil, Machala
Ghana	Accra
Nigeria	Lagos
Philippines	Maripipi
Trinidad	Port of Spain
Crillo-type beans	
Venezuela	Caracas, Maracaibo, Porto Cabello

red. Fermentation is carried out by covering the beans and allowing them to sweat for several days. This effectively kills the vegetative processes, softens the pulp and activates complex enzymatic reactions leading to the development of the characteristic flavor. The care with which this process is carried out has a significant effect upon the quality of the beans and their flavor when ultimately roasted prior to use. After fermentation the beans still contain about 30% moisture and have to be dried to about 8% prior to storage and shipment. As marketed, cocoa beans comprise about 14% outer shell and 85% kernel, the so-called nib.

Raw cocoa beans do not have the distinctive odor and flavor of cocoa, this is only developed during a roasting process. There are two main types of cocoa beans, Crillo and Forastero, which produce beans having distinctly different flavor characteristics, although this is not now so pronounced as in the past owing to considerable cross-breeding and hybridization of plants. Commercially only two categories are recognized—*bulk or basic beans* and *flavor beans*. The former have a strong, harsh flavor character whereas the latter are smoother and considerably more pleasantly aromatic. The process of roasting is described by Lee and Jackson (1975) but the optimum conditions are still a matter of expert judgment (Riedel 1974).

The Flavor of Cocoa

Considerable research has looked into the chemistry of cocoa flavor, the precursors responsible and the reactions involved at the various stages of production (Maga and Sizer 1973). These reactions are as follows:

a. *Fermentation* is when the flavor precursors are formed; the proteins present in the kernel are degraded and the level of free amino acids rises; sucrose is inverted to fructose and glucose which in turn are oxidized to alcohol and various acids; some theobromine and most of the tannins are lost (Rohan and Stewart 1967).

b. *Drying* reduces the moisture content to about 8% with loss of volatile acids and a rise in pH.

c. *Roasting* is when the precursors are converted into a complex mixture of some 300 aromatic compounds which qualitatively and quantitatively determine the profile of the roasted beans.

The following groups of compounds have been reported as present in roasted cocoa beans (Vitzthun *et al.* 1975; Yamanishi 1981):

Hydrocarbons (24)	Amines (6)	Pyrazines (32)
Alcohols (14)	Oxazoles (4)	Pyrroles (6)
Aldehydes (11)	Furans (1)	Sulfides and thiazoles (10)
Acids (14)	Ketones (10)	Quinoline and quinoxaline
Phenols (9)	Esters and lactones (49)	

The factors affecting the concentration of pyrazines in roasted cocoa beans have been investigated by Reineccius *et al.* (1972).

Cocoa beans are the source of three other important compounds: (a) cocoa butter (52–56%), a mixture of glycerides; (b) theobromine, an alkaloid related to caffeine but lacking its stimulant effect, and (c) cacao purple, an astringent coloring matter.

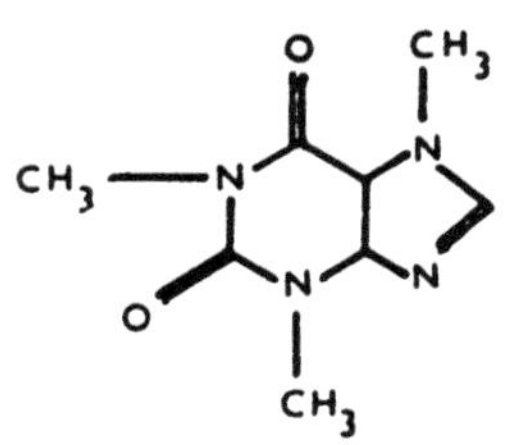

Caffeine

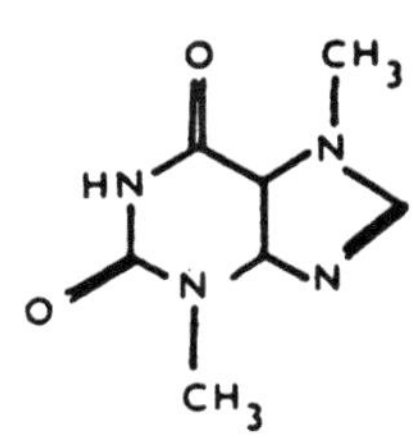

Theobromine

2. Chocolate

Chocolate is the product made by grinding freshly roasted and winnowed cocoa nibs. It contains 50–55% cocoa butter and when freshly made is liquid. This basic chocolate is usually cooled in molds to facilitate later handling. Commercial chocolate is prepared by blending the prime liquor with sugar, milk solids and flavorings, particularly vanilla.

3. Coffee

Coffee has a very long history as a stimulating beverage. It is derived from an evergreen tree or shrub of which there are two main species, *Coffea arabica* L. and *C. robusta* L. A third species, *C. liberica* L., is of little importance. Coffee is native to Ethiopia from where it has spread first to India and then to Indonesia, Brazil, Colombia, El Salvador and the Philippines. Presently the world crop is about 7.3 billion lb, the main commercial sources being Brazil (45%), Colombia (25%), and Mexico (6%). These tropical plants grow between 2000 and 6000 ft above sea level where the day temperature is about 70°C and the ground water is adequate. They bear clusters of small red, cherrylike fruits which contain two seeds. The fleshy berries are harvested almost continuously and, depending on the growing region, are either sun dried for 15 days prior to hulling or are hulled and then allowed to ferment prior to the separation of the greenish-yellow seeds. When dried these become the green coffee of commerce. As green coffee beans lack the characteristic odor and flavor normally associated with the beverage coffee, it is necessary to roast them. This induces a series of reactions resulting in significant changes to both the flavor and color of the resulting beans.

a. The Flavor of Coffee

The chemistry of coffee aroma and flavor has been extensively researched and an up-to-date review has been published by Yamanishi (1981). At least 500 aromatic compounds have been reported in the literature but the relative importance of each to the aroma and flavor profiles has yet to be established. The principal groups of compounds are listed in Table 6.27.

b. Caffeine

Coffee is a drink appreciated as much for its stimulating effect as for its attractive flavor. This is due to the presence of 1.2–1.9% of caffeine which exists in the raw beans and survives any subsequent processing. Caffeine is an alkaloid which is tasteless and is moderately soluble in hot water. To meet a demand for "caffeine-free" coffee, the alkaloid may be removed by extracting the dried beans using a nonpolar solvent. This extraction process has no detrimental effect on the flavor of the end product.

Table 6.27. Principal Constituents of Coffee Flavor[a,b]

Hydrocarbons	(49)	Alcohols	(19)
Aldehydes	(24)	Ketones	(83)
Acids	(22)	Esters	(29)
Lactones	(7)	Phenols	(21)
Amines	(4)	Pyrroles	(25)
Indoles	(3)	Pyridines	(7)
Quinolines	(2)	Pyrazines	(67)
Quinoxalines	(11)	Oxazoles	(25)
Thiazoles	(28)	Thiols	(5)
Sulfides	(17)	Thiphenes	(26)

[a] Source: Yamanishi (1981).
[b] A total of 528 compounds have been reported in coffee aroma.

4. Tea

The beverage which we call tea is a hot water infusion of the dried broken leaves of *Thea sinensis* (L) Sims. (= *Camillia sinensis*). This is an evergreen shrub which grows throughout the tropics from sea level up to about 6000 ft, the plants from higher altitudes producing what are considered to be the best quality leaves. In commerce, teas are designated by country of origin and often by growing region, but are also classified by the method of postharvest handling and preparation for market. Three main types are as follows:

a. Fermented teas (black tea)—India, Sri Lanka, Java, Sumatra, Pakistan, Japan, Taiwan, Malawi and Kenya. These represent about 80% of the total tea consumption.
b. Unfermented teas (green tea)—Japan, China, India and Indonesia.
c. Semifermented teas (Oolong or Pouchong tea)—Taiwan and Hong Kong.

a. Black Tea

The freshly harvested leaves are spread out on mats and allowed to dry out slowly and wither, after which they are mechanically rolled and broken to liberate the oxidase enzymes present in the cell sap and ensure their close contact with precursors. The rolled leaves are then fermented for several hours, usually in baskets or spread out under damp cloths. During this stage the colorless tea tannins are partially oxidized, changing to a reddish brown color and some essential oil is formed. This primary oxidation gives to the leaves a very agreeable odor and color and determines the flavor quality, strength, body and color of "tea" made from them. The process is stopped by "firing" at about 85°C. This induces a secondary oxidation, a reduction in moisture content to about 5% and the full development of the characteristic tea odor and color. The whole process demands skill and experience to achieve a consistent quality.

b. Green Tea

Green teas are not fermented. After harvesting, the fresh leaves are steamed to prevent fermentation and associated color changes. This is followed by partial drying and rolling under pressure. A secondary drying then takes place and the leaves are rolled and broken after which they are finally dried. The compounds present in the end product comprise tea tannins, largely in their original state. Very little essential oil is formed so that green teas lack the aromatic qualities of fermented teas.

c. The Flavor of Black Tea

The principal components of tea comprise (a) an essential oil (about 0.5%) formed during fermentation; (b) caffeine (1.8–5.0%) and (c) tannins (13–18%).

The enzymatic formation of the black tea aroma has been investigated and is reviewed by Yamanishi (1978, 1981) who established the involvement of the following biosynthetic pathways:

a. The main precursors are amino acids, carotenoids including β-carotene, lutein, neoxanthin and violaxanthin.

b. During fermentation a primary oxidation results in the significant reduction of carotenoids, that of β-carotene resulting in the formation of ionone and terpenoid carbonyls.

Table 6.28. Principal Constituents of Tea Flavor[a,b]

Hydrocarbons	(26)	Alcohols	(39)
Aldehydes	(48)	Ketones	(41)
Acids	(31)	Esters	(39)
Lactones	(15)	Phenols	(12)
Oxygenated compounds	(21)	Pyrroles	(4)
		Pyrazines	(20)
Pyridines	(8)	Other nitrogenous compounds	(12)
Thiazoles	(4)		
Sulfur compounds	(5)		

[a] Source: Yamanishi (1981).
[b] A total of over 300 compounds have been reported in the aroma of black, green and semifermented teas.

c. A secondary epoxidation takes place during "firing" resulting in the formation of dihydroactinidiole, an epoxyionone and two trimethyl-substituted cyclohexanones.

d. By similar reactions, other carotenoids give rise to ionone, linalool and substituted hydroxy- and epoxyionones (Sanderson and Graham 1973).

The total number of compounds identified in tea is over 300 and these are listed by Yamanishi. The principal compounds involved are given in Table 6.28.

H. Aromatic Vegetables

Although vegetables are mainly consumed for their nutritive value most contribute significantly to the texture of the end product. Their content of aromatic components, often formed during cooking, makes them a valuable source of distinctive and usually attractive flavors and since many are highly colored they contribute also to the attractive appearance of the end product. Most vegetables are not highly flavorful but some, notably the alliaceous vegetables (e.g., onion, garlic, leek and chives) are popular because of their strongly aromatic character. These have been used domestically since the earliest times and, in spite of handling and processing problems, still form a major commodity in modern food processing. An increase in convenience food products has led to the extensive use of dehydrated forms (e.g., onion chips and powder) and of distilled essential oils. More recently, the introduction of very concentrated fluid extracts, which have the full aromatic profile associated with the freshly cut vegetable and are water miscible, has widened the field of application.

1. Onion

Onion, *Allium cepa* L. is one of the oldest known and most popular of the alliaceous vegetables. Considerable crops are raised in the United States (California, Texas and New York), Egypt and Japan and to a lesser extent in most European countries, Mexico and Brazil.

The plant is variably a biennial or perennial although normally grown as an annual either from seed or small bulbs called sets. It has 4–6 aromatic, cylindrical, hollow leaves. When flowers are present, they are greenish-white, small and in rounded umbels appearing in late summer. Egyptian onion (var. *vivaparum*) produces small bulbs or top-sets in the flower head. The leaves die down at the end of the growing season leaving a hard-fleshed bulb having a thin outer scaly layer. The shape may be oblong, flat, globular or oblate and the color brown, white or red depending on the variety and source.

Numerous cultivars exist and are classified according to their dry-solids content and pungency, most being raised to satisfy a particular market demand (i.e., for domestic use, pickling, canning, freezing, dehydration, etc.).

Onions have almost no smell until the tissues are cut or bruised, but once this occurs the strong and characteristic flavor is produced very rapidly by the action of an enzyme on odorless precursors which coexist in the cell sap. The resulting complex mixture of sulfides and other sulfur-containing compounds, can be recovered as so-called onion oil by distillation although the amount recovered is only in the order of 0.02–0.03% and frequently much less than this.

a. The Flavor of Onion

The biochemistry of onion flavor is now well known and has been extensively reviewed by Shankaranarayana *et al.* (1974). The chemistry involved has been covered in Part I.

The distinctive flavor of onion and other alliaceous vegetables is due to *S*-alk(en)yl derivatives of L-cysteine sulfoxide which are rapidly hydrolyzed by the enzyme alliinase to give an unstable sulfenic acid derivative together with pyruvic acid and ammonia. The sulfenic acid compound breaks down and rearranges to form the relatively more stable thiopropanal *S*-oxide, which has lachrymatory properties, or it reacts with other compounds to produce a complex mixture of di- or trisulfides which ultimately characterize the product.

In the case of onion, the principal substrate is *trans*(+)*S*-(l-propenyl)L-cysteine sulfoxide which, when acted on by alliinase, produces a distinctive odor profile, a marked pungency, a transient lachrymatory effect, a bitter aftertaste and a pink discoloration. The reaction involved is described in an earlier section.

The characteristic aroma of freshly cut onion is due to thiosulfinates but these are soon changed into disulfides and thiosulfonates.

The stoichiometric formation of pyruvic acid has been used as a measure of onion pungency and odor intensity (Schwimmer *et al.* 1964). More recently Freeman and Whenham (1975) have evolved a spectrophotometric method for the determination of thiopropanal *S*-oxide which is a more reliable basis for judging the flavoring quality of fresh onion.

b. Onion Oil

The distillation of onion pulp which has been allowed to stand for 12 h results in a dark brown oil the composition of which has been studied by Brodnitz and Pollack (1970). Later work by Galetto and Bednarczyk (1975) established that three components (starred in following listing) are of significance in determining the profile. The principal components are as follows:

Methyl-1-propyl disulfide*	di-l-Propyl trisulfide
3,4-Dimethyl thiophene	*cis*-l-Propyl-propenyl disulfide
cis-Methyl-1-propenyl disulfide	Methyl-1-propyl trisulfide*
trans-Methyl-1-propenyl disulfide	di-1-Propyl trisulfide*

No allyl compounds were found to be present.

A more complete listing has been published by Shankaranarayana *et al.* (1974). The flavor of onion oil is approximately 4000 times stronger than that of fresh onions and is obviously far too strong to be added directly to food products. Diluted versions, both liquid and dry, are available and it is necessary to establish the relative flavoring power of these products from the supplier. It should be appreciated that, although such products provide an excellent source of onion flavor, they do not contribute anything of the textural quality provided by the raw vegetable. Onion powder may significantly influence the water-holding capacity of some food mixes.

c. Dehydrated Onion

In many food and snack products it is necessary to use dehydrated ingredients and considerable tonnages of dehydrated onion and garlic are produced annually to satisfy this demand. Onions grown especially for dehydration are not generally the same as those used domestically. Optimally these should have a high dry-solids content and a good level of flavor and pungency. Those grown in California are special strains of white globe onions. These are grown intensively, are mechanically harvested and are inspected in the fields prior to dispatch to the processing factory. Here they are flame peeled, washed to remove the burnt outer skin and then mechanically sliced onto the perforated drying belt. Drying is carried out in a tunnel drier using hot air which circulates through the carrier belt. Initially the fresh onions have a moisture content of about 80% and this is reduced to about 4% by the end of the drying cycle. An alternative fluid-bed technique is employed.

The dehydrated product may be marketed as "kibbled" to various mesh sizes or "powdered." All of these products are prone to absorb atmospheric moisture and must be packed in well-closed containers and stored in an ambient temperature of above 15°C.

The above process results in some loss of the fresh odor and flavor and the relative flavor strength of fresh and dehydrated onion is difficult to determine owing to the differences in profiles. For general purposes it is reasonable to assume that dehydrated onion is about eight to ten times stronger than fresh onion. Missing top-notes may be added in the form of the fluid extract or onion oil.

2. Garlic

Like onion, garlic has a very long history and has been cultivated in Eastern countries for centuries. It is the compound bulb of *Allium sativum* L. which comprises 8–20 small bulblets, having a silky white to pink to mauve colored skin and generally known as *cloves*, enclosed within a white membranous outer casing. The plant is a hardy perennial having long flattened, pointed leaves arising from a crown. The unbranched stem is about 3–4 in. (7.5–10 cm) long and carries an apical dense umbel of rose–white to greenish flowers often replaced by sterile pink colored bulbils.

The United States is the biggest producer of garlic but significant quantities are also available from Egypt, Bulgaria, Hungary and Taiwan.

a. The Flavor of Garlic

Like onion, garlic is without odor until the tissues are cut or bruised, but once this occurs the odor is intensely strong and obnoxious.

Garlic contains about 0.1–0.25% of volatile constituents, formed enzymatically when garlic cloves are crushed in a manner similar to that of onion. This complex of aromatic sulfides can be recovered as so-called garlic oil by distillation.

The chemistry of garlic flavor parallels that of onion and Freeman and Whenham (1976) established that any differences were due to qualitative and quantitative differences in the precursors present. The principal substrate is *S*-(2-propenyl)1-cysteine sulfoxide which when acted upon by alliinase produces allyl thiosulfinate (allicin) which ultimately degrades into allyl disulfide, allyl thiosulfonate and trace quantities of allyl trisulfide.

The principal components of garlic oil are as follows (Freeman and Whenham 1976): allyl-(1-propyl) disulfide (2%), allyl-methyl disulfide (13.4%), and allyl-(2-propenyl) disulfide (84.6%).

The chemistry of garlic oil has been reviewed by Whitaker (1976).

The overpowering flavor of garlic is not to everyone's liking. The odor is very persistent and when ingested may taint the breath for a long period afterwards.

b. Dehydrated Garlic

About 8000 tons of garlic powder are used annually in food processing. It is a creamy-white powder prepared by the dehydration of selected garlic cloves but, like onion powder, it is hygroscopic and must be packed in well-closed containers. For general use a concentration of about 5:1 against fresh garlic is a good starting point though it should be recognized that the flavor profile lacks much of the fine quality associated with fresh garlic.

References

ANON. 1983. Radiation preservation of foods. Food Technol. *37*(2), 55–60.

ARCTANDER, S. 1960. Perfumery and Flavor Materials of Natural Origin. Stephen Arctander, Elizabeth, N.J.

BRAVERMAN, J. B. S. and SOLOMIANSKY, L. 1957. Separation of terpeneless essential oils by chromatographic methods. Perf. Ess. Oil Record *48*, 284.

BRICOUT, J., FONTS, J-C., and MERLIVAT, L. 1974. Detection of synthetic vanillin in vanilla extracts by isotope analysis. J. Assoc. Off. Analyt. Chem. *57*, 713–715.

BRODNITZ, M. H. and POLLACK, C. L. 1970. Gas chromatographic analysis of distilled onion oil. Food Technol. *24*(1), 78–80.

BROGLE, H. 1982. Carbon dioxide as a solvent; its properties and applications. Chem. Ind. (June 19), 385–389.

CALAME, J. P. and STEINER, R. 1982. CO_2 extraction in the flavour and perfumery industries. Chem. Ind. (June 19), 399–402.

COHODAS, A. M. 1969. Spice grinding process. Canadian patent, 808644 (Mar. 18).

CORNELIS, J. C. 1982. Food Irradiation Now. Martinus Nijhoff/Dr. W. Junk, Publishers, The Hague, Netherlands.

COWLEY, E. 1973. Vanilla and its uses. Proc. Tropical Prod. Inst. Conf. London 1972.

FENAROLI, G. 1975. Fenaroli's Handbook of Flavor Ingredients, T. E. Furia and N. Bellanca, eds. CRC Press, Boca Raton, FL.

FORSS, D. A. 1981. Sensory characterization. *In* Flavor Research, R. Teranishi, R. A. Flath and H. Sugisawa, eds, pp. 125–174. Dekker, New York.

FREEMAN, C. G. and WHENHAM, R. J. 1976. Nature and origin of volatile flavour components of onion and related species. Int. Flavours Food Additives *7*, 222–227, 229.

GALETTO, W. G. and BEDNARCZYK, A. A. 1975. Flavor contribution of individual volatile components of the oil of onion. J. Food Sci. *40*, 1165–1167.

GILDERMEISTER, E. and HOFFMAN, F. 1965–1967. Die Aetherischen Oele, 4th edition. Akademie-Verlag, Berlin. (German.)

GUADAGNI, D. G., BUTTERY, R.G. and OKANO, S. 1966. Odor intensities of hop oil constituents. J. Sci. Food Agric. *17*, 142.

GUADAGNI, D. G., BUTTERY, R. G. and TURNBAUGH, J. G. 1972. Odor thresholds and similarity ratings of some potato chip components. J. Sci Food Agric. *23*, 1435.

GUENTHER, E. 1948–1952. The Essential Oils, Vols. 1–6. Van Nostrand, New York.

GREEN, R. J. 1975. Peppermint and spearmint production in the United States: Progress and problems. Int. Flavours Food Additives *6*, 246–247.

HARPER, R., LAND, D. G., BATE-SMITH, E. C. and GRIFFITHS, N. M. 1968. A glossary of odour stimuli and their qualities. Perf. Ess. Oil Record *59*(1), 22–27.

HEATH, H. B. 1978. Flavor Technology: Profiles, Products, Applications. AVI Publishing, Westport, CT.

HEATH, H. B. 1981. Source Book of Flavors. AVI Publishing, Westport, CT.

HEATH, H. B. 1983. Spices and aromatic extracts, influence of technological parameters on quality. *In* Int. Symp. Food Flavors, Paris, Dec. 1982, pp. 138–176. APRIA, Paris.

HODGE, W. H. 1975. A survey of flavour-producing plants. Int. Flavours Food Additives *6*, 244–245.

HOFFMAN, P. G. and SALB, M. 1979. Isolation and stable isotope ratio analysis of vanillin. J. Agric. Food Chem. *27*(2), 352–355.

KLINES, I. and LAMPARSKY, D. 1976. Vanilla volatiles—a comprehensive analysis. Int. Flavours Food Additives *7*, 272–273, 291.

KRISHNASWAMY, M. A., PATEL, J. D., PARTHASARATHY, N. J. and NAIR, K. K. S. 1973. Microbiological quality of certain spices. J. Plantation Crops *1*(Suppl.), 200.

LAWRENCE, B. M. 1976 onwards. Progress in Essential Oils. Monthly in Perf. Flavorist, Allured Publishing, Wheaton, IL.

LAWRENCE, B. M. 1978. Essential Oils 1976–1977. Allured Publishing, Wheaton, IL.

LEE, R. and JACKSON, E. B. 1975. Sugar and Chocolate Manufacture. Leonard Hill Books, London, England.

MAARSE, H. 1983. Recent advances in the methodology of flavour research. *In* Int. Symp. Food Flavors, Paris, Dec. 1982, pp. 3–26.

MAGA, J. and SIZER, C. E. 1973. Cocoa flavor. CRC Crit. Rev. Food Technol. *4*, 39.

MALINGRE M. 1982. Chemotaxonomy of spice plants. Perf. Flavorist *6* (Dec./Jan.), 15–22.

MARTIN, G. E., ETHRIDGE, M. W. and KAISER, F. E. 1977. Determining the authenticity of vanilla extracts. J. Food Sci. *42*(6), 1580–1583, 1586.

MARTIN, G. E., ALFONSO, F. C., FOGERT, D. M. and BUGGRUFF, J. M. 1981. Determination of the authenticity of vanilla extract. J. Assoc. Off. Analyt. Chem. *64*(5), 1149–1153.

MOYLER, D. A. 1984. Carbon dioxide extracted ingredients for fragrances. Perfum. Flavorist *9*(2), 109–114.

MURALIDHARANA, A. and BALAGOPOL, C. 1973. Studies in curing vanilla. Indian Spices *10*(3), 3–4.

NAGASWARA, T., UMEMOTO, T., TSUNEYA, T. and SHIGA, M. 1974. A component of peculiar odor in Japanese spearmint oil. Koryo *108*, 45–50.

NAGY, S., SHAW, P. C. and VELDHUIS, M. K., eds. 1977. Citrus Science and Technology, Vols. 1–2. AVI Publishing, Westport, CT.

NELSON, P. E. and TRESSLER, D. K. 1980. Fruit and Vegetable Juice Processing Technology, 3rd edition. AVI Publishing, Westport, CT.

OLIVER, R. 1973. Methods for the study of the aromatic components of vanilla extracts. Proc. Conf. Spices, London, April 1972, pp. 55–60. Tropical Products Institute, London.

PEYRON, L. 1984. Carbon dioxide, liquid and supercritical. Perfum. Cosmet. Aromes *55*(Feb./Mar.), 109–114. (French.)

PROVATOROFF, N. 1972. Some details of the distillation of spice oils. Proc. Tropical Prod. Inst. Conf., London, 1972.

PRUTHI, J. S. 1980. Advances in Food Research: Spices and Condiments. Academic Press, New York and London.

PURSEGLOVE, J. W., BROWN, E. G., GREEN, C. L. and ROBINS, S. R. J. 1981. Spices, Vols. 1–2. Longmans, London and New York.

QUARRÉ, J. 1975. Cultivation ot mint in the U.S.A. Parfum, Cosmet, Aromes. *2* (Mar./Apr.), 27–30, 33–35. (French.)

REINECCIUS, G. A., KEENEY, P. G. and WEISSBERGER, W. 1972. Factors affecting the concentration of pyrazines in cocoa beans. J. Agric. Food Chem. *20*, 202.

RIEDEL, H. B. 1974. Effects of roasting on cocoa beans. Conf. Prod. *40*, 193–194.

ROGERS, J.A. *et al.* 1971. Eugenol-Sources, production and usage. Papers 161st Natl. Meet. Amer. Chem. Soc., Los Angeles., CA.

ROHAN, T. A. and STEWART, T. 1967. The precursors of chocolate aroma. J. Food Sci. *32*, 395–398.

ROSSENBAUM, E. W. 1974. Vanilla extracts, synthetic vanillin. *In* Encyclopedia of Food Technology, A. H. Johnson and M. S. Peterson, eds. AVI Publishing, Westport, CT.

SALZER, U.-J. 1975. Analytical evaluation of seasoning extracts (oleoresins) and essential oils from seasonings. Int. Flavours Food Additives *6*, 151–157, 206–210, 253–258.

SANDERSON, G. W. and GRAHAM, N. H. 1973. The formation of black tea aroma. J. Agric. Food Chem. *21*(4), 576–585.

SCHWIMMER S., VESTROM, D. W. and GUADAGNI, D. G. 1964. Relation between pyruvate content and odor strength of reconstituted onion powder. Food Technol. *18*(8), 121–124.

SHANKARANARAYANA, M. L., RAGHAVEN, B., ABRAHAM, K. V. and NATARAJAN, C. P. 1974. Volatile sulfur compounds in food flavors. CRC Crit. Revs. Food Technol. *4*, 395–435.

SCHULTE-ELTE, K. H. *et al.* 1978. Vitispiranes, important constituents of vanilla aroma. Helv. Chim. Acta *61*(3), 1125–1133.

SMITH, D. M. and LEVI, L. 1961. Treatment of compositional data for the characterizing of essential oils. Determination of geographical origins of peppermint oils by gas chromatographic analysis. J. Agric. Food Chem. *9*, 230–244.

STAHL, W. H. 1972. Oleoresin quality analysis—fact or fancy. Proc. Tropical Prod. Inst. Conf., London, 1972.

THEODOSE, R. 1973. Traditional methods of vanilla preparation: Improvement of these techniques at the Antalaha station. Proc. Tropical Prod. Inst. Conf., London, 1972.

TRESSL, R. and SILWAR, R. 1981. Investigation of sulfur-containing components of roasted coffee. J. Agric. Food Chem. *29*, 1078–1082.

USP XX. 1980. U.S. Pharmacopoeia, 20th edition.

VITZTHUN, O.G., WERKHOFF, P. and HUBERT, P. 1975. Volatile components of roasted cocoa. J. Food Sci. *40*, 911–916.

WARNER, M., JONES, M. and SAAL, H. 1981. Citrus processor extracts juice, recovers peel oil at same time. Food Process. *42*(2), 94–95.

WHITAKER, J. R. 1976. Development of flavor, odor and pungency in onion and garlic. Adv. Food Res. *22*, 73–133.

WILD-ALTAMIRANO, C. 1969. Enzymatic activity during growth of vanilla fruit: I. Proteinase, peroxidase and polyphenoloxidase. J. Food Sci. *34*, 235–238.

YAMAGUCHI, K. and SHIBAMOTO, T. 1981. Volatile constituents of green tea, Gyokuro (*Camellia sinensis* L. var. *Yabukita*). J. Agric Food Chem. *29*, 366–370.

YAMANISHI, T. 1978. The aroma of various teas. *In* Flavor in Food and Beverages, G. Charalambous and G. E. Inglett, eds. Academic Press, New York and London.

YAMANISHI, T. 1981. Tea, coffee, cocoa and other beverages. *In* Flavor Research, Recent Advances, R. Teranishi, R. A. Flath and H. Sugisawa, eds. Dekker, New York and Basel.

YOUNG, R. A. 1984. Capturing the flavour. Food FIPP. *6*(5), 43–45.

7

Flavoring Materials Made by Processing

The term *processed flavor* is used in different contexts to describe (a) products where the natural raw materials are virtually flavorless and the desired aromatic profile is achieved only by deliberate processing (e.g., cocoa/chocolate); (b) flavors created as a result of Maillard and other related reactions between amino acids and sugars (e.g., imitation meat flavors); (c) flavors resulting from controlled enzymatic reactions (e.g., cheese, butter); (d) products made by fermentation (e.g., wines, vinegar) and (e) the application of biotechnology in the production of aroma chemicals.

Other well-known flavors are produced as a result of a combination of these processing methods (e.g., coffee and cocoa which involve fermentation and roasting; tea which may or may not be fermented before roasting and vanilla which must be fermented and cured). These products have already been discussed in Chapter 6, and here the relationship between the processing parameters encountered and the changes in the product chemistry will be considered, particularly as these affect the nature and quality of the flavoring materials available.

A. Natural Products Made by Roasting: Cocoa/Chocolate

The production of cocoa and chocolate from cocoa beans involves three prime processes: (a) fermentation, (b) drying and (c) roasting, each of which has a significant effect upon the final flavor profile. The chemistry involved is complex (Yamanishi 1981) and the quantitative mix of aromatic constituents resulting from the processing is determined by the precise conditions employed. The processing may be summarized as follows:

a. Fermentation enables the beans to be separated from the surrounding pulp and shell, inactivates the seeds, initiates the aromatic precursors and color development and significantly reduces bitterness. During this stage the sucrose and proteinaceous constituents are hydrolyzed, any polyphenolic compounds are oxidized and glucose is converted into alcohol and then oxidized to acetic and lactic acids. Beans which have not been fermented do not develop a chocolate flavor when roasted.

b. Drying removes moisture to about 15%.

c. Roasting develops the desired flavor profile and completes the color changes, further reduces the moisture content to about 5%, conditions the shell facilitating its removal and makes the "nib" more friable. During roasting the initial fermentation products are converted, by Maillard reactions, Amodori rearrangements and Stecker degradations, into a complex of volatile carbonyls, pyrazines, etc., which characterize the aroma and flavor of the final chocolate. The pyrazine content has been shown to vary according to the roasting temperature employed (Reineccius *et al.* 1972). There is some loss of volatile acids and up to 10% of acetic and propionic acids are released from the shell during roasting (Lees and Jackson 1973). Free amino acids present in the fermented beans also undergo a Strecker degradation yielding carbonyls which then react to give aldo condensation products. This probably accounts for the formation of 5-methyl-2-phenyl-hexenal which is significant to the flavor profile (Reymond and Rostagno 1978).

The temperature conditions encountered during the roasting process depend on the type of roaster employed but they usually lie within the range 115°–140°C (239°–284°F). The precise conditions are still one of skill and judgment based on a knowledge of the raw materials used and the ultimate product being made.

Production of Cocoa Powder

Having roasted the beans, it is necessary first to separate the shell from the kernel by passing through rollers and then to separate the broken shell and the cocoa nibs by winnowing, using graded screens and air elution. Final separation is carried out to ensure that the nibs are completely free from shell (in the United States shell is limited to 1.75%) prior to the next stage of processing.

Cocoa powder is manufactured by grinding the press cake after partial removal of cocoa butter. This product demands high roasted beans to give the end product a better flavor and darker color. Subjection of the separated nibs to a pressure of 7000 $lb/in.^2$ leaves a residual 20–28% of cocoa butter in the press cake. On leaving the press, the cake is broken to form small pieces of about 1-in. size which are then subjected to further milling under carefully regulated temperature conditions and cooled rapidly to 20°–23°C (68°–74°F) immediately on discharge, to ensure a properly tempered cocoa butter content, the correct particle size and an acceptable color.

Cocoa powder contains about 24% fat, 18% nitrogenous matter, 5.5% fiber, 10.5% starch and about 4% moisture. When permitted, commercial products may contain additives such as sucrose esters, tripolyphosphate, disodium sulfosuccinate and alginates in order to improve its wettability as well as vanillin and other flavorings to give it a more rounded profile.

a. The Dutch Process

Dutching or alkalizing is carried out to modify the color and flavor of cocoa. Depending on the degree of alkalization, the color becomes darker and the flavor milder and much less harsh. The process involves the addition of alkali to the milled chocolate liquor accompanied by efficient mixing and the removal of water by heating.

b. Chocolate

Chocolate is a suspension of cocoa particles and sugar held in cocoa butter, the manufacture of which entails four unit operations:

a. Mixing is done with the correct proportions of sugar, cocoa butter and cocoa liquor in a special plant known as a melangeur. This consists of two heavy rollers mounted over a heated rotating steel or granite bed, although other designs for continuous automatic operation are widely used so that mixing and flavor development are optimized. The mixing operation is carried out at 50°–60°C (122°–144°F) and usually takes 20–25 min but in an automatic plant the dwell time may be only about 5 min. The aim of this stage is to reduce the particle size of the sugar and cause loss of unwanted volatile acids. The product is a soft, plastic, pliable, nonflowing mass.

b. Refining involves considerable sheet forces which produce the desired particle size. This is carried out in a mill fitted with multiple rollers (3–5) and capable of being water cooled. Chocolate is usually refined to a particle size of 50–114 μm (micrometers) of which less than 10% are smaller than 5 μm and not more than 20% are larger than 22 μm. It is the distribution of particle sizes which determines the eating quality of the resulting chocolate (Allerton 1974).

c. Conching is necessary to ensure full flavor development by the elimination of the remaining volatile acids; it also removes moisture, smoothes down angular sugar crystals, modifies the viscosity and alters the color of the product. The specially designed plant produces efficient agitation and aeration to ensure good mixing and exposure of the product to air over a period of between 1 and 5 days at a controlled temperature below 100°C (212°F) depending on the type of chocolate being made. During conching, further oxidative changes occur together with some caramelization.

d. Tempering is necessary to induce crystallization of the cocoa butter in a stable form in the fluid chocolate mass. Cocoa butter can crystallize in four forms α, α′, β, β′, each with a different melting point. Only β fat crystals melting at 94°C (200°F) are suitable in chocolate. Incorrect tempering results in "blooming." The conditions, again, depend on the composition of the chocolate mix, particularly the percentage of fat present. Viscosity is affected by the state of temper and influences the ease of enrobing, but this quality is also affected by the amount of water present, the level of sugar, particle size distribution and the presence of additives such as lecithin and/or sucrose esters.

The use of cocoa and chocolate products as flavoring materials is described by Minifie (1980) and the question of cocoa substitution in beverages, confectionery and traditional chocolate products is discussed by Grab *et al.* (1978).

B. Reaction Flavors: Imitation Meat Flavors

Enzymatic and nonenzymatic browning reactions can produce very many volatile compounds having strong and distinctive odors depending on the reactants and the reaction conditions (Hodge *et al.* 1972). The classical Maillard reaction and the closely associated Strecker degradation have already been discussed in Chapter 3. Although these are only a part of the total chemical changes occurring during the heat treatment of foods, they probably generate more characteristic flavoring effects in processed foods than any other. The course of this reaction can be simply illustrated (Fig. 7.1) although its actual progress is highly complex. The nature of the end products depend on the following:

a. The precise nature and ratio of the α-amino acids/proteins and sugars present.

b. The time/temperature conditions of the reaction, which in turn determines the degree of secondary reactions involved.

c. The nature of any thermal degradation reactions which contribute to the overall profile (Morton and Macleod 1981).

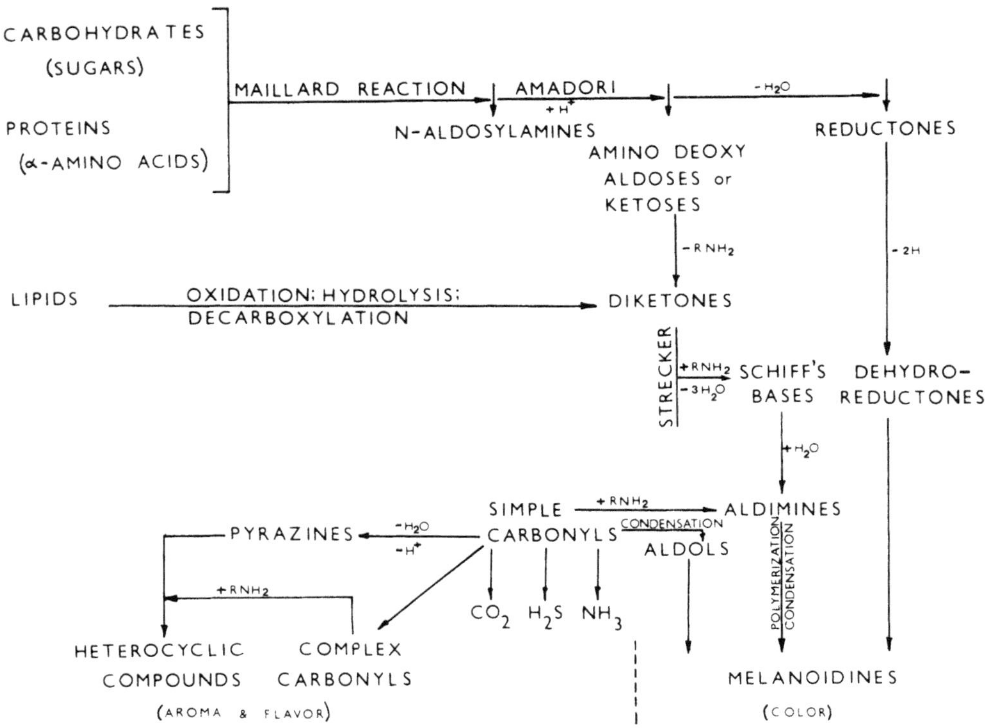

Fig. 7.1. The Maillard and associated reactions in flavor formation.

The highly desirable flavor of roast meat is probably a combination of the thermal degradation products of naturally occurring sugars, amino acids and polypeptides, the complex effects of the Maillard and related reactions coupled with the aromatic products arising from lipid oxidation. Of these, the Maillard reaction is of greatest significance in foods, particularly meat, heated above 90°C. The precise precursors involved are difficult to identify, but Hornstein (1967) considered that the low levels of free amino acids in the raw meat were probably more important for the browning reaction than the quantitatively greater peptide-bound amino acids.

The flavor chemistry of raw and cooked meats has been the subject of considerable research over the past 30 years (Flament *et al.* 1977; Katz 1981), and this has provided the flavor industry with invaluable data for the recreation of imitation meat flavorings. From the current knowledge of the aromatics produced by Maillard-type reactions, flavorists have been able to formulate innumerable products having meaty, nutty, bready and roasted characteristics based on synthetics identified in roast meats, although it is recognized that flavorings based on the duplication of research findings are not necessarily successful. In the case of meatlike flavors, this is probably due to the fact that meat flavor does not contain a single predominant characterizing volatile compound (Heinze *et al.* 1978). The direct application of the Maillard reaction based on model systems (Tressl 1982) offers a much better start-point for the making of authentic meatlike flavors most of which are based on either (a) the hydrolysis of vegetable proteins or, preferably, (b) the direct controlled reaction of selected amino acids and reducing sugars.

Hodge *et al.* (1972) classified the Maillard reaction products into four groups depending on their aroma type, chemical structure, molecular shape and processing parameters. His groupings are as follows:

a. nitrogen heterocyclics—pyrazines, pyrroles, pyridines, alkyl- and acetyl-substituted saturated *N*-heterocyclics. These compounds are responsible for corny, nutty, roasted and breadlike odors.
b. cyclic enolones—maltol, isomaltol, dehydrofuranones, dehydropyrones, cyclopentenolones; responsible for typically caramellike odors.
c. polycarbonyls—2-furaldehydes, 2-pyrrole aldehydes, C_3–C_6 methyl ketones.
d. Monocarbonyls.

These last two groups contain many very volatile compounds which tend to give specific nuances and supplementary aroma notes to the flavor profile rather than provide the main flavor impact.

The specific application of these aromatics in the formulation of meatlike flavors has been covered by many patents (Pintauro 1976; Heath 1983). The patent literature has many examples of the application of the Maillard reaction to the production of meatlike flavors (Dwivedi 1974; Heath 1981) and since the 1970s these products have come to be known as *reaction flavors* or *process flavors*. The nature of these products is now posing problems of categorization under the various systems of flavor legislation. The International Organization of the Flavour Industry (IOFI) has drafted guidelines for such products which it defines as the following:

> A product or a mixture prepared for its flavouring properties and which is produced from ingredients or mixtures of ingredients which are permitted for use in foodstuffs, or are present naturally in foodstuffs, or are permitted for use specifically in process flavourings by a process used for the preparation of foods for human consumption.

The ingredients for reaction or process flavors can include (a) a protein nitrogen source, (b) a carbohydrate source, (c) a fat or fatty acid source, and (d) other ingredients including herbs and spices; sodium chloride;

polysiloxane; acids, bases and salts as pH regulators; water; the salts and acid forms of thiamine, ascorbic, citric, lactic, inosinic and guanylic acids; esters or amino acids; inositol; sodium and ammonium sulfides and hydrosulfides; diacetyl and lecithin.

The processing conditions should not exceed 15 min at 180°C or proportionately longer at lower temperatures. The pH should not exceed 8.0.

In the United States guidelines are under study by the expert committee of FEMA.

Heating single amino acids with individually selected sugars, under precisely controlled conditions of time/temperature results in a wide spectrum of aromatic effects. When L-cysteine is used, the resulting aroma is markedly that of beef. Choice of sugars has been shown to be very important; pentoses need less reaction time than hexoses but the resulting flavor is inferior. Within this broad definition, meatlike aromas and flavors, closely resembling those of roasted beef, have been produced by heating together mixtures of L-cysteine and thiamine with a 6-desoxyhexose. A good roast chicken flavor has been produced by heating L-cysteine hydrochloride and glycine with a mixture of L-arabinose and dextrose. Variants of the flavor profile are obtained by the substitution of either L-cysteine or L-cystine. A synthetic ham or bacon flavor can be made by using the same ingredients but heated at a controlled pH. The substitution of D-glucose for dextrose gives a more hamlike note. Blanket patents are not uncommon and one manufacturer has covered a whole range of starting materials to make reaction flavors having profiles from beeflike to turkeylike. The starting materials are mainly mercaptans, heated at 300°F (148°C), in the presence of glutamic acid, L-cysteine hydrochloride, glycine, β-alanine and thaumine.

Imitation Meat Flavors

Meat flavorings are required in the manufacture of many convenience and processed foods which are increasingly forming a large proportion of the normal diet. They are needed for the acceptable flavoring of meat analogs based on textured vegetable protein which, although not matching up to predicted demands, are now widely used as a nutritional supplement to more conventional sources of protein, not only to lower the price but also to contribute to a lowering of lipid and cholesterol intake. Consumers still demand the real flavor of roast meats so that any fabricated product must have a good authentic meaty character. With these market requirements in mind, the flavor industry has developed a range of imitation meat flavors, although the specificity of the flavor profile (e.g., roast beef, roast chicken) has proved much more difficult to achieve. Obviously, flavor development progress has followed closely the research findings into the nature of meat flavor (Macleod and Seyeddain-Ardebili 1981), but its application poses many problems for the flavor manufacturer trying to reproduce these very complex aromatic mixtures as a stable flavoring product on a viable, commercial scale (Wilson and Katz 1974). Most complete beeflike flavors are achieved by using the reaction products of L-cysteine and other amino acids with dextrose, blended with various other flavorants to enhance specific notes. It has been found that hydrolyzed plant proteins are very effective protein source for the reaction products and are valuable and cheap start materials for imitation meat flavors.

Recent development work has resulted in the use of tripeptide glutathione as a replacement for expensive L-cysteine and of glyceraldehyde as a replacement for part of the sugars.

Most modern patents for reaction flavors designate (a) the amino acid source—almost invariably including *S*-compounds (e.g., L-cysteine or organic thiols); (b) the reducing sugar or carbonyl compounds to be used and (c) the precise processing conditions.

A review of the patent literature displays permutations of all possible start materials and reaction conditions and indeed there has been a steady fall-off of new patents for these products over the past 3 years. It is likely that future generations of imitation meat flavors will be based on defined reaction products blended with compounds isolated from roasted meats. Or, they may be based on meat flavor precursors which will yield their flavor during the heat processing of the end product. The quality characterization of meatlike flavorings has been studied by Rothe *et al.* (1981). Much work has still to be done in this area of biotechnology.

a. Hydrolyzed Vegetable Protein—HVP

The production of high quality hydrolyzed plant proteins and yeast autolysates has for a long time been the basis of most commercial imitation meat flavorings (Blake 1982). These are offered as single products for direct use in food mixes or blended with flavor potentiators (e.g., MSG, ribonucleotides) and, often, spices. Currently, there is a wide range of such hydrolyzates, produced in significant tonnages, each claiming specific color and flavoring characteristics, which in turn are dependent on the source protein (e.g., wheat gluten, maize gluten, soybean flour, etc.). By far the biggest proportion of HVP is based on soybean protein (Prendergast 1974).

The three available industrial methods are (i) enzymatic—a slow and not entirely effective process which often results in the formation of bitter peptides (Roozen 1979); (ii) alkaline hydrolysis—which, though satisfactory, results in unacceptable flavor profile and an unbalanced amino acid content, and (iii) acid hydrolysis—the most preferred method which is cost effective and yields a range of good flavors.

Complete hydrolysis of the protein is desirable using a high hydrochloric acid-to-protein ratio to ensure minimal amino acid degradation and a better flavor profile. Large excess of acid has to be avoided as its neutralization would result in an abnormally high salt content in the finished product. Charcoal filtration of the neutralized aqueous solution results in the partial removal of certain amino acids as well as some unwanted impurities. The degree of filtration has to be carefully controlled as this determines the level of color in the end product.

If the start material is high in glutamic acid, then the resulting HVP will have marked flavor potentiating properties. Some commercial HVPs are essentially free from reducing sugars so that further heating as part of a food system does not significantly enhance any Maillard reaction which may take place. Others are high in polysaccharides and give a decided roast meat or even burnt character on further heat processing. These differences underline the need to evaluate HVPs under actual processing conditions so as to determine their precise flavor contribution. The manufacturer's literature and specifications should be carefully studied before using these flavoring ingredients.

b. Autolyzed Yeast Extract

This is the product of the self-digestion of yeast cells, achieved by heating an aqueous slurry of active yeast at 45°C which effectively kills the cells without damaging any enzymes present. These enzymes are mostly proteases which attack the cell components so partially solubilizing them. The resulting low molecular weight compounds include peptides, amino acids and nucleotides. The degree of autolysis depends on the temperature/time conditions and the pH of the medium. The process is terminated by raising the temperature in order to inactivate the enzyme system. The degree of protein and carbohydrate breakdown can be varied to satisfy particular flavor and handling characteristics.

The flavor of the end product is determined, to a large extent, by the choice of yeast and the nature of the substrate. Aerobically grown *Saccharomyces cerevisiae* gives a much stronger, more beeflike flavor than anaerobically grown *Saccharomyces carlsbergensis*.

Hydrolysis with hydrochloric acid at 100°C is a more effective method of solubilizing yeast than autolysis. The conditions used are similar to those employed for the production of HVP. When the hydrolysis has reached the desired level the reaction liquor is neutralized to pH 5.5, clarified to remove all solid matter and partially concentrated. In some grades of yeast extract the cell wall tissue is retained, the higher glycan content making these effective emulsifiers and thickeners. Further purification involves using either adsorption or ion-exchange techniques to remove undesirable components. The purified liquor is then finally concentrated and may be spray dried.

Autolyzed yeast and hydrolyzed yeast extracts are excellent savory flavoring agents as well as contributors to the nutritive value of an end product.

C. Enzymatically Derived Flavorings: Butter, Cheese

Fresh milk, which is a balanced system of lipids emulsified in a continuous phase comprising aqueous solutions of salts, lactose and lactalbumin together with a colloid suspension of proteins and calcium phosphate, has a variable composition depending on the breed of cattle, feed used and the period of lactation. It has a very low level of flavor but a quite distinctive odor, both of which are significantly altered by (a) heat, (b) microbiological action, (c) enzymatic reactions and (d) oxidation. The flavor profile of any dairy product depends not only on the sensory qualities of the start milk but on the nature and severity of the processing parameters.

The nature of the components responsible for the odor and flavor of fresh milk include (a) carbonyls—particularly acetaldehyde, acetone and formaldehyde (Harper and Huber 1956); (b) dimethyl sulfide (Bassette *et al.* 1966) and (c) free fatty acids—obviously the presence of lipids is of significance particularly during storage (Webb *et al.* 1974; Atherton and Newlander 1977).

Most flavors in milk cannot be attributed to any one compound but rather to the combined, almost subliminal, effects of many (Price and Manning 1983).

It is the enzymatic modification of the basic flavor of milk when converted into butter or cheese with which we are concerned in this section. Enzymes are complex proteins which act as catalysts in various biochemical interactions. Their precise mechanism in enabling or facilitating certain, often quite specific, reactions is not yet well understood. They generally function at ambient temperatures in a neutral pH medium and are effective at very low concentrations. Reactions involving enzymes are often rapid and, in food products, usually associated with hydrolysis involving the breakdown of complex molecules into simpler moieties.

1. Butter

In the United States butter must contain not less than 80% by weight of milk fat, but its composition is variable depending on the start material and the method of manufacture. Most butters contain about 16% of water, 1% of protein and 0.4% lactose. The salt content is adjusted to suit market requirements. The natural color of butter is due to carotenoids and other fat-soluble pigments present in the milk lipids and originating in the feed—the yellower the milk the deeper the color of the butter. Where permitted, butter is often additionally colored by the use of natural coloring matters such as annatto, obtained from the fruit of Bixa orellana.

The commercial production of butter consists of first separating the fresh sweet cream from milk followed by a churning process to break the emulsion (i.e., a fat in water emulsion becomes a water in fat emulsion). The precise operation is well described in the literature (Harper and Hall 1976) and may be either batch or continuous.

In most modern factories butter is made continuously in cylindrical agitators, the process involving (a) heating the cream to 52°–65°C (126°–150°F); (b) centrifugal separation; (c) removal of the skim milk; (d) pasteurization of the fats; followed by (e) controlled solidification and crystallization of the butter in a suitable plant.

a. The Flavor of Butter

The flavor of butter is produced by the fermentation of bacteria in the cream. Natural souring gives a good full flavor but the use of selected pure cultures or "starters" permits the better control of flavor and minimizes the development of off-flavors. The most widely used organisms are *Streptococcus lactis, Leuconostoc citrovorus* and *L. paracitrovorus* (*dextranicus*). Initially, *Streptococcus cremoris* was used but this has now largely been replaced. A vigorous aroma producer *S. lactis* subsp. *diacetylactis* is also of interest. These organisms are added to the pasteurized cream and allowed to ferment under controlled conditions prior to churning. This produces numerous flavor components predominant among which are diacetyl, acetoin and acetaldehyde. What we call "butter flavor" is open to wide interpretation, ranging from almost bland, for butters made from sweet cream, to those having a positive impact made from ripened cream. In the United States, cultured butter is no longer made and only that from sweet cream, in which there has been no significant bacterial growth, is now available. The flavor is pleasing but it lacks the distinctive character of butter made from mature cream or that containing added bacteria.

It has long been recognized that the lactic cultures used in making butter produce volatile acids and that diacetyl, which is present at about 0.05 to 2 ppm, is the key flavoring component, but the reactions involved obviously result in a highly complex mixture of compounds the relative organoleptic contribution of which is not well understood. One might expect butyric acid to predominate but this compound is generally not present or at most in very low concentrations well below its threshold.

The formulations of imitation butter flavors usually include a high percentage of diacetyl, modified by various butyrates, vanillin and some butyric acid.

b. Enzymatic Production of Butter Flavors

Apart from the traditional use of specific nature-identical organics in the compounding of imitation butter flavors, used extensively in the manufacture of margarine and in baking, the industry has been extending its use of selected enzyme systems for the production of natural butter flavors under carefully controlled conditions.

c. Butter Oil

This is a commercial product made from melted butter by centrifuging or directly from cream by a patented process by which the unwanted curd is removed by running the nearly churned cream into hot water. This causes the curd to precipitate and the butter oil can then be recovered by centrifuging. Butter oil can be used directly in many food products or may be treated with selected organisms which bring about lypolysis. The resulting very flavorful product is known as enzyme-modified butter oil (LBO).

The nature of the lipid fraction of butter depends to a large extent on that of the raw milk used. Yu *et al.* (1982) reported that the neutral fraction contained triglycerides, 60.5%; 1,3-diglycerides, 13.2%; 1,2-diglycerides, 7.8%; and free fatty acids, 6.2%. They found that the polar fraction comprised cephalin, 22.9%; lecithin, 17.3%; lysophosphatidic acid, 16.4%; and sphingomyelin, 11%. In the product they were examining the ratio of saturated to unsaturated compounds was 2:1.

The commercial production of enzyme-modified butter oil has been described by Andres (1981) and the making of a dry-powdered version by low temperature, spray-mat, drying is discussed by Farnham *et al.* (1981). These products offer richer, more intensified butter flavor profiles having wide applications, particularly in baked goods and spreads. Very little information has been published on the precise organisms or enzymes employed but the use of the microorganisms *Trametes sanguinea* has been patented (Hori and Shimagono 1967). The efficacy of *Streptococcus lactis* subsp. *diacetylactis* in the production of the butter flavor has been discussed by Piatkiewicz *et al.* (1981), and Olson *et al.* (1981) have described the microencapsulation of multienzyme systems. Aqueous cellfree extracts of *S. lactis* subsp. *diacetylactis* are emulsified in milk fat and extruded into cold water to produce the capsules. It is claimed that these products are effective in significantly increasing, up to eightfold, the production of diacetyl and acetoin in young cheese.

Commercially, butter oil is modified by controlled enzyme action which releases the volatile and flavorful short chain fatty acids, C_4–C_{10} (e.g., butyric, caprylic and capric), and also the nonvolatile long chain fatty acids. The enzyme system is heat inactivated when the process has reached the desired end point. Such products contain about 48% fat.

The function and effect of LBOs in food processing vary according to the rate of application. They are of particular value in sugar confectionery (0.1–0.3%) to enhance buttery character in soft caramel and toffee and to reduce the sweetness in products having a high sugar content, also in margarine (0.05–0.15%), or in bakery margarines (0.2–0.6%), to give a rich buttery flavor profile similar to that of real butter. It is usual for these products to be "tempered" at 20°–25°C for 12 h prior to use to make them easier to incorporate into food mixes, and in order to allow flavor equilibrium to be achieved, end products containing them should not be submitted to sensory assessment for a period of 24–48 h after manufacture. Manufacturers recommend that LBOs should be stored at 2°–6°C and be used within 6 months from date of manufacture.

2. Cheese

The making of cheese is a convenient, easy way of converting milk constituents into a palatable product which has good keeping qualities, is transportable, and can be standardized or modified to suit market needs. There are at least 1200 named varieties of cheese worldwide, of which about 500 have distinctive characteristics. From a flavor point of view, the spectrum of profiles is enormous, but taking texture into account they can be classified into six main groups as given in Table 7.1. A list of the principal cheeses, their names and descriptions, is given by Tressler and Sultan (1975).

Cheese is made by coagulating milk by the enzymatic action of rennet, cutting the resulting curd into small pieces followed by removal of the whey by gentle heating then pressing, molding and ripening to enable full flavor development. The start milk may be derived from a variety of animals (e.g., cows, sheep, goats). Acidity is one of the major factors in the formation of a workable rennet clot and to achieve this "starters" of specially prepared cultures of *Streptococcus* spp., selected for their activity in producing lactic acid from the lactose present in the milk, are

Table 7.1. Types of Natural Cheese Classified by Distinguishing Differences in Processing[a]

Distinctive Processing	Distinctive Characteristics	Typical Varieties
Curd coagulated primarily by acid[b]	Delicate, soft curd	Cottage cheese, cream, Neufchatel
Curd particles matted together	Close texture[c], firm body	Cheddar, Cheshire
Curd particles kept separate	More open texture	Colby, Monterey, Edam, Gouda
Presence of small amount of copper from copper cheese kettle or vat	Granular texture; brittle body	Grana—hard grating types: Parmesan or Reggiano, Romano or Sardo
Stretched curd	Plastic curd; threadlike or flaky texture	Provolone, Caciocavallo, Mozzarella
Bacteria ripened throughout interior with formation of eyes[d]	Gas holes or eyes throughout cheese	Emmentaler or Swiss (large eyes); Gruyere, Asiago (small eyes)
Mold ripened throughout interior	Visible veins of mold Typical piquant, spicy flavor	Blue, Roquefort, Stilton, Gorgonzola
Surface ripened principally by mold	Edible crust; soft, creamy interior Typical pungent flavor	Camembert, Brie
Surface ripened principally by bacteria and yeast	Surface growth; soft, smooth, waxy body Typical mild to robust flavor	Bel Paese, Brick, Limburger, Muenster, Port du Salut
Protein of whey or whey and milk coagulated by acid and high heat	Sweetish flavor of whey	Whey Cheese: Gjetost, Sap Sage, Mysost, Primost, Ricotta

[a] From Hall *et al.* (1971).
[b] In constrast to coagulation by acid and enzymes or in whey cheese, by acid and high heat.
[c] Close texture means no mechanical holes within the cheese; open texture, considerable mechanical holes.
[d] In contrast to ripening by bacteria throughout interior without eye formation.

added and the temperature raised to about 29°C (85°F) and held for several hours. The level of acidity is critical and is an essential factor in determining the quality of the resulting product. The use of rennet enzymes is centuries old and although calf rennet is still recognized, it is a shrinking natural source and has largely been replaced by an alternative chymosin enzyme preparation developed for this purpose. Such products are well known to the dairy industry and suppliers provide specifications and usage data. The congealed curd is cut and cooked by gentle heating to express the whey. The warm curd still contains the bacteria and enzymes necessary to produce the desired flavor character in the end product. The characteristic cheese flavor results from the action of the residual bacterial flora and enzymes on lactose, lipids and protein, mainly casein, present in the start milk. The primary compounds formed contribute very little to the ultimate flavor profile and it is the sequential reactions which take place slowly during the ripening and maturation periods which have the significant effect (Dumont and Adda 1978; Adda *et al.* 1978). The formation of flavor in cheese is a dynamic system involving several reactions which proceed until they are in equilibrium, this depends on the conditions employed throughout the process (Adda *et al.* 1982).

a. Cheese Flavor

Each named cheese is distinguished not only by its physical characters but by its unique flavor which is determined by the start milk, the processing parameters, the degree of ripening and storage conditions. Recent work (Forss and Sugisawa 1981) indicates that the typical cheese flavor does not appear to depend on any one key component but originates from various compounds present in the protein, lipid and carbohydrate fractions of the milk. Individually such compounds have profiles which are quite different from that of cheese but collectively, in the right proportions, they may broadly be described as "cheesey."

In some of the more highly flavored cheeses, some specific compounds do have a significant influence on the profile, e.g., blue cheese: methyl ketones, arising from the enzyme systems in *Penicillium roqueforti*; Swiss Emmentaler: alkanols and propionates; cheddar: 2-butanone and 2-butanol. The chemistry of these complex reactions has been presented, schematically, by Dumont and Adda (1978).

Two very well-known types of cheese—cheddar and blue—are worth closer study:

b. Cheddar Cheese Flavor

The flavor of freshly prepared cheddar cheese is bland and only develops its sweetly aromatic, slightly nutty character during ripening. The profile is well rounded with no outstanding attributes. As the cheese ages it develops a distinct sharpness or bite which may leave the palate unpleasantly tingling.

The number of compounds isolated from cheddar cheese is about 150 (Kristoffersen 1978) but as yet their relative importance to the profile has not been established. After a lengthy investigation, Kristoffersen (1973) concluded that (a) amino acids individually had no definable relationship to the flavor but in total may affect the ultimate profile; (b) ammonia and hydrogen sulfide are significant degradation products of the amino acids; (c) free fatty acids appear to be unrelated to the flavor character; (d) carbonyls do not relate directly to the profile; (e) there is a direct relationship between the levels of free fatty acids and hydrogen sulfide which may be used as a quality indicator.

Scarpellino and Kosikowski (1962) suggested that the significant amounts of 2-butanone and 2-butanol found in cheddar cheese arise from acetoin which itself is a degradation product of pyruvic acid. Keen *et al.* (1974) concluded that this reaction was induced by the presence of *Lactobacillus plantarum* and *L. brevis* which are organisms present in the cheese during ripening.

c. Blue Cheese Flavor

Blue cheese is applied to Roquefort, Goronzola, Stilton and Wensleydale cheeses in which the blue–green mold, *Penicillium roqueforti*, grows in the cracks and is an important ripening agent. The use of molds to produce this family of cheeses is lost in antiquity. The mold may be added to the milk after the addition of the rennet or to the drained curd. The cheese is then formed with a relatively loose texture to provide the right growth environment and 2.3–3% salt is added.

When freshly made these cheeses have little flavor and the distinctive appearance and characteristic flavor are only produced during the 60–90-day ripening period. Good quality Roquefort cheese has a sharp peppery note but is still smooth to the palate. The chemistry of the ripening process has been studied and it is established that there is a continuous lipolysis of triglycerides with the simultaneous liberation and conversion of fatty acids to methyl ketones. Forss (1972) has reviewed the formation of the methyl ketones, which contribute significantly to the flavor, and suggested that they were formed from either the enzymatic breakdown of α-ketoacids or from their thermal decomposition. Enzymes originating in the *Penicillium* mold hydrolyze the milk fat to lactic and other fatty acids which are then converted by α-oxidation and carboxylation into 2-alkanones, thence by reduction into secondary alcohols. The compounds isolated from blue cheese are broadly similar to those present in cheddar cheese.

d. Enzyme-Modified Cheese (EMC)

Cheese flavor is very important in many savory food and snack products. In most instances an adequate flavor level can be achieved by the careful selection and use of natural cheeses, using the powder form if necessary, but in some instances the percentage needed to attain the correct flavor is not technologically acceptable and frequently is uneconomical. In such products, cheese flavor concentrates are required and the industry has produced these by the application of enzyme-induced, controlled lipolysis. An article in Food Engineering (Anon. 1982) describes the production of EMCs by the acceleration and enhancement of the aging process in young cheddar, Swiss and Romano cheeses by treatment with selected Streptococcal multi-enzyme systems. In Europe, other cheese bases are employed and products based on Emmentaler, Gouda and mozzarella cheeses are available.

Close attention to the processing conditions ensures consistency of flavor strength and quality. The enzyme systems used are completely inactivated by heat treatment of the final proudcts. These have the consistency of a smooth to slightly grainy textured paste with an off-white to cream color. The flavor intensity depends on the type of cheese used and the precise modification process, but products in the ranges 10–15, 20–25 and 25–30 times the flavor of mature cheese are manufactured. Most of them contain about 25–30% fat and 22–27% protein.

It is claimed that the use of EMCs can result in considerable raw material cost savings, as up to 50% of cheese solids in many product formulations can be replaced by a small quantity of EMC. Their use also enables food processors to adjust any variable flavor in natural cheese included in the product. They are of particular value in (a) processed cheese, (b) cheese powders, (c) cheese snacks, (d) pizza sauce, (e) cheese biscuits, (f) cheese spreads, (g) quiche, (h) dips, and others. The rate of addition is directly related to the profile of the natural cheese present and the strength required in the end product. On average, the recommended use rate is 1–3% of the total recipe weight. As with LBO, the EBCs should be stored at 2°–6°C and used within 6 months from date of manufacture.

3. Lactic Acid Fermentation—Yogurt

Lactic acid fermentation is extremely important to the food industry and is applied in food preservation, pickle manufacture and in the formation of characteristic flavors in some products of which sour cream, buttermilk and yogurt are the best known. Sour cream and buttermilk depend for their flavor on the combined action of strains of *Streptococcus lactis* and *S. cremoris* to produce lactic acid and on *Leuconostoc cremoris* and *S. diacetylactis* for flavor production.

Yogurt is the product of two dissimilar organisms—*Streptococcus thermophilus* and *Lactobacilus bulgaricus*—which grow simultaneously to give the desired characteristics in the end product. Yogurt is made in two types: (a) firm bodied and (b) viscous or semifluid. Firm yogurt is produced from milk to which is applied nonfat dry milk or skim milk concentrate to give a total milk-solids–not-fat content of 12%. This is warmed to 60°C (140°F), homogenized and then heated to 82°C (180°F) at which temperature it is maintained for 30 min followed by cooling to 45°C (114°F). It is then inoculated and blended with 2% of the starter organisms and allowed to incubate at 43°C (110°F) for 45 min after which it is packaged. The packages are allowed to stand for 3 h prior to cooling to 7°C (44°F).

Semifluid yogurt is generally incubated in bulk vats and its processing differs only in the temperature and rate of incubation. After coagulation, the product is chilled to 10°C (50°F) at which stage it is practicable to incorporate fruit and/or flavorings.

Alternatively, the product may be processed using a UHT technique. After prewarming the milk it is subjected to UHT at 149°C (300°F) for 3.3 sec followed by rapid cooling to 43°C (110°F). Subsequent procedures are similar to those already described.

a. Yogurt Flavor

It is generally accepted that lactic acid, acetic acid and acetaldehyde are important in the flavor of natural yogurt and that symbiotic action of *S. thermophilus* and *L. bugaricus* are necessary to obtain the full flavor profile and correct acid balance. A pH of between 4.5 and 4.7 has always been thought to be optimum but a lower range of 4.0 to 4.4 is now increasingly preferred.

The flavor chemistry involved is complex and the number of compounds identified in yogurt suggests that many may be artifacts. Workers in this field have variously suggested that (a) alcohols may significantly contribute to the flavor profile by the formation of esters with any fatty acids present; (b) acetaldehyde is of major importance although a low concentration of this compound does not necessarily mean an unacceptable product; (c) the presence of esters gives to the product a desirable "fruity" character and (d) ketones may modify the flavor profile depending on the degree of preheat treatment given to the milk. Sensory flavor analysis and GC examination of yogurt prepared by UHT and normal vat processing reveal no significant differences in flavor profile or sensory attributes (Labropoulos *et al.* 1982).

b. Flavorings for Yogurt

A very wide range of natural and imitation flavorings is available commercially. These may be either entirely natural, natural with added synthetic topping notes or entirely synthetic. The natural products are often packed with whole or cut fruits so as to facilitate handling and incorporation into the yogurt mix. Flavorings may be added in one of three ways: (a) mix blending—the flavoring, sugar and coloring matter being added to the milk prior to fermentation; (b) in package—the fruit and flavoring being filled directly into the package prior to the addition of the prepared yogurt, which is then stirred before eating and (c) bulk mixing—the flavoring, sugar and coloring matter being added to the product after preparation and prior to filling.

Flavored yogurts are increasing in popularity because of their more acceptable level of acidity and reduced astringency compared with natural yogurt.

Because of the inherent acidic quality of the yogurt base, the most acceptable flavorings are those of the citrus fruits and other acidic fruits such as raspberry, strawberry, cherry and blueberry where the acid contributes to and enhances the flavor. Although not very acidic, banana and coconut are also popular flavors, particularly with children. Some savory flavors have been tried but without success. Flavors such as chocolate are not compatible with yogurt.

Marshall (1982), in an article on flavor compounds in fermented milks, reviews the different products achieved by fermentation with (a) *Streptococcus* and *Leuconostoc* spp., (b) *Lactobacillus* spp., (c) *Streptococcus* and *Lactobacillus* spp., and (d) *Streptococcus* and *Lactobacillus* spp. with yeasts. The author exemplifies the several products commercially available, inlcuding (i) buttermilk; (ii) Scandinavian fermented milks (e.g., filmjolk, Latfil, Langfil); (iii) acidophilus milk (e.g., yakult); (iv) yogurt and numerous much less well-known milk products. Tables show the nature of the organisms used, their metabolic products and other characteristics. The comparative flavors of such products and their quality assessment are also discussed.

D. Flavors Made by Fermentation

The word fermentation—from the Latin *fermentare* meaning to cause to rise—was first applied to the foaming observed during the making of wines and beer, a process whose origins are lost in history. It is now used to describe enzyme-catalyzed reactions in which compounds such as the carbohydrate sugars are broken down anaerobically by bacterial or yeast enzymes leading to the formation of pyruvic acid. In the case of yeast, this intermediate is split into carbon dioxide and acetaldehyde which in turn is converted into ethanol. Industrially, fermentation is carried out, using carefully selected microorganisms appropriate to the substrate, under precisely specified conditions. The products are wide ranging and include alcohol, glycerol, butyl alcohol, acetone, as well as lactic, acetic, citric, gluconic and glutamic acids, many of which are important in food flavorings.

The whole process of yeast fermentation, whatever the precise nature of the substrate, has been clearly stated by Drawert and Rapp (1965).

1. Yeasts

Yeast is a microscopic, uncellular organism of which there are several hundred species, only a very few of which are of commercial importance, namely, *Saccharomyces cerevisiae, Saccharomyces carlsbergensis, Candida utilis* (formerly *Torula utilis*), *Candida tropicalis*, and *Kluyveromyces fragilis*.

Selected *Saccharomyces* spp. are used for the production of wines and other alcoholic beverages.

The choice of yeast depends on the nature of the substrate in which it is to be grown. *S. cerevisiae* and *S. carlsbergensis* utilize glucose, galactose, maltose and sucrose; *Candida utilis* (the so-called torula yeast) prefers pentose sugars such as arabinose and xylose and is widely used in the fermentation of molasses and sulfite waste liquors; *Kluyveromyces fragilis* grows on lactose contained in dairy by-products.

2. Vinegar/Actetic Acid

Vinegar has a long and useful history both as a condiment and as a food preservative. It is the product of the alcoholic and acetic fermentation of a sugar-containing solution without any intermediate distillation except in the production of spirit vinegar. Numerous vinegars are available for use in food processing. These include (i) wine vinegar, the alcoholic fermentation being from grapes; (ii) malt vinegar, made from malted barley; (iii) spirit vinegar, prepared by the acetic fermentation of distilled alcohol, itself made by fermentation; (iv) fruit wine vinegars, made from grape and nongrape fruit wines; (v) cider vinegar, made from fermented apple juice and (vi) distilled vinegar, obtained by the distillation of malt vinegar. The flavor characters of these vary considerably.

Two entirely different processes are involved in the manufacture of vinegar:

a. Fermentation of sugar by yeast:

$$\underset{\text{Hexose}}{C_6H_{12}O_6} \rightarrow \underset{\text{Ethanol}}{2CH_3CH_2OH} + \underset{\text{Carbon dioxide}}{2CO_2}$$

b. Conversion of alcohol to acetic acid by the action of acetobacter organisms:

$$\underset{\text{Ethanol}}{2CH_3CH_2OH} + \underset{\text{Oxygen}}{2O_2} \rightarrow \underset{\text{Acetic acid}}{2CH_3COOH} + \underset{\text{Water}}{2H_2O}$$

The process is usually carried out within the temperature range 24°–27°C (75°–80°F). The alcoholic fermentation is induced by a strain of *Saccharomyces cerevisiae* var. *ellipsoides* which has been grown in the presence of a minimum level of sulfur dioxide to eliminate extraneous organisms.

Using a substrate of 8–20% fermentable sugars, the primary fermentation is carried out in open vats fitted with a cooling system to regulate the temperature. The added yeast, during its normal reproduction, converts the sugars to alcohol and carbon dioxide and, in the case of malt vinegar, similarly converts dextrins obtained from the malted barley. Fermentation takes about 7 days to completion at which stage the product contains 6–7% alcohol which represents about a 90% of theoretical conversion of the sugars. After sedimentation of the yeast cells, the supernatant alcoholic liquor is filtered and is ready for the second acetic oxidative fermentation.

This secondary fermentation requires considerable oxygen and is carried out in one of two ways:

a. Packed generator systems. The traditional acetifier comprises a tall vat packed with an inert material (usually birch twigs or wood shavings) which act as the carrier for the acetobacter culture. The alcoholic substrate is sprayed on the top of the column and allowed to flow down against a stream of air, introduced from the bottom of the vat, into a false bottom and is recirculated for several days until most of the alcohol has been oxidized to acetic acid.

b. Fring's acetator. This is a submerged culture fermentation involving vigorous stirring of the substrate to induce adequate aeration. Several purpose-designed acetators are available but in each of the acetobacter organisms are introduced into the substrate and grow in a suspension of fine air bubbles produced by the considerable upwards turbulence. The process is continued until the desired level of acetic acid is achieved. This type of plant requires less space than the traditional vats and gives a more rapid conversion but is more energy consuming. The resulting vinegar is filtered and pumped into storage vats where it is allowed to mature for several months during which it develops a delicate and characteristic bouquet.

In the United States, Federal standards require vinegar to contain not less than 4 g of acetic acid per 100 ml.

Distilled vinegar is prepared by distilling the acetic acid and volatile constituents from a normally fermented vinegar. It is colorless and retains much of the characteristic aroma and flavor of the original vinegar. It is of particular use in products where color is an important attribute (e.g., pickles, mayonnaise, tomato ketchup).

Concentrated vinegar, containing up to about 40% acetic acid, may be prepared by freeze concentration using a scraped-surface Votator heat exchange unit operating at −8°C (17°F). The water is removed as ice and most of the aromatics remain in the concentrated product.

3. Wines

The growing of grapes and the fermentation of the expressed juice to produce alcoholic beverages or wines has a very long history. It is known that the European grape, *Vitis vinifera* L., has been grown in the Mediterranean countries for at least 5000 years. This species was found to be so good for wine production that it was introduced to the Americas, South Africa and Australia during the seventeenth century and specificially to California by the middle of the eighteenth century. The native American vine is another species, *Vitis labrusca*, and, in order to produce strains resistant to disease, insect and fungal attack and climatic conditions, a hybrid between these two species was soon developed. Today there are literally thousands of varieties of *Vitis* spp., which differ from each other in color, size and shape of the berry, juice composition, ripening time, resistance to disease and notably flavor potential on fermentation.

Presently, out of a total world crop of about 25 million acres, there are in the United States some 600,000 acres under cultivation of grapes yielding about 380 million gallons of wine. The European production amounts to about 70% of the total.

a. Quality Factors

The principal quality determinant is the grape used, particularly its species and variety, the climatic conditions prevailing during its growth and ripening, its method of culture and resistance to infestion, particularly fungal and/or insect attack.

Because of its high natural sugar content (18–24%) and the relatively low acidity (0.5–1.5%) of its juice at maturity, *Vitis vinifera* L. is by far the major source of commercial wine. Although much lower in sugar content (10–20%) and with a higher acid content (1–2.5%), some production is carried out in the United States from domestic grapes, *V. labrusca* and *V. rotundifolia*, but wines, particularly from the "Concord" varieties of the eastern United States, have a characteristic profile with a strong methyl anthranilate note often described as "foxy" (Nelson *et al.* 1977). In all the regions, numerous hybrids and strains have been developed to produce optimum juice composition, to induce resistance to diseases and to condition the plants to regional climatic conditions.

Wines produced from such cultivars contain sufficient sugar to give an alcohol content of about 10% or higher on fermentation. An acidity of less than 1%, calculated as tartaric acid, and a pH of 3.1–3.7 make grape juice ideal among the fruit juices for the making of wine, although very potable wines can be made from other fruit juices.

The type of wine is significantly influenced by the skin pigmentation which displays a range of colors from greenish yellow through yellow–brown to pink, red, reddish violet to deep blue–black. Although the extracted juice is almost colorless, fermentation in the presence of the skin produces wines having very distinct and often characteristic profiles and colors.

The composition of the grape juice is influenced by climate, particularly hours of sunshine, which are probably the greatest source of differences between wines from different growing regions. For grapes to attain full maturity they must have an average daily temperature in excess of 10°C for a period of about 180 days. If this "heat summation" of 1800 is not achieved it may be necessary to adjust the sugar content of the juice prior to fermentation.

b. Wine Making

The making of wine falls naturally into the following stages:

a. Harvesting the grapes. Timing is critical in order to achieve the correct sugars/acid balance. Hand gathering is still widely practiced in Europe but mechanical harvesters are being used increasingly in California, New York state and in Australia. The gathered grapes are bulked and transferred to the winery.

b. Crushing. Mechanical crushing and pressing is now widely used and replaces the traditional treading methods. The pressed pulp passes through a very coarse cylindrical sieve to remove the stems.

c. Separation of the juice. White wine is normally fermented from must after removal of the skin and seeds. This is done immediately after crushing but, if additional flavor and some color is required, this operation may be delayed by up to 24 h. Separation is achieved by mechanical pressing using one of several designs. Red wine is fermented from juice, skin and seeds together and any solids are removed after the fermentation process.

d. Settling the must. Fresh pressings are normally cloudy and the fine suspended matter is allowed to settle as a sludge. To save valuable time the must may be clarified by centrifuging.

e. Fermentation. Although simple, the fermentation stage needs care and control, particularly in the following. (i) Selection of the yeast strain: Strains of *Saccharomyces cerevisiae* are preferred as they allow final fermentation liquor to contain up to 18% ethanol, if conditions are right. (ii) Adequate nutrients: A well-balanced must usually contains adequate nitrogenous constituents to encourage yeast growth. (iii) Temperature control: Optimum temperature for fementation is 25°C (77°F) but, as the process produces heat, temperature control is necessary to prevent the fermenting mass from reaching 30°C (86°F) at which temperature yeast growth is seriously impaired. (iv) Resistance to oxidation: This is only a problem in small fermentation units where the use of air traps is necessary. At the end of fementation 50–100 ppm of sulfur dioxide or ascorbic acid may be added as an antioxidant. (v) Control of the "cap": In red wine, the skins and solid pulp float to the top of the fermenting juice and this must be submerged at least twice a day to prevent it from drying out and to give better flavor and color to the product. In bigger wineries this is achieved by recycling the must as a spray over the "cap."

f. Postfermentation. Fermentation comes to a halt when the fermentable sugars fall below about 0.1%. Under normal conditions this stage may be reached in from 10 to 30 days, after which the yeast coalesces and settles as a sludge, known as *lees*, leaving an opalescent supernatant wine. The yeasts are removed by a process known as *racking*.

g. Secondary malolactic fermentation. In some instances it may be desirable to leave the wine in contact with the lees as this encourages secondary fermentation brought about by lactic acid bacteria which attack any malic acid present converting it to lactic acid and carbon dioxide. The process is inhibited by the presence of 70–100 ppm of sulfur dioxide.

h. Clarification. A process dictated by conditions and involves, as necessary, the use of finings (e.g., Bentonite), filtration with or without filter aids, centrifugation, temperature adjustment particularly refrigeration between −5° and −7°C (23° and 17°F).

i. Aging. This stage calls for considerable skill and experience to ensure that the wine attains an optimum stable profile.

The preceding is of necessity only an outline of the techniques employed in wine making. The whole process calls for considerable practical experience and the literature abounds with information applicable to wines from the various producing regions worldwide (Amerine *et al.* 1980).

c. Flavor Constituents of Grapes and Wine

The list of identified volatile aromatic components in grapes and wines is in excess of 350 (Rapp *et al.* 1978; Ribereau-Gayon 1978). Many of these are present at subthreshold levels and may indeed be artifacts of the analytical techniques employed. The constituents demonstrably affecting the flavor of wines originate from five sources, namely: (a) unchanged constituents from the raw grape juice, (b) products arising from the alcoholic fermentation of the must by wine yeast; (c) products derived from microbiological reactions other than the wine yeast; (d) chemical transformations occurring within the wine during processing and subsequent aging and (e) extraneous aromatic compounds derived during processing.

Obviously, fermentation is only one of the biochemical processes taking place during the production of wine, and biotechnological aspects of flavor in fermented beverages have been reviewed by Schreier (1978). The following summarizes the main considerations which apply to the five flavor sources just listed.

a. Unchanged constituents. The prime juice constituents largely unaffected by the alcoholic fermentation process are the organic acids, principally tartaric, malic and citric acids. High acid levels in the grape juice are undesirable as this leads to a sharp wine, but juices having too low a level of acids give wines which lack character, are flat and lifeless and usually do not keep well. Some aromatic components also survive the fermentation. The terpene alcohols (e.g., linalool, geraniol, α-terpineol) and the terpene oxides (e.g., neroli oxide) readily characterize the prime grape juice and can also be found in the finished wine, probably accounting for its decided roselike nuance. The complex composition of the grape must, the variety and specificity of the enzyme systems present in the selected yeast strain and the processing parameters result in innumerable compounds which may or may not contribute significantly to the ultimate profile of the wine. Although present at parts-per-billion levels the terpen-, furan- and pyran-derivatives are sufficiently significant to enable their presence to be used as a basis for typing musts (Schreier 1978) and have been identified in the finished wine (Schreier 1979).

b. Products resulting from fermentation. During fermentation, *Saccharomyces cerevisiae* anaerobically metabolizes the sugars present in grape juice to yield ethanol and carbon dioxide but other metabolites may also be found and persist in the finished wine (e.g., glycerol, acetaldehyde, pyruvic acid). As amino acids are present in the must, these form a substrate for a further fermentation by wine yeast resulting in the formation of fusel oils. Being highly aromatic alcohols these have a significant effect on the bouquet of the wine. The alcohols resulting from the fermentation undergo esterification with the fruit acids present in the wine (mainly acetic acid) giving rise to a complex mixture of esters which may significantly modify the profile. Schreier (1978) concludes that the influence of yeast fermentation is not crucial to the character of any given wine variety as this is determined more by the original constituents of the grape must which remain unaltered. Secondary malic–lactic acid fermentation may also result in the decarboxylation of malic acid into pyruvic acid and, by reduction, to lactic acid. A sufficient number of aromatic compounds may accompany this reaction to introduce distinct nuances. The production of lactic acid results in a wine having a smoother palate.

c. Extraneous microbiological reactions. Naturally grapes carry a heavy load of microorganisms including several types of yeasts, bacteria and molds. When freshly prepared a must may contain up to 10 yeast cells per milliliter, but its low pH ensures that most are inhibited leaving only certain acid-resistant yeasts and acetic and lactic acid bacteria to develop. Spontaneous fermentation of the must is not initiated by *Saccharomyces cerevisiae* but by other yeasts present on the grape skin. It is the rapid growth rate of *Saccharomyces* which ensures its predominance in the later stages of the fermentation. Extraneous fermentation can only be prevented by pasteurization of the must followed by seeding with a pure strain of the wine yeast. It is spontaneous fermentation which is thought to account for the formation of phenylethyl alcohol, fusel oils and acetaldehyde, the later oxidation of which by the action of micrococci leads to the formation of acetic acid. Other metabolites of this complex fermentation sequence are acetoin, diacetyl and certain amines all of which are detrimental to the quality of the finished wine (van Straten *et al.* 1978).

d. Chemical transformations. Such reactions as esterification and polymerization are very slow and only take place during long storage or aging. One of the most important of these reactions is the polymerization of tannins leading to a reduction in the astringency of the wine and also changes in its anthocyanin content. This latter is directly related to changes in the color of the product. Not all these changes are advantageous and overaging can result in a wine which lacks character or in extreme cases becomes "vinegary."

e. Extraneous odors and flavors. Wines readily pick up aromatics from materials with which they are in contact (e.g., wooden casks, corks). Modern technology dictates the use of stainless steel for the processing plant but not everybody agrees that the resulting wine has a fully developed and acceptable profile. Work by Tanner *et al.* (1982) has shown that the compound responsible for "corking" in wines is 2,4,6-tichloranisole which can make a wine unacceptable at a level of only 30 ppt. Whereas the "musty" note is due to tetra and pentachloroanisole derived from the wood of casks.

Like its source materials, techniques of production, storage, bottling and presentation, wine is a product displaying infinite variety. Its chemistry, though simple in concept, is highly complex in practice and for many it is sufficient merely to enjoy the results of the vintner's art whether as an ingredient in a food product or as a superb supplement to a meal.

4. Dried Inactive Yeast Powder

An inevitable and commercially important by-product of fermentation processes is yeast. This can be recovered, dried and inactivated to form a valuable food ingredient. Although the brewing industry is the major source, the demand is sufficient to warrant the growing of selected yeasts primarily for this purpose. Henry (1982) has described the production of dried Torula yeast resulting from the fermentation of the waste sulfite liquors from paper production.

The yeast slurry, containing about 1% solids, is continuously separated from the substrate by centrifuging, the collected yeast cells are washed to remove all adhering chemicals and then dried to a moisture content of max 7%, in one of several ways, to give the commercial product. This is characterized by a very high protein content (min. 45%) and a substantial content of Group B vitamins. The USF XIII confines the source to either *S. cerevisiae* or *Candida utilis*.

Dried inactive yeasts are used not only as valuable nutrient and vitamin supplements but in the flavoring of dry savory seasonings, dip mixes, gravy and sauce mixes, snack products, etc., to which they give a rich, full-bodied flavor character. Becauses of the many variables which affect the flavor profile of inactivated yeast, the products of different manufacturers may have quite distinctive flavoring properties. The profile may be characterized by such nuances as "sweet," "caramellike," "roasted," "beefy," "cheesey," "creamy," and this implies that such products must be individually assessed prior to use.

E. Biotechnology: Production of Aroma Chemicals

Biotechnology is one of the latest words currently being used to describe the application of microbiological cultures, their endogenous enzyme systems as well as commercially produced enzymes to a whole spectrum of sophisticated techniques which can directly affect the processing, quality and availability of raw materials and food products and other industrial processes. Of course, much of the basic technology has been used in food production for most of this century. Fermentation, using numerous bacterial, fungal and yeast cultures has been of great service in food and beverage manufacture, in the making of dairy produce, sausages, baked goods, etc. But, the whole subject of utilizing the powerful enzyme systems present in microorganisms has been given far greater impetus by the need to economize on energy use, to satisfy a growing need for improved nutrition for a growing world population and to find natural alternatives to the synthetic aroma chemicals widely used as flavorants.

These factors, coupled with the recently acquired knowledge of how to carry out genetic engineering involving the modification of gene patterns, has given researchers a challenge to restructure organisms either to synthesize enzymes having specific actions on designated substrates or to remove some of the uncertainty associated with naturally occurring enzyme systems, some of which may produce unwanted attributes or even be hazardous to health.

This increased interest in biotechnology has had a direct impact on research and development into the application of induced microbiological enzyme reactions under controlled conditions for the commercial production of aroma chemicals which truly imitate nature. Presently not all the biosynthetic pathways of flavor formation in aromatic plant materials have been studied or elucidated so that demands for organic flavorants continue to be met by classical chemical syntheses even though many of these involve complex multistage reactions. But, although progress is as yet slow, there is an increasing move towards microbial processes which provide certain advantages over the more traditional chemical methods. Reactions induced microbiologically are often single stage; the whole complicated series of transformations taking place under very gentle conditions of fermentation. The reactions can often be studied and optimized by control of the substrate and fermentation conditions. Enzymatic reactions are generally fast and clearly specific; they are adaptable and may be used as a stage in a more complex chemical process and, as they are often stereospecific, are frequently of value in producing flavoring chemicals having unique flavor profiles. Microorganisms have long been known as a rich source of enzymes which themselves are of value in flavor production and/or enhancement in foods. This makes the whole subject of biotechnology of vital interest to food producers and we are likely to see tremendous strides in its applications during the next decade.

To date, success has been achieved in the use of microbial metabolism in many industrial applications, including sweeteners (e.g., fructose, aspartame); conversion of milk waste, etc; flavor potentiators (e.g., 5′-nucleotides); starch hydrolysates; tenderizers for meats; improved pickle production; clarification of fruit juices and production of aroma chemicals.

Very many more products are likely to be developed commercially as our knowledge of enzyme systems in biosynthesis is increasingly improved, but there are many practical difficulties to be overcome in converting laboratory results into viable commercial production. New food products arising from this technology are likely to attract legislative control (Jones 1985).

Microorganisms in Flavor Formation

Comprehensive reviews of the part played by microorganisms in flavor formation have been published by Kempler (1983) and Schreier (1978). These articles reveal that already well-established microbiological methods are available for the production of the following.

a. Terpenes. Terpenes are produced by the submerged culture of *Ceratocystis* (Collins and Halim 1977); *C. monoliformis* (Lanza and Palmer 1977); *Trametes odorata* and *Phellinus* spp. (Collins and Halim 1972) and *Kluveromyces lactis* (Drawert and Bartos 1978). The conversion of l-menthone into l-menthol as the source of natural menthol other than from *Mentha* spp., results from the culture of *Pseudomonas putida* (Nakajima *et al.* 1978). Alternatively, the hydrogenation of thymol to four isomeric forms of menthol using cultures of *Absidia hyalospora, B. subtilis* var. *niger, Geotrichum candidum* or *Trichoderma* spp. is of real commercial interest and is already patented in the United States. A Japanese patent involves the use of *Gleophyllum* or *Schizophyllum* cultures for the hydrolysis of carboxylic esters to l-menthol. Doubtlessly these are only the start of many other similar applications of commercial value.

b. Lactones. Yeasts (*Candida* spp. and *Saccharomyces cerevisae*); molds (*Penicillium notatum, Cladosporum butyri* and *Cl. suaveolens*); a fungus (*Trichdema viride*) and bacteria (*Sarcina lutea*) have all been shown to be capable of converting keto acids into lactones in yields up to 85% in certain instances (Collins and Halim 1972; Tahara *et al.* 1972). This area of aroma chemical production is of significant interest in view of the wide application of lactones in flavorings.

c. Esters. Fruity odors from cultures of *Pseudomonas fragi* have been recognized since the 1950s and, more recently, fruity notes associated with cheese production have been confirmed as due to ethyl butyrate and hexanoate produced by the lipolytic activity of this organism (Bills *et al.* 1965; Reddy *et al.* 1968). Other organisms (*S. lactis* subsp. *diacetylactis, S. cremoris*) are also known to be involved in ester formation (Hosono and Elliott 1974).

d. Diacetyl (2,3-butadiene). The biochemical pathways of the enzyme-induced reactions in cheese production have been closely studied (Forss and Sugisawa 1981). Joensson *et al.* (1980) produced starter distillates by the steam distillation of the culture of any or all of the following species of bacteria: *Streptococcus lactis, S. cremoris, S. lactis* subsp. *diacetylactis, Leuconostoc citrovorum* and *L. dextranicum* grown on a medium of skim milk in the presence of 0.1% citric acid. All these organisms convert α-acetolactate to diacetyl and the distillate contains about 2% of flavor compounds of which 80–90% is diacetyl. Such distillates can be used directly in butter flavorings and their use has been given FDA approval.

Numerous other examples are quoted in the reviews listed here. Many of them are still only laboratory trials and have yet to be translated into commercial production. This whole area of biotechnology is now receiving very close attention by the food processing and related industries and is likely to see significant progress as legislation recognizes this advancing technology and costings become more realistic. An overview in *Food Technology* (Oct. 1985) gives an updating of this topic.

F. Flavors Made by Pyrolysis: Smoke Flavors

1. The Smoking of Foods

The process of curing and smoking food, which has evolved slowly over thousands of years, was probably discovered by accident when it was observed that meats and fish hung up to dry in a smoky atmosphere remained edible for longer periods and also acquired a pleasant taste. For many hundreds of years the process has been carried out as a means of preserving both meat and fish, although the emphasis has gradually changed from preservation to the imparting of a light, very pleasant flavor to the food product.

Initially, the smoking process was carried out in a kiln with little or no control over the smoking process. Today, smoke generators are designed to ensure the burning of hard wood sawdust under controlled conditions of forced air recycling and temperature. Even so, the slow burning of wood demands considerable skill and may give rise to products of variable flavor quality. The direct use of wood smoke is relatively inefficient, results in messy tarry deposits throughout the smoke generating plant and often results in the deposition of unsightly particulate matter on the surface of the smoked product. In spite of these problems the smoking of meats and fish is still extensively used. But, it is for these reasons that the use of alternative natural liquid smoke flavors have gained wide acceptance. In the United States, some 1.8 million kg of liquid smoke flavorings are used annually directly in foods and an additional 8 million kg are applied by atomization and showering. These use figures are steadily increasing.

2. Natural Liquid Smoke Flavorings

The Code of Practice issued by the International Organization of the Flavour Industry (IOFI) specifies smoke flavors as follows.

> Smoke flavours are concentrated preparations, not obtained from smoked food materials, used for the purpose of imparting smoke type flavour to foods. They are prepared according to one or more of the following methods:
>
> 1. By subjecting various untreated hardwoods to processes described as (a) controlled burning, or (b) dry distillation at appropriate temperatures, usually between 300° and 800°C, or (c) treatment with superheated steam at temperatures usually between 300° and 500°C, and condensation and capturing of those fractions which have the desired flavour potential;
> 2. By applying further isolation techniques to the fractions obtained under 1. [item 1] in order to retain only the flavourwise important fractions or components;
> 3. By compounding chemically defined flavouring substances. Smoke flavours should contribute no more than 0.03 ppb 3,4-benzopyrene to the final food product.

3. Pyroligneous Acid

When wood is heated without an excess of air to temperatures in excess of 250°C, it decomposes into charcoal and a volatile fraction which condenses on cooling. The aqueous phase of this condensate is known as pyroligneous acid. The chief component is water containing about 10% of acids, mainly formic, acetic and propionic.

4. Smoke Condensates

These are prepared by drawing the distillation and combustion products from damp hardwood sawdust into a condenser fitted with an electrostatic precipitator. This removes all particulate matter from the condensate which is collected and settled to allow separation into aqueous and tarry phases. The clear aqueous phase is the desired product.

Modern generators produce smoke under carefully controlled conditions of combustion temperature and air flow. Sawdusts of hickory, maple and oak woods are used as the source material. The smoke passes through a settling chamber to remove heavy tars and wood ash and is then directed upwards through a column of inert ceramic material countercurrent to a steady flow of water. This is recycled as necessary until the titratable acidity is 5–12%, as acetic acid. The product is then filtered and set aside to age, during which time it loses its unpleasantly sharp flavor character and additional tarry matter deposits. A U.S. patent (Tolin 1961) describes an alternative use of yeast, wheat and soybean hydrolysates in place of water to give greatly improved flavor pick-up.

The latest plant consists of an inclined rotary calciner. The sawdust is fed continuously into one end of the inclined tube and discharged at the other end as charcoal. The vapors are handled as already described. The rate of feed, inclination of the tube, and rate of air flow are optimized to give a product with the best smoke flavor and the lowest content of polycyclic aromatics.

5. Chemistry of Smoke Flavors

There are three important and interrelated aspects which determine the nature and character of liquid smoke flavors. These are the following.

a. Flavor Chemicals

Kim (1974) identified some 150 compounds present in oak smoke condensate and Daun (1979) determined that of these the most important flavorants are guaiacol,4-methyl guaiacol and 2,6-dimethyloxyphenol. These compounds alone do not give a satisfactory "smoke" flavor and numerous other constituents must play a part in the characteristic profile.

b. Color Compounds

The nature of the compounds responsible for the brown color in smoke condensates has been investigated (Ruiter 1979). They are thought to be due to a Maillard-type reaction between carbonyls produced in the smoke vapor phase and amino acids on the food itself. The most important carbonyls present in smoke condensates are dicarbonyls and hydroxycarbonyls, being derived from glycolic aldehyde, glyoxal and methylglyoxal precursors (Watts 1974).

c. Polycyclic Aromatics

These are of particular importance in view of their carcinogenic nature. Levels of up to 50 ppb of benzo(a)pyrene (3,4-benzopyrene) have been reported in certain food products exposed to excessive smoking (Potthast 1980) but figures in excess of 1 ppb are considered to be unacceptably high. The level recommended by IOFI is only 0.03 ppb. Reduced levels of polycyclic aromatics can be achieved by the use of lower temperatures of pyrolysis. Cold-smoked products have less benzo(a)pyrene than hot-smoked products. Smoke flavors can now be consistently produced under carefully controlled conditions and these commercially available flavorings contain no detectable levels of benzo(a)pyrenes (Hollenbeck 1979).

6. Methods of Application

The four methods used for the application of liquid smoke flavors, are direct addition, dipping, showering and atomization (Wistreich 1979; Schneck 1979).

a. *Direct addition* applies mainly to the production of comminuted meat products such as frankfurters and pate but liquid smoke flavorings are also useful in many snack products, pizza toppings, etc., at about 0.1%.

b. *Dipping* is used as a means of application to small meat and fish products where the flavor is only required on the surface.

c. *Showering* is really very similar to dipping but is applied to the continuous processing of sausages such as frankfurters. The liquid smoke flavoring is recycled. Such products may contain up to 0.5% of smoke flavor.

d. *Atomization* is the most widely used method of applying smoke flavors. This consists of exposing the food product to a fine spray of the liquid smoke. The use of electrostatic deposition considerably increases the efficiency of this process and a good color is achieved by preheating to dry the surface of the product and heating again following the smoke application. Using these techniques bacon may be flavored with about 0.5% of liquid smoke.

Products flavored with liquid smoke must be labeled "smoke flavoring added."

Most flavor manufacturers offer smoke flavors based on known synthetic GRAS components. These have the advantage of a known and entirely safe composition and a standardized flavoring and coloring effect.

References

ADDA, A. J., ROGERS S. and DUMONT, J-P. 1978. Some recent advances in the knowledge of cheese flavor. *In* Flavor of Food and Beverages, pp. 65–74, G. Charalambous and G. E. Inglett, eds. Academic Press, New York and London.

ADDA, A. J., GRIPPON, J. C. and VASSAL, L. 1982. The chemistry of flavor and texture generation in cheese. Food Chem. *9*, 115–129.

ALLERTON, J. 1974. Chocolate and cocoa products. *In* Encyclopedia of Food Technology, pp. 195–215, A. H. Johnson and M. S. Peterson, eds. AVI Publishing, Westport, CT.

AMERINE, M. A. 1974. Wine. *In* Encyclopedia of Food Technology, pp. 961–968, A. H . Johnson and M. S. Peterson, eds. AVI Publishing, Westport, CT.

AMERINE, M. A. *et al.* 1980. Technology of Wine Making, 4th edition. AVI Publishing, Westport, CT.

ANDRES. C. 1981. Butter flavor variations meet specific usage demands. Food Process. *42*(5), 62–63.

ANON. 1982. Products taste 'cheesier' . . . with less cheese. Food Eng. *54*(11), 64.

ATHERTON, H. V. and NEWLANDER, J. A. 1977. Chemistry and Testing of Dairy Products. AVI Publishing, Westport, CT.

BASSETTE, B., BAWDON, R. E. and CLAYDON, T. J. 1966. Volatile components in blood, milk and urine of cows fed silage grain, bronegrass pasture and hay-grain test meals. J. Dairy Sci. *49*, 811.

BILLS, D. D., MORGAN, M. E., LIBBEY, L. M. and DAY, A. E. 1965. Identification of compounds responsible for fruity flavor defect of experimental Cheddar cheeses. J. Dairy Sci. *48*, 1168–1173.

BIRCH, G. G., BLAKEBROUGH, N. and PARKER, K. J. 1981. Enzymes in Food Processing. Elsevier, Barking, Essex, England.

BLAKE, A. 1982. The world of meat flavours. Food Manuf. *57*(9), 65–67, 71.

COLLINS, E. B. 1972. Biosynthesis of flavor compounds by microorganisms. J. Dairy Sci. *55*(7), 1022–1028.

COLLINS, R. P. and HALIM, A. F. 1972. Characterization of the major aroma constituents of the fungus Tricoderma viride PGRS. J. Agric. Food Chem. *20*, 437–438.

COLLINS, R. P. and HALIM, A. F. 1977. Essential oil composition of *Ceratocyctis virescens*. Mycologia *69*, 1129–1136.

DAUN, J. H. 1979. Interaction of wood smoke components and foods. Food Technol. *33*(5), 66–71, 83.

DRAWERT, F. and BARTOS, H. 1978. Biosynthesis of flavor compounds by microorganisms: 3. Production of monoterpenes by the yeast *Kluyveromyces lactis*. J. Agric. Food Chem. *26*, 765–766.

DRAWERT, F. and BERGER, R. 1981. Possibilities of the biotechnological production of aroma substances by plant tissue cultures. *In* Flavor '81, pp. 509–527, P. Schreier, ed. de Gruyter, Berlin and New York.

DRAWERT, F. and RAPP, A. 1965. Biotechnological sequences in wine making. Vitis *5*, 27.

DUMONT, J. P. and ADDA, J. 1978. Flavour formation in dairy products. *In* Progress in Flavour Research, pp. 245–262, D. G. Land and H. E. Nursten, eds. Appl. Sci. Publ., London.

DWIVEDI, B. K. 1973. The role of enzymes in food flavors: Part 1. Dairy products. CRC Crit. Rev. Food Technol. *3*, 457–478.

DWIVEDI, B. K. 1974. Meat flavor—a review. CRC Crit. Rev. Food Sci. Nutr. *5*, 487–535.

EMBERGER, R. 1981. Aspects of the development of industrial flavor materials. *In* Flavor '81, pp. 245–262, P. Schreier, ed. de Gruyter, Berlin and New York.

FARNHAM, J. G., JORDAN, R. and SAAL, H. 1981. Spray/mat effectively dries tough enzyme modified butter-oil, cheese. Food Process. *42*(3), 104, 106.

FLAMENT, I., SONNAY, P. and OHLOFF, G. 1977. The flavor of grilled beef. Helv. Chim. Acta. *60*(6), 1873–1883.

FORSS, D. A. 1972. Odor and flavor compounds from lipids. *In* Progress in the Chemistry of Fats and Other Lipids, Vol. 13, p. 177, R. T. Holman, ed. Pergamon Press, Oxford.

FORSS, D. A. 1981. Sensory characterization. *In* Flavor Research: Recent Advances, pp. 126–174, R. Teranishi, R. A. Flath and H. Sugisawa, eds. Dekker, New York and Basel.

FORSS, D. A. and SUGISAWA, H. 1981. Brief review of dairy and soy products. *In* Flavor Research: Recent Advances, pp. 355–371, R. Teranishi, R. A. Flath and H. Sugisawa eds., Dekker, New York and Basel.

GRAB. W., BRUGGER, W. and TAYLOR, D. 1978. Cocoa substitution. *In* Flavor in Foods and Beverages, pp. 81–90, G. Charalambous and G. E. Inglett, eds. Academic Press, New York, San Francisco and London.

HALL, C. W., FARRALL, A. W. and RIPPEN, A. L. eds. 1971. Encyclopedia of Food Engineering. AVI Publishing, Westport, CT.

HARPER, W. J. and HALL, C. W. 1976. Dairy Technology and Engineering. AVI Publishing, Westport, CT.

HARPER, W. J. and HUBER, R. M. 1956. Some carbonyl compounds in raw milk. J. Dairy Sci. *39*, 1609.

HEATH, H. B. 1981. Source Book of Flavors. AVI Publishing, Westport, CT.

HEATH, H. B. 1983. Developments in flavors and flavorings. Perf. Flav. *8* (June/July), 53–60.

HEINZE, R. F., INGLE, M. B. and REYNOLDS, J. F. 1978. Flavoring vegetable protein meat analogs. *In* Flavor in Food and Beverages, pp. 43–56, G. Charalambous and G. E. Inglett, eds. Academic Press, New York and London.

HENRY, B. S. 1982. The production, properties and applications of inactive yeast powders and extracts. Proc. Symp. British Soc. Flav., London, June.

HODGE, J. E., MILLS, F. D. and FISHER, B. E. 1972. Compounds of browned flavour derived from sugars–amine reactions. Cereal Sci. Today *17*, 34–38.

HOLLENBECK, C. M. 1979. Liquid smoke flavoring—status of development. Food Technol. *33*(5), 88–89.

HORNSTEIN, I. 1967. The flavor of red meats. *In* Chemistry and Physiology of Flavors, pp. 228–250, H. W. Schultz, E. A. Day and L. M. Libbey, eds. AVI Publishing, Westport, CT.

HOSONO, A. and ELLIOTT, J. A. 1974. Properties of crude ethylester-forming enzyme preparations from some lactic acid and psychrotrophic bacteria. J. Dairy Sci. *57*, 1432–1437.

HOSONO, A., ELLIOTT, J. A. and McGUGAN, W. A. 1974. Production of ethyl esters by some lactic acid and psychrotrophic bacteria. J. Dairy Sci. *57*, 535–539.

IOFI. Code of Practice. International Organization of the Flavour Industry, Geneva.

JACOBS, L. C. 1981. Biotechnology: Creates chemicals, synfuels from renewable resources. Food Prod. Develop. *15*(2), 38–39, 50, 53.

JOENSSON, H., PETTERSON, H. E., ANDERSSON, K. and JOHANSSON, S. 1980. The production of starter distillates. Belgium patent. 883,752.

JONES, D. D. 1985. How the federal government will oversee food products of emerging biotechnologies. Food Technol. *39*(6), 59–70.

KATZ, I. 1981. Recent progress in some aspects of meat flavor chemistry. *In* Flavor Research: Recent Advances, pp. 217–230, R. Teranishi, R. A. Flath and H. Sugisawa, eds. Dekker, New York and Basel.

KEEN, A. R., WALKER, N. J. and PEBERDY, M. F. 1974. The formation of 2-butanone and 2-butanol in Cheddar cheese, J. Dairy Sci. *41*, 249.

KEMPLER, G. M. 1983. Production of flavor compounds by microorganisms. (Manuscript unpublished.)

KIM, K. 1974. Identification of flavor constituents in carbonyl, noncarbonyl, neutral and basic fractions of aqueous smoke condensates. Agric. Biol. Chem. *338*, 55–63.

KONRAD, G., LEWERENZ, H. J., LIESKE, B., SCHLIEMANN, J. and WOLM, G. 1982. Manufacture of a meat flavour concentrate. German Democratic Republic Patent DD 154,764. (In German.)

KRISTOFFERSEN, T. 1973. Biogenesis of cheese flavor. J. Agric. Food Chem. *21*, 573–575.

KRISTOFFERSEN, T. 1978. Biogenesis of cheese flavor. *In* Encyclopedia of Food Science, pp. 68–71, M. S. Peterson and A. H. Johnson, eds. AVI Publishing, Westport, CT.

LABROPOULAS, A. E., PALMER, J. K. and TAO, P. 1982. Flavor evaluation and characterization of yogurt as affected by ultra-high temperature and vat processes. J. Dairy Sci. *65*, 191.

LANGHUS, W. L. 1974. Cheese. *In* Encyclopedia of Food Technology, pp. 178–183, A. H. Johnson and M. S. Peterson, eds. AVI Publishing, Westport, CT.

LANZA, E. and PALMER, J. K. 1977. Biosynthesis of monoterpenes by *Ceratocystis moniliformis*. Phytochem. *16*, 1555–1560.

LAW, B. A. 1981. The formation of aroma and flavour compounds in fermented dairy products. Dairy Sci. Abstr. *43*, 143–154.

LAW, B. A. 1982. Flavour compounds in cheese. Perf. Flavorist *7*(5), 9–21.

LAW, B. A. and WIGMORE, A. S. 1982. Microbial proteinases as agents for acceleration of cheese ripening. J. Soc. Dairy Technol. *35*(2), 75–76.

LAWRIE, R. A. 1982a. Flavour of meat and meat analogues. Food FIPP. *4*(2), 11–13.

LAWRIE, R. A. 1982b. Flavour of meat and meat analogues. Food FIPP. *4*(3), 9,11,15.

LEES, R. and JACKSON, E. B. 1973. Sugar Confectionery and Chocolate Manufacture. Leonard Hill Books, Intertext Publisher, Aylesbury, Bucks.

MACLEOD, G. and SEYEDDAIN-ARDEBILI, M. 1981. Natural and simulated meat flavors (with particular reference to beef). CRC Crit. Rev. Food Sci. Nutr. *14*(4), 309–439 (439 refs.).

MAGA, J. S. and SIZER, C. E. 1973. Pyrazines in foods. A review. J. Agric. Food Chem. *21*, 22–30.

MARSHALL, V. M. 1982. Flavour compounds in fermented milks. Perf. Flavorist *7*(2), 27–34.

MINIFIE, B. W. 1980. Confectionery Science and Technology, 2nd edition. AVI Publishing, Westport, CT.

MORTON, I.D. and MACLEOD, A. J. 1981. Food Flavours: Part A. Introduction. Elsevier, Amsterdam.

MOSKOWITZ, G. J. AND LaBELLE, G. G. 1981. Enzymatic flavor development in foods. *In* Quality of Foods and Beverages, Vol. 2, pp. 21–35, G. Charalambous and G. E. Inglett, eds. Academic Press, New York.

NAKAJIMO, O., IRIYA, R. and HAYASHI, T. 1978. Microbiological transformation of terpenoids: 14. Conversion of (−)-menthone by *Pseudomonas putida*. Nippon Nogei Kagaku Kaishi. *52*, 167–174.

NELSON, R. R., ACREE, T. E., LEE, C. Y. and BUTTS, R. M. 1977. Methyl anthranilate as an aroma constituent of American wine. J. Food Sci. *42*(1), 57–59.

OLSON, N. *et al.* 1981. Microencapsulation of multi-enzyme systems. *In* Advances in Biotechnology: Vol. III. Fermentation Products, pp. 361–366, A. Mizrahi and A. L. van Wezel, eds. A. R. Liss, New York.

PIATKIEWICZ, A. *et al.* 1981. Selection and mutation of *Streptococcus diacetylactis* for butter production. *In* Advances in Biotechnology: Vol. II. Fuels, Chemicals, Foods and Waste Treatment, pp. 491–495, A. Mizrahi and A. L. van Wezel, eds. A. R. Liss, New York.

PINTAURO, N. D. 1976. Food Flavoring Processes. Noyes Data Corp., Park Lodge, NJ.

POTTHAST, K. 1980. Recent figures for benzo(a)pyrene contents in meat products. Fleischwirt., *60*(11), 1940–1941, 1945–1949, 2032–2035. (In German.)

PRENDERGAST, K. 1974. Protein hydrolysis—a review. Food Trade Rev. *44*, 14–21.

PRICE, J. C. and MANNING, D. J. 1983. The flavor of milk. Perf. Flavorist *8*(Apr./May), 50–58.

RAPP, A., HASTRICH, H., ENGEL, L. and KNIPSER, W. 1978. Possibilities of characterizing wine quality and vine varieties by means of capillary chromatography in Flavor. *In* Foods and Beverages, pp. 391–417, G. Charalambous and G. E. Inglett, eds. Academic Press, New York, San Francisco and London.

REDDY, M. C., BILLS, D. D. and LINARDS, V. 1968. Ester production by *Pseudomonas fragi*: I. Identification and quantification of some esters produced in milk culture. J. Dairy Sci. *51*, 656–659.

REINECCIUS, G. A., KEENEY, P. G. and WEISSBERGER, W. 1972. Factors affecting the concentration of pyrazines in cocoa beans. J. Agric. Food Chem. *20*, 202.

REYMOND, D. and ROSTAGNO, W. 1978. Flavor aspects of chocolate. *In* Flavor in Foods and Beverages, pp. 75–79, G. Charalambous and G. E. Inglett, eds. Academic Press, New York, San Francisco and London.

RIBEREAU-GAYON, P. 1978. Wine flavor. *In* Flavor in Foods and Beverages, pp. 355–380, G. Charalambous and G. E. Inglett, eds. Academic Press, New York, San Francisco and London.

ROBINSON, R. K., ed. 1981. Dairy Microbiology. Elsevier, Barking, Essex, England.

ROOZEN, J. P. 1979. The bitterness of protein hydrolysates. *In* Progress in Flavour Research, pp. 321–326, D. G. Land and H. E. Nursten, eds. Appl. Sci. Publ., London.

ROTHE, M., SPECHT, M. and ERHARDT, V. 1981. Studies on the quality characterization of meat-like flavourings. *In* Flavor '81, pp. 203–210, P. Schreier, ed. de Gruyter, Berlin and New York.

RUITER, A. 1979. Color of smoked foods. Food Technol. *33*(5), 54–63.

SCARPELLINO, R. and KOSIKOWSKI, F. V. 1962. Evolution of volatile compounds in ripening raw and pasteurized milk Cheddar cheese observed by gas chromatography. J. Dairy Sci. *45*, 343–348.

SCHNECK, J. C. 1979. Liquid smoke application to cured meats. Proc. Annual Recip. Meat Conf. Amer. Meat Sci. Assoc., pp. 101–103.

SCHREIER, P. 1978. The role of microorganisms in flavour formation. *In* Progress in Flavour Research, pp. 175–196, D. G. Land and H. E. Nursten, eds. Appl. Sci. Publ., London.

SCHREIER, P. 1979. The flavor composition of wines: A review. CRC Crit. Rev. Food Sci. Nutr. *12*(1), 59–111. (398 refs.)

SEITZ, E. W. 1981. Microbial formation of flavors. *In* Flavor '81, P. Schreier, ed. de Gruyter, Berlin and New York.

SHIPE, W. R. *et al.* 1978. Off-flavors of milk: nomenclature, standards and bibliography. J. Dairy Sci. *61*, 855.

VAN STRATEN, S., JONK, G. and VAN GEMERT, L. 1978. Alterations in a wine distillate during aging. *In* Flavor in Food and Beverages, pp. 381–390, G. Charalambous and G. E. Inglett, eds. Academic Press, New York, San Francisco and London.

TAHARA, S., FULIWARA, K., ISHIZAKA, H., MIZUTANI, J. and OBATA, Y. 1972. Decalactone—one of the constituents of volatiles in cultured broth of *Sporobolomyces odorus*. Agric. Biol. Chem. *36*(13), 2585–2587.

TALBOTT, L. L. and McCORD, C. 1981. The use of enzyme modified cheeses for flavoring processed cheese products. Proc. 2nd Bienn. Marschall Int. Cheese Conf., pp. 81–92.

TAMINE, A. Y. 1981. Microbiology of "starter cultures." *In* The Microbiology of Milk Products, Vol. 2. Elsevier, Barking, Essex, England.

TANNER, H., ZANIER, C. and WUNRDIG, G. 1982. Analytical differentiation between "musty" and "corky" flavors of wine. Weinwirtschaft *1118*(1), 15–16.

THOMAS, E. L. 1981. Trends in milk flavors. J. Dairy Sci. *64*, 1023–1027.

TOLIN, S. 1961. Smoke flavored yeast hydrolyzates. United States patent 3,000,743 (Sept. 19), assigned to Vico Products Co.

TRESSL, R. 1982. Formation of flavor components from proline and hydroxyproline with glucose and maltose and their importance to food flavor. *In* Flavor '81, pp. 397–416, P. Schreier, ed. de Gruyter, Berlin and New York.

TRESSLER, D. K. and SULTAN, W. J., eds. 1975. Food Products Formulary: Vol. 2. Cereals, Baked Goods, Dairy and Egg Products. AVI Publishing, Westport, CT.

VINE, R. P. 1981. Commercial Winemaking, Processing and Controls. AVI Publishing, Westport, CT.

WATTS, B. M. 1974. Smoking process. *In* Encyclopedia of Food Technology, pp. 802–804, A. H. Johnson and M. S. Peterson, eds. AVI Publishing, Westport, CT.

WEBB, B. H., JOHNSON, A. H. and ALFORD, J. A. 1974. Fundamentals of Dairy Chemistry. AVI Publishing, Westport, CT.

WESLEY, P. 1981. Biotechnology in fermentation—Foods and the future. Food Product Development *15*(2), 22–23, 26–27, 30–37.

WILSON, R. A. and KATZ, I. 1974. Synthetic meat flavours. Flav. Ind. *5*(1/2), 30–35, 38.

WISTREICH, H. E. 1979. The smoking process—application of liquid smoke. Food Technol. *33*(5), 90.

YAMANISHI, T. 1981. Tea, Coffee, Cocoa and Other Beverages. *In* Flavor Research: Recent Advances, R. Teranishi, R. A. Flath and H. Sugisawa, eds., pp. 231–304. Dekker, New York and Basel.

YU, J. H., KIM, Y. I. and CHUNG, E. Y. 1982. Studies on the lipids in domestic butter. Korean J. Dairy Sci. *4*(1), 55–64.

8

Synthetic Flavoring Materials

A. Imitation Flavorings: Matching Nature

Over the past two decades and increasingly since the introduction of such powerful analytical techniques as coupled GC–MS computer data handling and HPLC, the chemical composition of most natural flavoring substances has been revealed. Hundreds of organic chemicals, embracing the whole spectrum of chemical structure, have been isolated, identified and, in many cases, successfully synthesized. Natural biochemical processes, mostly associated with plant metabolism, have been researched (Buttery 1981) and considerable progress has been made in reproducing these under controlled conditions using single-cell cultures. Such reactions result in highly complex mixtures of organics which provide the distinctive odor and flavor profiles which are usually diagnostic of the source material. Unfortunately, the use of flavorings processed only from natural sources has definite limitations, particularly when the flavoring is derived from fruits. Of particular concern to the food manufacturer is seasonal availability. In many modern processed foods their flavoring performance is often less than adequate and for many years food processors have resorted to the use of imitation flavorings, the composition of which can be precisely controlled, their aromatic profiles and physical form designed to meet specified consumer requirements or manufacturing parameters and their consistent supply assured. In the past, the flavor chemist or flavorist studied individual organics, memorized their odor and flavor characteristics and used this knowledge to form imitations of a natural target flavor mostly by a process of trial and error (Seidman 1979). Before the days of sensitive and reliable instrumental analytical techniques, some of the blends so produced were nothing less than inspired artistic creations. Today most of the guesswork has been eliminated and the flavorist is guided in his/her formulations by a wealth of research data, to which must still be applied a degree of intelligent interpretation to achieve a really satisfactory imitation flavoring having the correct nuances and capable of withstanding often severe processing conditions. Throughout, the aim of the flavorist is to recreate, using permitted flavoring chemicals, the natural image as closely as possible, in a very concentrated form suitable for direct application in a food product, beverage, etc. Wherever possible this is done using organics which are known to occur in the natural target flavor but this does not preclude the use of other permitted compounds to achieve a particular odor or flavor nuance.

Imitation flavorings are mixtures comprising, as necessary, one or more of the following types of components: (a) natural flavoring materials and products made from them by physical means (e.g., herbs and spices, essential oils, extracts); (b) isolates—organic chemicals obtained from natural sources by physical and/or chemical means (e.g., menthol from Japanese or Chinese mint oil by freezing, linalool from Bois de Rose oil by fractional distillation); (c) nature-identical compounds—synthetic organics identical in chemical composition with those found in nature, although not necessarily from the named source of the flavor (e.g., vanillin in an imitation strawberry flavor); (d) artificial compounds—synthetic organics which have not as yet been found in nature but recognized as safe for use in foods and of proven value as flavorants (e.g., ethyl vanillin) and (e) approved solvents (e.g., water, ethanol, propylene glycol) or carriers (e.g., salt, maltodextrin, dextrose) as required to produce the flavoring in a form and concentration for use by the food processor.

1. Synthetic Organics

Researchers, using the optimum analytical techniques available, are now able to separate and identify most of the compounds responsible for odor and flavor of natural materials (Rapp *et al.* 1979; Nursten 1979)—their synthesis poses quite different and sometimes insoluble problems. Nature is able to achieve what the organic chemist cannot without the use of very unusual start materials, without processing conditions calling for high energy input or high pressures and without the prospect of obtaining only a negligible yield. To be of any value to the flavorist or flavor manufacturer a synthetic organic chemical must be capable of ready synthesis from available start materials in sufficient yield to make it economically viable. The product must be adequately pure and stable to achieve consistency of effect in the end product and have physical characters which enable its quality to be controlled within acceptable limits.

The purity of synthetic organic flavorants is a very important quality attribute. Chemicals are rarely so pure that they have simple odor and flavor profiles. Most contain sufficient trace impurities, stemming either from the start materials or processing conditions used in their manufacture, to display a spectrum of perceptible character notes. Sometimes it is the presence of these which characterize the compound and make it far more acceptable as a flavorant than the pure material itself. It is the complex odor and flavor characters which are recognized as normal and which lead to the chemically pure material being rated as unacceptable. From a quality point of view, the purchasing of synthetic organic flavoring materials calls for precise analysis of their physical characters, particularly the refractive index and specific gravity, coupled with a critical appraisal of their odor and flavor under controlled conditions against an acceptable reference sample. This latter evaluation is of far greater value in determining suitability for use than insistence on rigid compliance with specified physical standards—the two methods of assessment should always be considered together and the organoleptic results given the greater emphasis in deciding acceptance or otherwise.

2. Quality Control

The evaluation of organic compounds to be used in flavorings calls for considerable care owing to their extremely powerful aromatic effects. Under no circumstances should any such chemical be tested in an undiluted condition. Care should also be taken in sniffing a chemical sample directly from a bottle; not that the "neck odor" is any reliable indication of quality in any case. The odor profile should first be determined in a manner similar to that used for essential oils, using a suitable smelling strip for liquids. In the case of solid chemicals the closed vessel headspace technique is preferable. For the assessment of the flavor profile, the chemical should first be diluted as follows: Dissolve 0.1 g (or 0.1 ml if liquid) of the sample in 5 ml of 95% ethanol; add 0.02 ml of this solution to 100 ml/g of the test medium (equivalent to 4 ppm). Any comparative reference sample should be similarly diluted at the same time. The comparison should then be carried out by one of the accepted sensory test methods described in the literature. It must not be assumed that the comparison of odors is a simple matter. To decide how similar or dissimilar is the odor quality, or what is the nature of any differences between two samples of the same chemical requires both patience and expertise. Whatever comparison method is used it must be meaningful in terms of acceptability for use of the material under test and reproducible on successive occasions (Dravnieks *et al.* 1978; Forss 1981).

Highly aromatic materials can be evaluated by sensory methods without difficulty so long as the following procedure is observed: (a) The material is first diluted to a level at which the nose and palate are not swamped causing temporary anosmia. (b) Sufficient time is allowed between repeat tests for the senses to recover their initial acuity. (c) The tests are repeated in reverse order of presentation to eliminate, or at least minimize, any carry over effects.

3. Consumer Attitudes toward Synthetic Chemicals

Not surprisingly, consumers express considerable concern about the indiscriminate use of "chemicals" in foods although Stofberg and Stoffelsma (1980) have shown that the amounts used and consumed as flavorings is infinitesimal compared with the input from natural sources the safety of which is not questioned. Nevertheless, in most countries, the laws governing safety of foodstuffs recognizes this consumer attitude and the Expert Panel of the Flavor and Extract Manufacturers' Association (FEMA) in the United States has, over the past 20 years, carried out a very comprehensive survey of materials currently in use or likely to be used in the formulation of flavorings. The work of this committee has established the generally recognized as safe (GRAS) status of flavoring materials in accordance with the Food Additives Amendment to the U.S. Federal Food, Drug and Cosmetic Act. The procedure adopted entailed the development and application of criteria for the safety evaluation of such substances (Oser and Hall 1977). The basis for deciding safety in use involves a detailed consideration of (a) toxicological data using both long and short term studies; (b) biochemical and metabolic data; (c) occurrence in nature, qualitatively and quantitatively; (d) chemical structure and physical characters; (e) proposed use levels and average daily intake (ADI); (f) prior use in foods and (g) toxicological significance of continuous ingestion.

To date 14 lists of GRAS substances have been published by FEMA as a result of this work. Details of these listings can be found in the references (Hall 1960; Hall and Oser 1961–1972; Oser and Ford 1973–1979; Oser *et al.*, 1984, 1985).

Parallel with this work, the FDA has also carried out an extensive program of toxicological testing, the results of which appear in the Federal Register from time to time giving approval for specified materials to be classed as GRAS and safe for use in foodstuffs in the United States.

A similar, but not identical, listing of flavoring substances has been published by the International Organization of the Flavour Industry (IOFI) in the form of a Positive List of artificial flavoring substances the use level data for which is available (IOFI 1982).

These listings have been consolidated into a single alphabetical list by Heath (1981) and the Chemical Sources Assoc. (1981). These are up to date to the time of publication but, as this is a continuing process of assessment, updatings are to be expected.

B. Classification of Flavorants by Molecular Structure

Naturally occurring aroma and flavoring components embrace the whole spectrum of organic chemistry and the most important groups of compounds are listed in Table 8.1. Parliament (1980) has pointed out the wide range of structure, threshold and character of aroma chemicals present in any one plant source. Although the functional group of an organic compound has a quite significant affect on its aroma and hence flavor profile there is not much correlation between chemical structure and organoleptic properties and there are literally thousands of different

Table 8.1. Chemical Nature of the Major Aromatic Compounds Used in Flavorings[a]

Hydrocarbons	Esters
Aliphatic	Aliphatic
Aromatic	Aromatic
Cyclic terpenes (mono- and bicyclic terpenes)	Terpene esters
	Ethers
Sesquiterpenes (mono-, bi- and tricyclic sesquiterpenes)	Heterocyclic compounds
	Ketals
Alcohols	Ketones
Aliphatic (saturated, unsaturated and terpenic alcohols)	Aliphatic
	Aromatic
Aromatic	Cyclic terpene ketones
Cyclic terpene alcohols	Ionones
Sesquiterpene alcohols	Irones
Aldehydes	Lactones
Aliphatic	Oxides
Aromatic	Phenols
Cyclic terpene aldehydes	Phenol ethers
Heterocyclic aldehydes	Compounds containing nitrogen and sulfur
Acetals	
Acids	Amines
Aliphatic	Amino compounds
Aromatic	Imino compounds
Anhydrides	Sulfides
	Pyrazines

[a] From Heath (1978).

odor qualities. Some compounds of similar structure do have broadly similar odors but there are very many exceptions and even where the odors are similar the intensities may vary considerably. Macleod (1980), in reviewing the chemistry of odors, gave examples of the following categories of aromatic chemicals: (a) compounds of similar structure with similar odors (Fig. 8.1); (b) compounds of different structure with similar odors (Fig. 8.2); (c) compounds of similar structure with different odors (Fig. 8.3) and (d) stereoisomers with different odors (Fig. 8.4).

These examples show that structural relationships are not sufficiently precise to allow predictability of odor or flavoring effects either qualitatively or quantitatively. Numerous attempts have been made to evolve a theory which will explain structure–activity relationships but most of these apply only to a narrow range of examples and Macleod (1980) concluded that the most important factors affecting this relationship would seem to be overall shape and size of the molecule, its detailed stereochemistry and certain chemical properties (e.g., polarity, nature of functional groups) and physical properties (e.g., volatility and solubility) of the compound. Such considerations are of little importance when a characterized aromatic component is being synthesized but are of real concern to the organic chemist aiming to create an organic compound having a defined odor and/or flavor character. At present, organic chemists have to work on an empirical basis observing the effects of substitution, etc., by sensory analysis as the work progresses.

Certain relationships are worthy of comment. Within very broad limits, the flavoring strength of an organic compound, as opposed to its aromatic quality, is inversely related to its vapor pressure. The lower the vapor pressure, the greater the aromatic impact. For this reason, branched chain substitution is likely to result in compounds having a higher flavoring value than the equivalent straight chain compound (e.g., isoamyl butyrate has a considerably lower threshold and hence greater flavoring strength than *n*-amyl butyrate). Unfortunately the effect is not quantitatively predictable. The substitution of groupings within the molecule may also have a profound effect on its aromatic attributes. The degree of effect depends to some extent on the size of the molecule; in small molecules the substitution of one functional group for another may radically change its whole chemistry whereas in larger molecules, where functional groups form a much lesser part of the total structure, the effect is so pronounced.

The modification of simple molecules such as CH_3OH, CH_3SH and CH_3NH_2 demonstrates this effect whereas in the much larger vanillin molecule the substitution of an ethoxy group for the normal methoxy group results in a compound having three times the flavoring power and only a subtle change in the aromatic profile.

Lemonlike odor

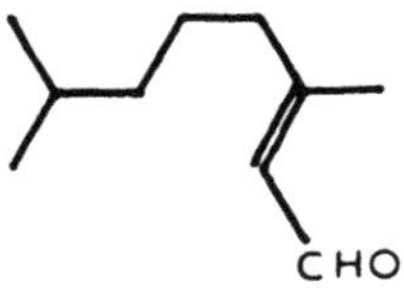

Citral

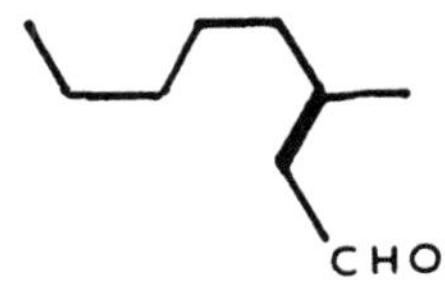

3-Methyloct-2-enal

Cocoalike odor

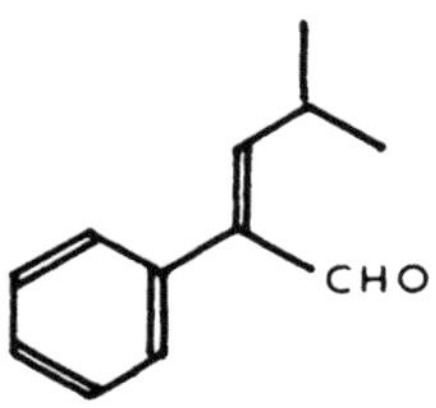

5-Methyl-2-phenyl-2-hexenal

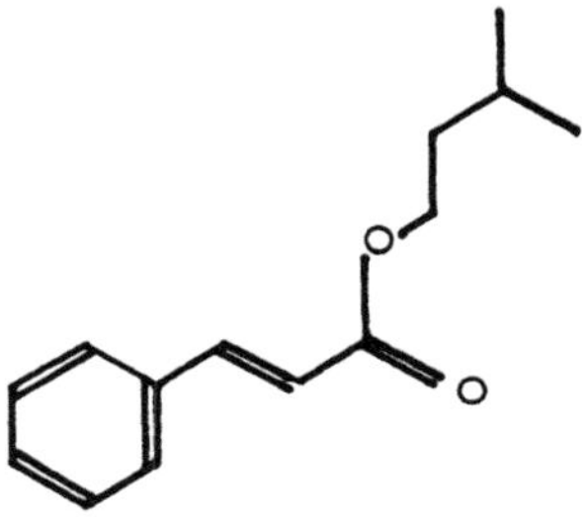

Meatlike odor

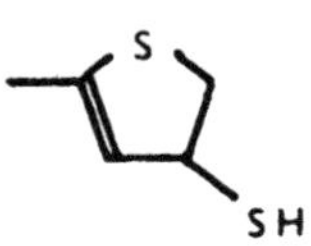

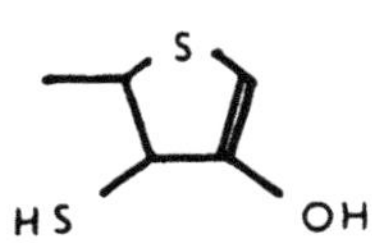

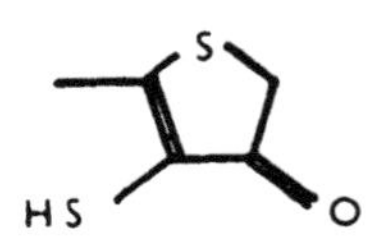

Mustardlike odor

$CH_2{:}CH{\cdot}CH_2{\cdot}N{:}CS$

Allyl isothiocyanate

$CH_3{\cdot}CH_2{\cdot}N{:}CS$

Ethyl isothiocyanate

Fig. 8.1. Compounds having similar structures and similar odors.
From Macleod (1980)

Musklike odor

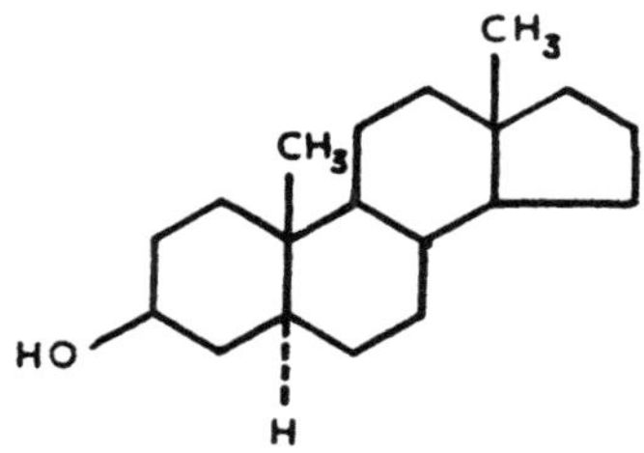

Homoandrostan-3-ol

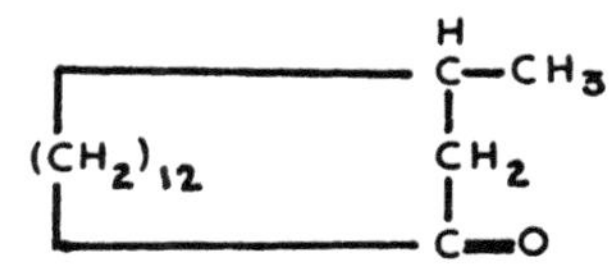

3-Methyl-1-cyclopentadecanone

Camphoraceous odor

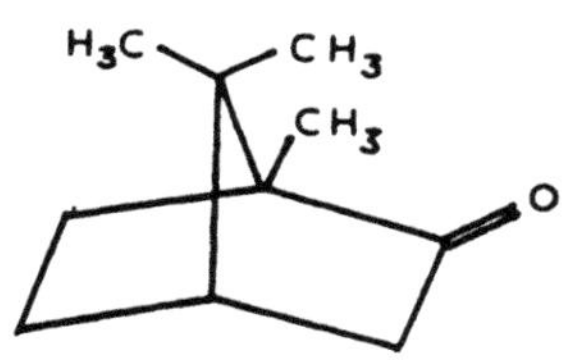

Camphor

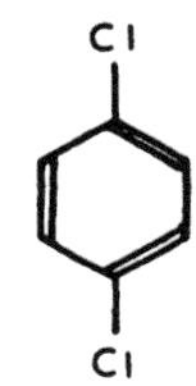

Dichlorobenzene

Fig. 8.2. Compounds having different structures and similar odors.
From Macleod (1980)

Grapefruitlike odor

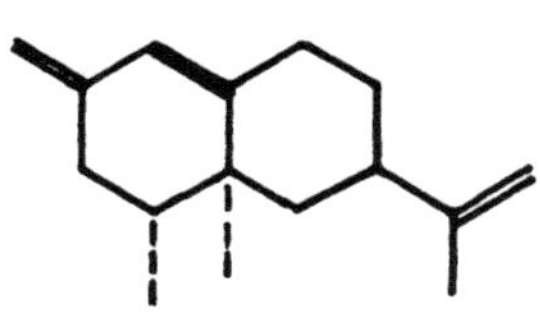

Nootkatone

Woody odor

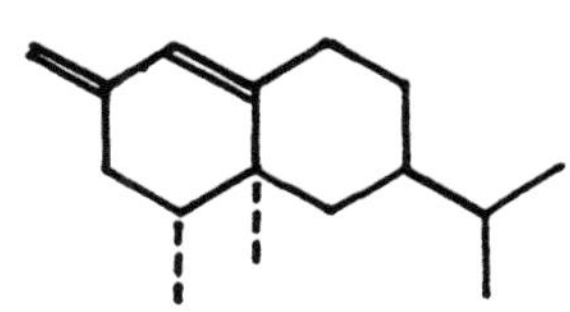

11,12,-Dihydronootkatone

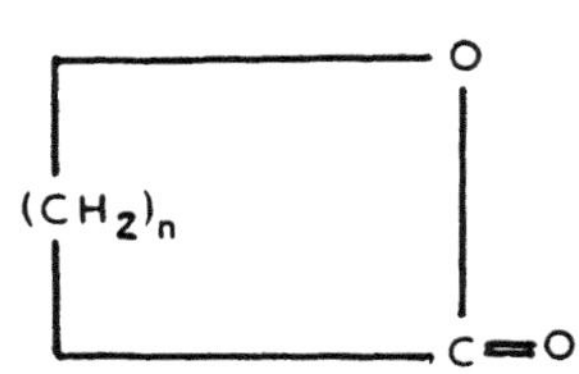

Lactones where n = 5, fruity; n = 14, musklike; and n = 17, civetlike

Fig. 8.3. Compounds having similar structures and different odors.
From Macleod (1980)

Similar odors—citral

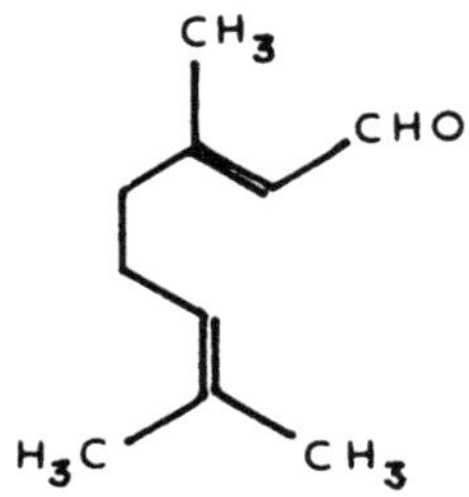

2-*trans*-3,7-Dimethyl-2,6-octadien-1-al (geranial), strongly lemonlike

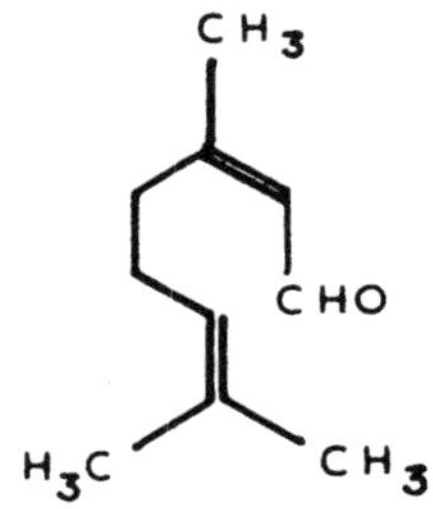

2-*cis*-3,7-Dimethyl-2,6-octadien-1-al (neral), strongly lemonlike

Different odors—norandrostanol

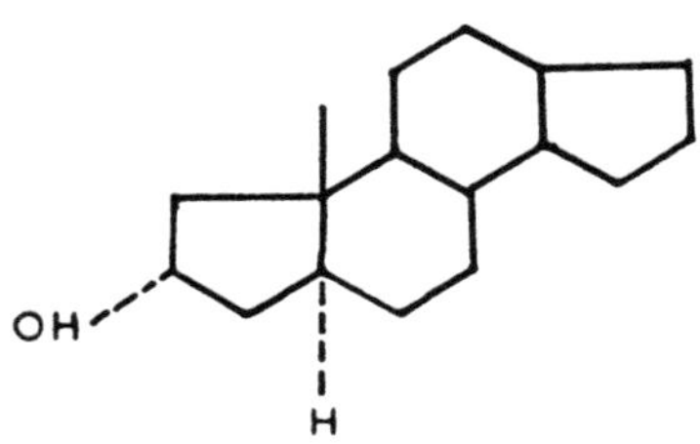

Norandrostan-2α-ol, musklike

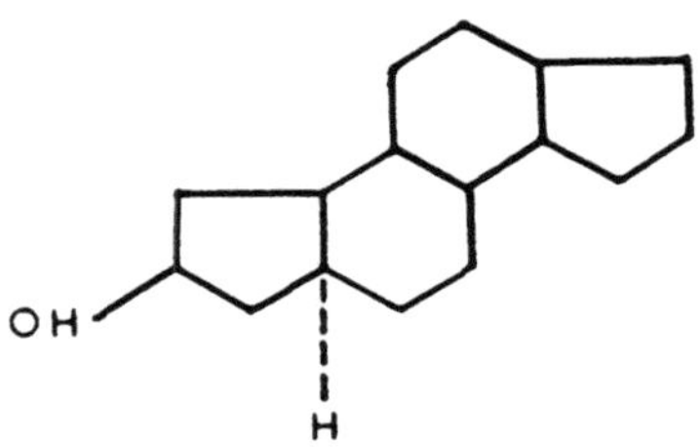

Norandrostan-2β-ol, cedarlike

Fig. 8.4. Isomers and sterioisomers.
From Macleod (1980)

Structural relationships between synthetic organics used as flavorants are listed in Table 8.2.

C. Threshold Values

One of the principal characters of an organic compound which is to be used in flavorings is its threshold value under precisely defined conditions. The American Society for Testing and Materials (ASTM) has published a listing of available threshold data (Stahl 1973) and gives various threshold definitions—sometimes based on those proposed by the British Standards Institution (BSI)—in the following ways: (a) Detection threshold—minimum physical intensity detectable where the subject is not required to identify the stimulus. (b) Difference threshold—smallest change in concentration of a substance required to give

Table 8.2. Structural Relationships of Flavoring Organics[a]

Class of Compound	Distinguishing Systematic Suffix or Prefix	Functional Group	Typical Structure: Aliphatic	Typical Structure: Aromatic
Alcohols	-ol	—O—H	R·OH	$C_6H_5R \cdot OH$
Phenols	-ol	—O—H		C_6H_5OH
Carboxylic acids	-ic acid or -oic acid	—C(=O)—O—H	R·COOH	C_6H_5COOH
Esters	-ate	—C(=O)—O—	R·COO·R′	$C_6H_5 \cdot COO \cdot R$
Aldehydes	-al	—C(=O)—H	R·CHO	C_6H_5CHO
Ketones	-one	—C(=O)—	R·CO·R′	$C_6H_5CO \cdot R$
Ethers	-ether	—O—	R·O·R′	$C_6H_5O \cdot R$
Lactones	-lactone or -ide	—C(=O)—O—	R / \ O—CO (ring)	
Amines	-amine or amino	—N(H)H	$R \cdot NH_2$	$C_6H_4(NH_2) \cdot R$
Nitro-compounds	nitro-	—N(O)O	$R \cdot NO_2$	
Sulfides	-ide	—S—	R·S·R′	
Disulfides	di-	—S—S—	R·S·S·R	
Thio-alcohols	-mercaptan	—S—H	R·SH	

[a] From Heath (1978).

a perceptible change. (c) Recognition threshold—lowest concentration at which a substance is correctly identified. (d) Terminal threshold—the concentration of a substance above which changes in concentration are not perceptible.

The preceding clearly stresses the need for an understanding of the conditions under which any threshold values are obtained, and although comparative results published in a single work may be valid their wider application may not be so significant. In a natural flavor the quantitative composition is not a true indication of its odor and/or flavor profile. In a few instances a single chemical may dominate and characterize the profile (e.g., 2-methoxy-3-isobutylpyrazine in bell peppers). More frequently very many compounds are isolated and it is then necessary to establish their relative contribution to the overall profile. Determination of their individual thresholds gives an indication of their odor intensities although these may not be so significant in a mixture of compounds. The factors affecting threshold values and their determination have been discussed by Forss (1981).

Guadagni *et al.* (1966) evolved a new concept for the evaluation of the relative importance of aromatic components to the odor and flavor profiles and defined what they called the *odor unit number*. This figure is the quotient of the concentration of a volatile component and the threshold concentration of that compound in water, i.e.,

$$C_\alpha/T_\alpha = U_\alpha$$

where C_α is the concentration of the component α; T_α is the threshold concentration of component α in water and U_α is the number of odor units of α. This was extended to mixtures of components by the following relationship:

$$U_1 + U_2 + U_3 + \cdots + U_n = U_m$$

where $U_1, \ldots, U_n$ are the odor units of the components and U_m that of the mixture. Obviously the value for U_m can be checked by determining the values for C_m and T_m in respect to the mixture. The authors also postulated that the relative contribution (%) of any component would be given by the value of

$$100 \times U_n/U_m$$

This concept is proving to be of increasing value in flavor research which had previously centered on identifying components without much concern about their contribution to the total aromatic profile of the mixture. [See Land (1984).] The method is not without some practical problems, and there are some doubts about the validity of the basic assumptions. These have been pointed out by Rothe (1976) and Frijters (1979), but the original concept is still of considerable value in understanding the complexities of natural aromatic compositions (Forss 1981).

Odor quality cannot so readily be quantified and at present there is no satisfactory way of predicting the odor quality of a mixture from a knowledge of its quantitative composition (Enns and Hornung 1985). Some generalizations are possible where the mixture itself is composed of complex components (e.g., the blending of herbs and spices into seasonings) by the extension of the concept of odor units to the individual spices used. When it comes to pure flavoring chemicals enhancement and/or masking may occur but rarely is a new aromatic note created. Various mathematical concepts have been applied to this problem (Moskowitz and Warren 1981) and multidimensional scaling has revealed some of the complex quantitative relationships which can occur in mixtures of chemicals (Schiffman *et al.* 1977): (a) In binary mixtures one chemical may completely suppress the other and totally dominate the mixture. (b) The power to suppress diminishes as the number of components increases. (c) The ability to mask the odor of another component is passed on (i.e., *A* suppresses *B*; *B* suppresses *C*, then *A* suppresses *C*).

D. Sensory Characters of Organics

In this section only the very basic organic chemistry is noted. From what has already been discussed, there is little value in studying long lists of compounds and, for those who are interested, the principal flavorants in each of the following groups have already been listed, together with their sensory characteristics by Heath (1978). Those directly concerned with flavoring compositions will have developed their own records and/or index system based on their personal evaluation of these chemicals. Those coming into this field are strongly recommended to start records of their own evaluations of odor and flavor characters as soon as possible and to maintain these as their involvement progresses. In addition, reference should be made to standard works such as "Fenaroli's Handbook of Flavor Ingredients" (Furia and Bellanca 1975) or Arctander (1969) in which flavor and fragrance chemicals are listed alphabetically with their physicochemical and sensory data and personal records supplemented as necessary.

1. Hydrocarbons

The simplest of the organic chemical groupings is the hydrocarbons comprising only carbon and hydrogen. The saturated paraffin hydrocarbons, having the general formula C_nH_{2n+2}, are of little importance in a flavor context although some of the higher members of the series (C_5–C_8) are of some value as extraction solvents.

Aromatic hydrocarbons, based on a single (benzene), double (naphthalene) or triple (anthracene) ring structure are more aromatic but, again, are of no direct use in flavorings.

Benzene

Naphthalene

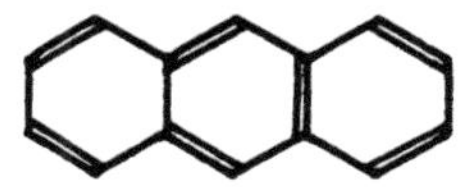

Anthracene

In flavor chemistry it is the terpenoid hydrocarbons which are of greatest importance. These compounds, which are found in so many of the essential oils, form a very distinctive group of chemicals which have carbon skeletons comprising isoprene units joined together in a regular head-to-tail configuration in accord with what is called the "isoprene rule."

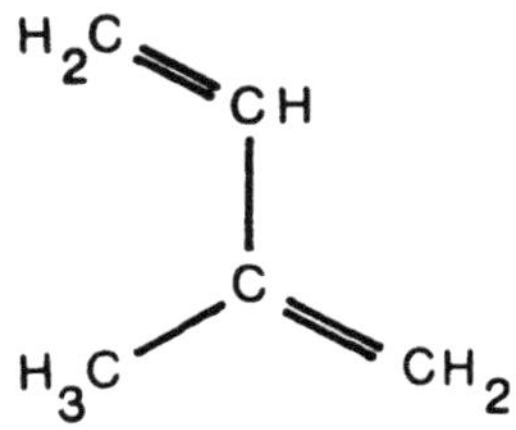

Isoprene

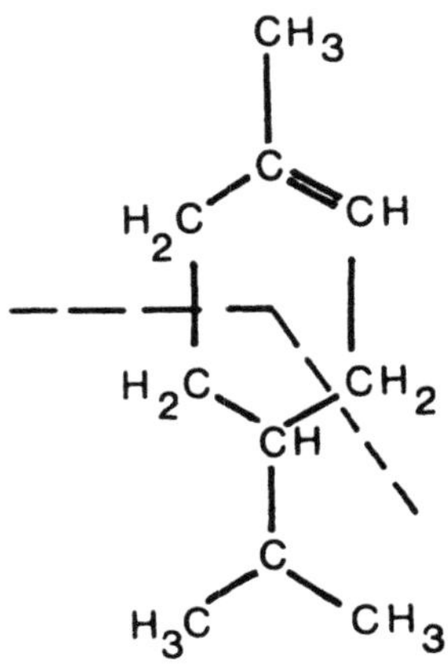

Dipentene

They may be open or closed chain or cyclic compounds of the general formula C_nH_{2n-4}. Where $n = 2$ they are called terpenes; where $n = 3$, diterpenes; and where $n = 4$, sesquiterpenes. The compounds may be saturated or unsaturated. Typical examples of these compounds are given in Table 8.3.

The complex relationships between structure and odor of the terpenes has been the subject of much study and it is concluded that the terpene hydrocarbons act rather as diluents and carriers for other, usually oxygenated, flavoring constituents although their presence adds a considerable degree of freshness to the odor and flavor profiles. Essential oils which have been deterpenated are flat and lack this fresh attribute (Heath 1978).

2. Carboxylic Acids

The carboxylic acids are characterized by having a carboxyl group attached to either an alkyl group (R·COOH) or an aryl group

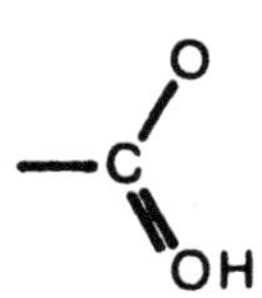

Table 8.3. Examples of Terpenoids[a]

Structure may be open chain, closed chain or cyclic, saturated or unsaturated, based on units of isoprene:

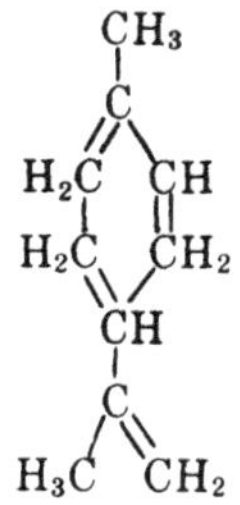

Typical Structures Include:

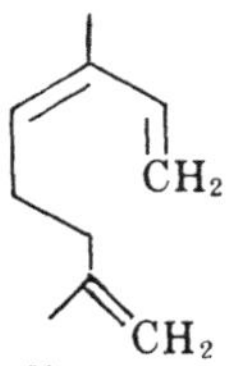

dipentene

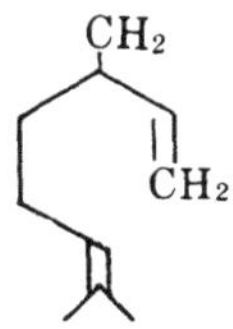

myrcene

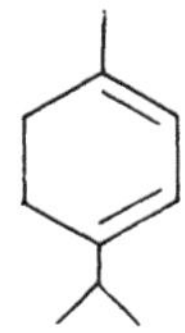

α-terpinene

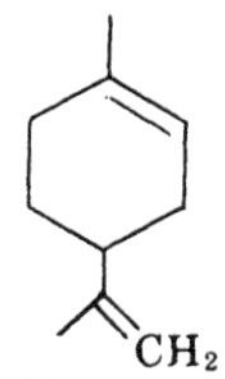

d-limonene

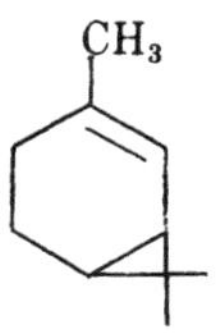

Δ^{4}-carene

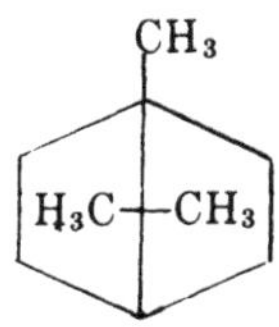

camphane

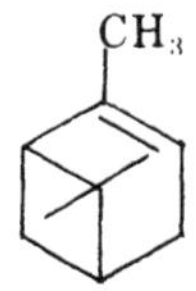

α-pinene

[a] From Heath (1978).

(Ar·COOH). The presence of the hydroxyl group tends to suppress the odor (e.g., propionic acid CH_3CH_2COOH has a pungent odor whereas lactic acid $CH_3CH(OH)\cdot COOH$ is odorless). The di- and tri-carboxylic acids are also odorless.

In foods, acetic acid CH_3COOH is by far the most important of the carobxylic acids and typifies the saturated fatty acids (C–C_3) which are generally unpleasant with strongly pungent, penetrating odors. In the range C_4–C_8 this character is progressively replaced by a rancid buttery/cheesey note and above C_9 the odor becomes unpleasantly fatty/aldehydic. Acids above C_{14} are waxy solids at room temperature and have little or no odor.

The lower members of the unsaturated oleic series are pungent and more acrid, their character changing, with increasing molecular weight, to a slightly spicy, aldehydic and tallowlike quality. Those above C_{11} are odorless.

The aromatic acids do not have a well-defined pattern of odors; all are faintly balsamic with various spicy, floral or even fruity overtones. Benzoic and cinnamic acids and the higher members of the series are odorless, though compounds based on a saturated cyclic structure (e.g., hexahydrobenzoic acid) have an unpleasant rancid character.

As a group, the organic acids impart a sharp sour taste to a product. The di- and tri-carboxylic acids are of particular value in this respect as they occur naturally in so many fruits. These compounds are odorless and are not used directly in flavoring formulations but citric, malic and tartaric acids are often used in end products to improve and even enhance added fruit flavorings which would otherwise be untypical of the natural fruit.

3. Acetals

Alcohols and glycols add to the carbonyl group of aldehydes in the presence of an anhydrous acid to form mixed ethers called acetals. It is suggested that, in alcoholic solution, aldehydes really exist in equilibrium with compounds called hemiacetals but that these are too unstable to be isolated. In the presence of acids the hemiacetals act as an alcohol and react to form the acetal. The reaction is reversible and, in the presence of water, acetals are readily hydrolyzed back to the parent aldehyde and alcohol.

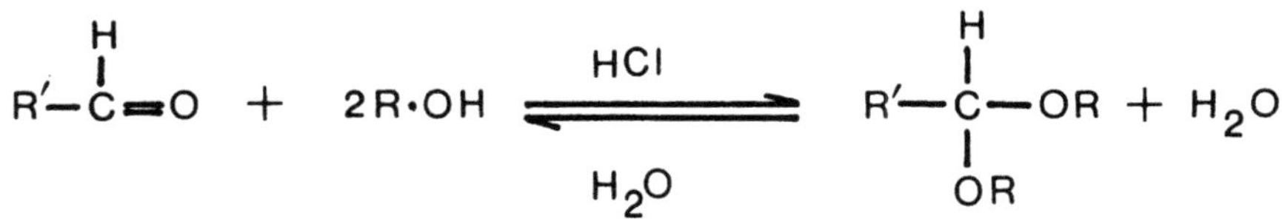

The acetals are liquids having a wide range of odors but are not extensively used in flavorings. Two are of particular interest: (a) acetaldehyde diethyl acetal—pleasantly nutty and (b) cinnamaldehyde ethylene glycol—warm, sweetly spicy.

The product of glycol with hydrotropic aldehyde (1-oxo-2-phenyl-2-methylethane) has a strong odor reminiscent of mushrooms.

4. Alcohols

Alcohols are compounds having the general formula R·OH where R is any alkyl or substituted alkyl group. Alcohols may be grouped in several ways, including (a) primary (R·CH OH), secondary (R·R′·CHOH), tertiary (R·R′·R″·OH); (b) open chain or cyclic; (c) saturated or unsaturated; (d) containing a halogen atom and (e) containing an aromatic ring.

In the aromatic series where the hydroxyl group (–OH) is substituted directly on the benzene nucleus the compounds are known as phenols. These differ significantly from the alcohols and are considered separately.

The alcohols as a group are among the most important of flavoring materials and are extensively found throughout nature. All contain the hydroxyl group which largely determines their charcteristics.

In the lower alcohols, where the hydroxyl group represents a large part of the molecule, they are polar solvents and readily miscible with water. The change in solubility with increasing carbon number is a gradual one and for normal primary alcohols those above C_5 are generally regarded as not water soluble.

Almost all alcohols can be distilled and this forms a convenient method for their preparation and purification. Alcohols of lower molecular weight have a sweet, spiritous odor which intensifies up the series C_9 which has an unpleasant oil note. They are excellent and widely used solvents in the preparation of liquid flavorings. With increasing molecular weight the physical form changes and they become viscous fluids, the pleasant odor being progressively replaced by a fatty and irritating odor character. The higher alcohols are waxy solids having almost no odor.

Unsaturated alcohols, in which the double bond is close to the hydroxyl group, have a harsh, penetrating and irritating odor. The aromatic alcohols are mostly pleasant and have well-rounded odor profiles.

The substitution of two or more hydroxyl groups removes the odorous character of the compound in both the aliphatic and aromatic series (e.g., glycerol).

The terpenoid alcohols represent a very distinct and important group of flavoring compounds found widely in essential oils. Two examples are of interest: (a) Geraniol (*trans*-3,7-dimethyl-2,6-octadien-1-ol)—which occurs in rose, lemongrass and geranium oils—has a characteristic roselike odor. (b) Linalool (3,7-dimethyl-1,6-octadien-3-ol)—which occurs widely in either its optically active or inactive forms—has a typical floral odor.

Aromatic alcohols in which the hydroxyl group is in the side chain have relatively weak but usually pleasant odor characters.

5. Carbonyls

These are compounds characterized by the presence of the carbonyl group (C=O) and may be either of the following:

a. Aldehydes of the general formula

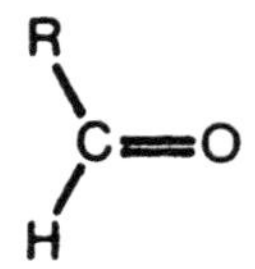

b. Ketones of the general formula

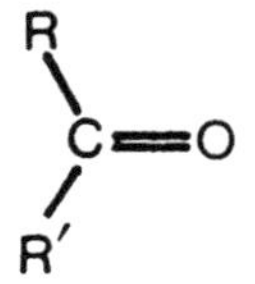

where R and R′ may be either the same or different aliphatic or aromatic groups.

Diketones contain the group

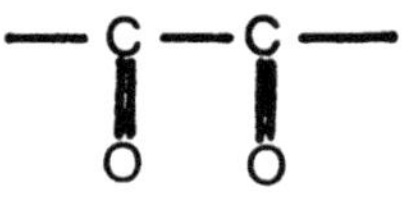

The aldehydes and ketones closely resemble each other in most of their properties but are sufficiently different to warrant their being discussed separately.

6. Aldehydes

The aldehydes are intermediate between the alcohols and acids, being more reactive than the alcohols because of their double bond linkage with oxygen. Aldehydes are particularly susceptible to oxidation to acids.

Aldehydes of lower molecular weight are characterized by their unpleasant, sharply pungent, irritating odors. With increasing molecular weight there is a progressive change in the profile to a pleasing fruity character though the penetrating nature of the odor remains apparent throughout the series hence the true aromatic profile of aldehydes is better appreciated in dilution where this attribute is mostly eliminated. Aldhydes C_8–C_{10} are markedly floral and most attractive on dilution. They are best used at low concentrations in compositions in which a floral nuance is required. As the molecular weight increases further the compounds lose their odor and become somewhat waxy.

Unsaturated aldehydes have particularly irritating odors and this is further increased by the presence of a triple bond. The aromatic aldehydes are powerful and display a wide spectrum of profiles depending on their complexity.

Misnamed "aldehydes." By long tradition certain compounds are commonly known by names such as "aldehyde C_{10}" or "strawberry aldehyde"; "aldehyde C_{14}" or "peach aldehyde." The sensory quality is accurate; the chemical designation is not. "Aldehyde C_{10}" is in fact ethyl 2,3-epoxy-3-phenyl-butanoate whereas "aldehyde C_{14}" is 4-hydroxyundecanoic acid, γ-lactone.

7. Ketones

The ketone molecule contains two substituent carbon radicals on the carbonyl group. These may be either similar or different. Those of low molecular weight are of little value as flavorants although acetone is used in the flavor industry as a valuable, if somewhat flammable, extraction solvent.

The first ketone of real value as a flavoring material is C_7-methyl amyl ketone which has a strong fruity character which persists through the next members of the series. With increasing molecular weight, the fruity character is replaced by floral notes but the effects are not sequential and the ketones display a wide range of aromatic profiles and strengths. Ketones are mostly used in the formulation of imitation fruit flavorings.

8. Esters

Esters are formed by the reaction of alcohols or phenols with acids and their derivatives.

$$R{\cdot}COOH + R'{\cdot}OH \rightleftharpoons R{\cdot}COO{\cdot}R' + H_2O$$

where R may be alkyl or aryl and where R′ is usually alkyl.

To the flavorist, esters form the principal source of a very wide spectrum of odor and flavoring effects, and there are over 200 of these compounds permitted for use in foods. Esters are widely distributed in the essential oils and in some instances represent the major constituent (e.g., oil of wintergreen is about 90% methyl salicylate).

Although some general structural/sensory relationships can be found within the group, particularly their overall "fruity" character, these are by no means sufficiently well defined to form the basis of a worthwhile classification. From the broad similarities which exist between them, it is preferable to consider esters according to their acid constituent (e.g., acetates, buyrates, isobutyrates, benzoates, etc.) rather than on their alcohol or phenol radical (e.g., methyl, ethyl, benzyl, eugenyl, etc.), although this moiety does have an increasingly significant effect on the profile as it becomes larger and/or more unsaturated. The profiles of the constituent acids and alcohols have almost no sensory relationship with that of the resulting ester.

Tabulations of permitted esters are of little value and the practical flavorist will almost certainly evolve his/her own system of recording their invididual profiles and flavoring effects and the sensory changes which occur when they are used in mixtures.

9. Ethers

Ethers are compounds of the general formula R–O–R; R–O–Ar or AR–O–Ar and are named from the two groups which are attached to the oxygen atom followed by the name *ether* (e.g., diethyl ether). Some have common names which are more widely used (e.g., anisole which is methyl phenyl ether). If one group has no simple name the compound may be called an alkoxy derivative (e.g., 2-ethoxyethanol).

Ethers of lower molecular weight are extremely volatile and have a sweet, spiritous odor whereas the higher ethers have very interesting profiles which may be ethereal and fruity. The ether group does not impose much sensory impact and most ethers have only light odors influenced more by the substituents in the molecule. There is little relationship within the group and individual characters must be experienced at first hand.

10. Heterocyclic Compounds

Heterocyclic compounds are those which contain a ring made up of more than one kind of atom. The most common from a flavor chemistry point of view are

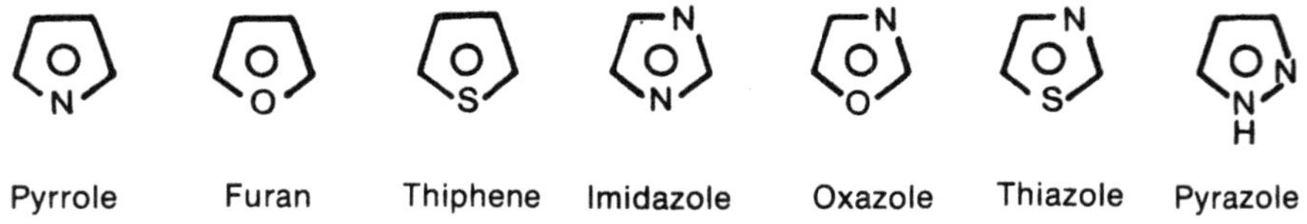

Aliphatic heterocyclic compounds are also found in some alkaloids (e.g., piperidine) and the anhydrides of succinic and glutaric acids also come into this category.

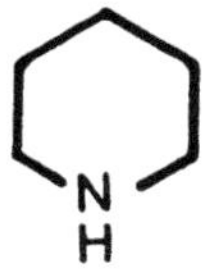

Piperidine

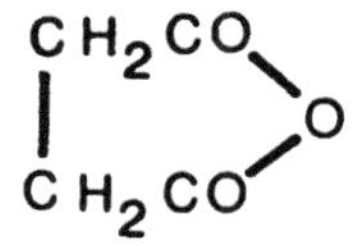

Succinic anhydride

Substituted compounds of these parent heterocycles are of great interest as they possess strong and characteristic aromatic effects which may be very unpleasant when concentrated but improve considerably on dilution (e.g., furfural).

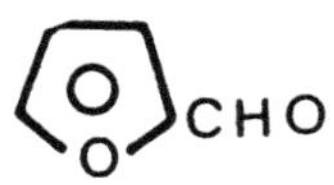

As in many other cases the individual profiles are unpredictable from a knowledge of their chemical structures.

11. Ketals

These compounds are analogous to the acetals and result from the reaction of alcohols with the carbonyl group of ketones. They are generally unstable but the more stable compounds are used mostly in perfumery.

12. Lactones

Internal esterification resulting in the loss of one molecule of water from the γ- and δ-hydroxyacids yields a class of cyclic compounds called lactones. The five-membered ring system is readily formed and the resulting γ-lactones are very stable.

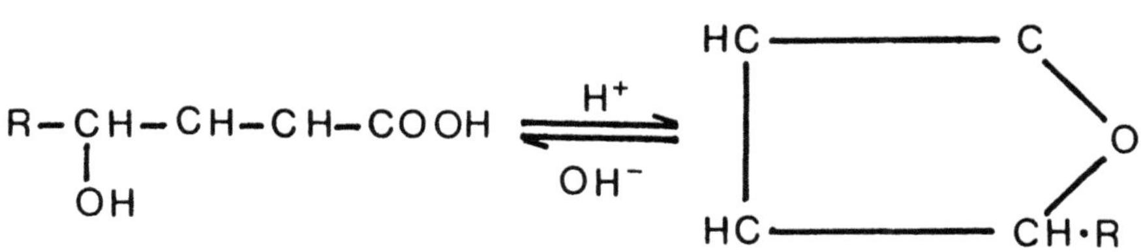

γ-Hydroxy acid γ-Lactone

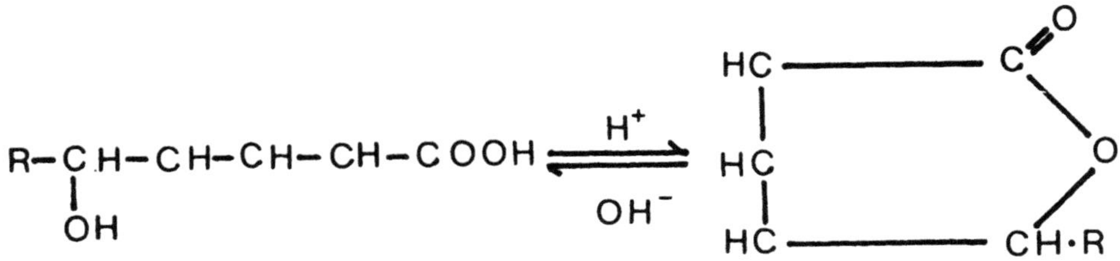

δ-Hydroxy acid δ-Lactone

As a group, the lactones are powerful and distinctive flavorants which are responsible for many of the characteristic notes of fruits and which give to peaches and apricots their distinctive profiles. A comparison of the odors of δ- and γ-lactones indicates that the increase in the size of the ring in the δ-compound results in an increased odor intensity but that the γ-lactones have in addition some "green" nuances not present in the δ-lactones. Many lactones have a strong musklike odor which is very persistent.

In the aromatic series the best known lactone is coumarin, but this is no longer permitted for use in foods.

The part played by lactones in foods has been reviewed by Maga (1976).

13. Nitrogen-Containing Compounds

a. Amines

The amines are substituted ammonia compounds having the general formula RNH_2 (primary); R_2NH (secondary) and R_3N (tertiary), where R is any alkyl or aryl group.

Aliphatic amines are named according to the alkyl group(s) attached to the nitrogen following these with the word *amine*. Sometimes the prefix *amino* is added to that of the parent chain. Aromatic amines in which the nitrogen is attached directly to the aromatic ring are named as derivatives of aniline.

All amines of low molecular weight have strong and very distinctive profiles which on dilution have an ammoniacal character. The profile becomes increasingly "fishy" in the secondary and teriary amines although this objectionable odor is gradually lost with increasing molecular weight.

In the aromatic amines an amino group substituted on the side chain results in a marked ammoniacal odor nuance whereas direct substitution of the amino group onto the nucleus results in a very weak and undefined odor character (e.g., anthranilic acid is odorless whereas methyl anthranilate has a strong orange-flower-like profile).

The amines are of very limited use in flavorings as most of them have an objectionable odor but they do play an important part in foods (Maga 1978b).

b. Amides

The hydroxyl group of carboxylic acids may be replaced by the amino group to give compounds having the general formula

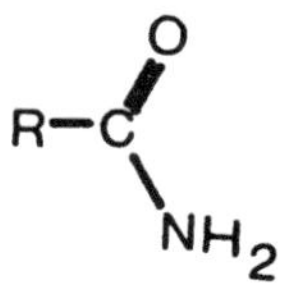

and called amines. These are chemically inert compounds, and as they are odorless are of no value as flavorants.

c. Imines

In flavor chemistry one sometimes comes across compounds called imines. These are usually only intermediates in complex reactions and have the general formula RCH=NH or R_2C=NH.

d. Amino Acids

Although not directly used as flavorants the amino acids, as the prime constituents of proteins, not only play an essential role in body nutrition but also enter into the Maillard reaction when heated with pentose sugars. This initiates a series of interrelated reactions resulting in a complex mixture of highly aromatic compounds and a characteristic brown color. The chemistry of this reaction is dealt with in greater detail elsewhere. The Maillard-type reaction, using selected amino acids and/or pentose sugars, is used as a source of meatlike flavoring materials which have very wide application in food processing.

e. Isothiocyanates

The isothiocyanates contain the –N:C:S group and are found in several pungent vegetables, particularly, (a) allyl isothiocyanate—produced by the enzymatic action of myrosin on the glycoside sinigrin in black mustard seed; (b) benzyl isothiocyanate—found in garden cress and (c) phenyl ethyl isothiocyanate—responsible for the sharp pungency of horseradish.

Of these only allyl isothiocyanate is permitted for use in foods in the United States.

14. Phenols

Phenols are compounds of the general formula ArOH where Ar is phenyl, substituted phenyl or one of the other aryl groups. They are characterized by having the hydroxyl group attached directly to the aromatic ring.

Phenol has a very distinctive odor long associated with disinfectants but substitution of methyl groups into the benzene nucleus reduces the sharp acidic character associated with phenol so that the cresols and xylenols have much weaker odors. Within the cresols, *meta* substitution relative to the hydroxyl group gives a lower odor impact than either the *ortho* or *para* positions. The introduction of two or more hydroxyl groups totally suppresses the odor character.

Phenols are important in food flavors and have been reviewed by Maga (1978a).

15. Sulfur-Containing Compounds

Sulfur and oxygen, being chemically related, form similar compounds:

R·OH	R·SH
Alcohols	Mercaptans
or alkyl thiols	

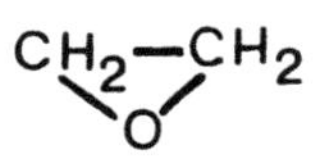

Ethylene oxide

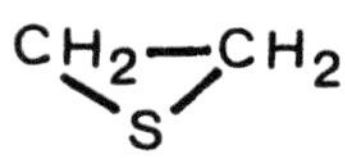

Dimethyl sulfide

The –SH group is known as mercapto-, thiol- or sulfhydryl-.

Sulfur-containing compounds play an important role in natural flavor chemistry as they are not only responsible for the objectionable odors associated with rotting vegetable matter but also contribute to such agreeable odors as in roast meats (Schulte 1973; Shankaranarayana *et al.* 1974; Maga 1975a,b).

Sulfides

Also known as thioethers, the sulfides, disulfides and polysulfides are the characterizing components of the alliaceous vegetables (e.g., onion, garlic, leek) although the more subtle attributes of these profiles are contributed by substituted thio- compounds.

The preceding section on sensory characters of organics covers only the major organic groupings widely encountered in flavor work. It is intended merely to draw attention to the enormous range of aromatic effects which can be derived from synthetic chemicals and in no way is a substitute for practical experience of the odor and flavor characteristics of the chemicals permitted for use in foods.

E. Solvents

Solvents must be considered under the two headings: (a) those used for extraction of flavoring materials (e.g., spices) and totally removed from the product prior to its use in foods and (b) those used in the extraction of certain aromatic materials (e.g., vanilla) or in the production of liquid flavorings and present when these are incorporated into foodstuffs.

Extraction Solvents

The solvents used in the recovery of highly concentrated flavoring components from plant material may be divided into the following categories:

a. Polar solvents. These contain hydroxy or carboxy groups characterized by their miscibility with water (e.g., methanol, ethanol, isopropanol, acetone).

b. Nonpolar solvents. These are generally inert hydrocarbons which are not miscible with water (e.g., hexane, pentane).

c. Organic compounds. Some esters, notably ethyl acetate, have very good solvent properties of particular value in the direct extraction of wet plant materials.

d. Chlorinated hydrocarbons. These have low boiling points and relatively high specific gravities, are not miscible with water and pose no fire hazard although they are increasinglyy suspect as a toxicological hazard.

e. Liquefied or supercritical gases. Liquid carbon dioxide acts as a nonpolar solvent but is miscible with all common liquid solvents. Its solvent power is not high compared with other traditional solvents.

The handling of organic solvents for the production of flavoring extracts and oleoresins calls for close attention to flammability and toxic hazards due to excessive concentrations of vapor in the processing area.

Although acetone is an excellent solvent, it poses a high fire and explosion risk calling for special plant precautions. The nonflammable but significantly more toxic chlorinated hydrocarbons are now widely used in place of acetone for spice extraction work but their continued acceptance is under question. As these solvents are not permitted in foodstuffs they must be completely stripped from the resulting extract or oleoresin before it is used or offered for sale. The removal technique is of considerable importance to the flavor quality of the end product as, in most cases, the solvent must be evaporated in the presence of other low-boiling constituents which are essential to the profile of the extract. To avoid undue losses, the distillation to recover solvent must be carried out as rapidly as possible at the lowest practicable temperature.

Regulations on residual solvents permitted in the United States under the Food Additives Regulations were published in the Federal Register (1968).

The limits are as follows:

	Maximum permitted (ppm)
Acetone	30
Methanol	50
Isopropanol	50
Hexane	25
All chlorinated solvents	30

These limits are considered desirable and have been widely adapted in other countries.

Of the solvents which may remain in flavoring preparations, ethanol is by far the best but in most countries this attracts high fiscal duties which may not be recoverable when used in flavorings. This may make ethanol-based flavorings unacceptably expensive. In consequence, some countries permit the use of isopropanol but its acceptance is by no means universal. Another important nonalcoholic solvent often used to replace ethanol is propylene glycol which is a colorless, practically odorless, hygroscopic, somewhat viscous liquid which is miscible with water and is a useful solvent for many organics. This subject has been discussed at greater length by Heath (1981).

F. Nomenclature of Organic Chemicals

One cannot read any research paper involving flavor chemistry without encountering innumerable compounds having almost impossibly long chemical names. These are generally systematic and give a direct indication of the molecular structure of the compound concerned but unfortunately some confusion may arise from the use of different naming systems. Trivial or common names are most likely to be used in a commercial laboratory (e.g., isopropyl alcohol not isopranol) and in some older references these designations are used almost exclusively. The International Union for Physics and Chemistry (IUPAC) system is by far the most preferred, and reference should be made to standard texts on organic chemistry to find the rules pertaining to the various types of organic compounds involved. In general, the system is based on the identification of the longest chain containing the substituent group and applying to it the name of the parent hydrocarbon or basic compound named on a strictly systematic basis. The position of the substituent group or groups is then indicated by sequential numbers using the lowest possible combination. It is a good practice to adopt the designation used in the official GRAS listings and to maintain an alphabetical listing of synonyms for ready reference (Heath 1981).

References

ARCTANDER, S. 1969. Perfume and Flavor Chemicals, Vols. 1 and 2, Steffen Arctander, Olyphant, PA.

BUTTERY, R. G. 1981. Vegetable and fruit flavors. Food Sci. (NY) *7* (Flav. Res., Recent Adv.), 175–216.

CHEMICAL SOURCES ASSOC. 1981. Flavor and Fragrance Materials—1981. Allured Publishing, Wheaton, IL.

DRAVNIEKS, A., BOCK, F. V. C., POWERS, J. J., TIBBETTS, M. and FORD, M. 1978. Comparison of odors directly and through profiling. Chem. Senses *3*(2), 191–225.

ENNS, M. P. and HORNUNG, D. E. 1985. Contribution of smell and taste to overall intensity. Chem. Senses *10*(3), 357–366.

FORSS, D. A. 1981. Sensory characterization. *In* Flavor Research: Recent Advances, pp. 125–174, R. Teranishi, R. A. Flath and H. Sugisawa eds. Dekker, New York and Basel.

FRIJTERS, J. E. R. 1979. Some psychophysical notes on the use of odor unit number. *In* Progress in Flavour Research, pp. 47–52, D. G. Land and H. E. Nursten, eds. Appl. Sci. Publ., London.

FURIA, T. E. and BELLANCA, N., eds. 1975. Fenaroli's Handbook of Flavor Ingredients, 2nd edition, Vols. 1 and 2, CRC Press Inc., Boca Raton, FL.

GUADAGNI, D. G. *et al.* 1968. Correlation of subjective–objective methods in study of odor and taste. ASTM STP440, American Society for Testing and Materials, Philadelphia, PA.

GUADAGNI, D. G., BUTTERY, R. G. and OKANO, S. 1966. Odor intensities of hop oil constituents. J. Sci. Food Agric. *17*, 142.

HALL, R. L. 1960. Recent progress in the consideration of flavoring ingredients under the Food Additives Amendment. Food Technol. *14*(10), 488.

HALL, R. L. and OSER, B. L. 1961. Recent progress in the consideration of flavoring ingredients under the Food Additives Amendment: 2.GRAS substances. Food Technol. *15*(12), 20.

HALL, R. L. and OSER, B. L. 1965. Recent progress in the consideration of flavoring ingredients under the Food Additives Amendment: 3.GRAS substances. Food Technol. *19*(2), 151.

HALL, R. L. and OSER, B. L. 1970. Recent progress in the consideration of flavoring ingredients under the Food Additives Amendment: 4.GRAS substances. Food Technol. *24*(5), 25.

HEATH, H. B. 1978. Flavor Technology: Profiles, Products, Applications. AVI Publishing, Westport, CT.

HEATH, H. B. 1981. Source Book of Flavors. AVI Publishing, Westport, CT.

IOFI. 1982. Code of Practice for the Flavour Industry—Positive List of Artificial Flavouring Substances. International Organization for the Flavour Industry, Geneva.

LAND, D. G. 1984. Flavour research in 1984—Where now and where next? *In* Progress in Flavour Research 1984, pp. 615–632, J. Adda, ed. Elsevier, Amsterdam.

MACLEOD, A. J. 1980. Chemistry of odors. Proc. Symp. Zool. Soc., London. *45*, 15–34.

MAGA, J. A. 1975a. Role of sulfur compounds in food flavor: II. Thiophenes. CRC Critical Rev. Food Technol. Nutr. *6*, 241–270.

MAGA, J. A. 1975b. Role of sulfur compounds in food flavor: III. Thiols. CRC Critical Rev. Food Technol. Nutr. *7*, 147–192.

MAGA, J. A. 1976. Lactones in foods. CRC Critical Rev. Food Technol. Nutr. *8*, 1–56.

MAGA, J. A. 1978a. Simple phenols and phenolic compounds in food flavor. CRC Critical Rev. Food Technol. Nutr. *10*, 323–372.

MAGA, J. A. 1978b. Amines in foods. CRC Critical Rev. Food Technol Nutr. *10*, 373–403.

MOSKOWITZ, H. R. 1977. Psychophysical and psychometric approaches to sensory evaluation. CRC Crit. Rev. Food Sci. Nutr. *9*(1), 41–79.

MOSKOWITZ, H. R. and WARREN, C. B. 1981. Odor quality and chemical structure. *In* ACS Symp. Ser. No. 148, Amer. Chem. Soc., Washington, D.C.

NURSTEN, H. E. 1979. Why flavour research? How far have we come since 1975 and where no? *In* Progress in Flavour Research, pp. 337–355, D. G. Land and H. E. Nursten, eds. Appl. Sci. Publ., London.

OSER, B. L. and HALL, R. L. 1972. Recent progress in the consideration of flavoring ingredients under the Food Additives Amendment: 5.GRAS substances. Food Technol. *26*(5), 35.

OSER, B. L. and FORD, R. A. 1973a. Recent progress in the consideration of flavoring ingredients under the Food Additives Amendment: 6.GRAS substances. Food Technol. *27*(1), 64.

OSER, B. L. and FORD, R. A. 1973b. Recent progress in the consideration of flavoring ingredients under Food Additives Amendment: 7.GRAS substances. Food Technol. *27*(11), 56.

OSER, B. L. and FORD, R. A. 1974. Recent progress in the consideration of flavoring ingredients under the Food Additives Amendment: 8.GRAS substances. Food Technol. *28*(9), 76.

OSER, B. L. and FORD, R. A. 1975. Recent progress in the consideration of flavoring ingredients under the Food Additives Amendment: 9.GRAS substances. Food Technol. *29*(8), 70.

OSER, B. L. and FORD, R. A. 1977. Recent progress in the consideration of flavoring ingredients under the Food Additives Amendment: 10.GRAS substances. Food Technol. *31*(1), 65.

OSER, B. L. and FORD, R. A. 1978. Recent progress in the consideration of flavoring ingredients under the Food Additives Amendment: 11.GRAS substances. Food Technol. *32*(2), 60.

OSER, B. L. and FORD, R. A. 1979. Recent progress in the consideration of flavoring ingredients under the Food Additives Amendment: 12.GRAS substances. Food Technol. *33*(7), 65.

OSER, B. L., and HALL, R. L. 1977. Criteria employed by the Expert Panel of FEMA for the GRAS evaluation of food flavoring substances. Food Cosmet. Toxicol. *15*, 457.

OSER, B. L., FORD, R. A. and BERNARD, B. K. 1984. Recent progress in the consideration of flavoring ingredients under the Food Additives Amendment: 13.GRAS substances. Food Technol. *38*(10), 65.

OSER, B. L., WEIL, C. S., WOODS, L. S. A. and BERNARD, B. K. 1985. Recent progress in the consideration of flavoring ingredients under the Food Additives Amendment: 14.GRAS substances. Food Technol. *39*(11), 108, 110, 112.

PARLIAMENT, T. H. 1980. The chemistry of aroma. Chem. Tech. (March), 284–289.

RAPP, U., HOHN, M. and DIELMANN, G. 1979. Optimizing coupling techniques for gas chromatography–mass spectrometry in flavour research. *In* Progress in Flavour Research, pp. 151–158, D. G. Land and H. E. Nursten, eds. Appl. Sci. Publ., London.

ROTHE, M. 1976. Aroma values—a useful concept? Nahrung *20*, 259–266.

SEIDMAN, M. 1979. Sensory methods in the work of the flavour chemist. *In* Progress in Flavour Research, pp. 15–26, D. G. Land and H. E. Nursten, eds. Appl. Sci. Publ., London.

SCHIFFMAN, S., ROBINSON, D. F. and ERICKSON, R. P. 1977. Multidimensional scaling of odorants: Examination of psychological and physiological dimensions. Chem. Senses Flav. *2*(3), 355–390.

SCHULTE, L. 1973. Precursors of sulfur-containing flavor compounds. CRC Critical Rev. Food Technol. Nutr. *4*, 457–506.

SHANKARANARAYANA, M. L., RAGHAVEN, B., ABRAHAM, NATARAJAN, C. P. and BRODNITZ, M. H. 1974. Volatile sulfur compounds in food flavor. CRC Critical Rev. Food Technol. Nutr. *4*, 395–435.

STAHL, W. H., ed. 1973. Compilation of odor and taste threshold values data. ASTM Data Ser. DS48. Amer. Soc. for Testing and Materials, Philadelphia, PA.

STOFBERG, J. 1980. The future of artificial flavoring ingredients. Perf. Flavorist *5*(Aug./Sept.), 17–22.

STOFBERG, J. and STOFFELSMA, J. 1980. Consumption of flavoring materials as food ingredients and food additives. Perf. Flavorist *5*(Dec. 80/Jan. 81), 19–22.

9

Flavor Potentiators

Flavor potentiators are compounds which have no flavor of their own at usage levels but yet intensify or enhance the flavor of a food. Compounds used in the food industry which fall into this category are table salt, monosodium glutamate (MSG) and certain nucleotides, e.g., guanosine 5′-monophosphate (GMP) and inosine 5′-monophosphate (IMP). Maga (1983) has published an excellent comprehensive review of this subject which is the basis for much of this chapter. The reader is encouraged to go to Maga's article for detail on this subject.

Monosodium glutamate and the 5′-nucleotides were originally identified by Japanese researchers as occurring naturally in food ingredients used to "enhance" the flavor of traditional recipes. Ikeda (1909) isolated MSG from sea tangle (*Laminiaria Japanica*) and Kodama (1913) identified the histidine salt of 5′-inosinic acid in dried bonito (type of fish). The MSG went into commercial production nearly immediately and as of 1978, worldwide usage was about 300,000–330,000 metric tons (Ebert 1978). It was not economically feasible to produce the 5′-nucleotides at the time they were first identified, and commercial production was not begun until the early 1960s. Annual production of IMP and GMP in Japan (major manufacturer) was reported to be about 4680 tons in 1974 (Kuninaka 1978).

Two other amino acids (tricholomic and ibotenic acids) have been identified as having flavor potentiator properties. Tricholomic and ibotenic acids were found originally in Japanese mushrooms (*Tricholoma muscarium* and *Aminita strobiliforms*, respectively) by Takemoto *et al.* (1964a,b,c) and Takemoto and Makajima (1964a,b). These amino acids were of particular interest since they were found to be effective in killing the common house fly. While both amino acids have lower sensory thresholds for potentiating effects than either MSG or the 5′-nucleotides, they have not found commercial application.

A. Chemical Properties

1. Structure

The structures of the naturally occurring flavor potentiators are presented in Fig. 9.1. The commonly available naturally occurring flavor potentiators are either 5-carbon L-amino acids or purine ribonucleoside 5′-monophosphates having a 6-oxy group.

It is of interest that flavor potentiating effect is very dependent upon structure of the molecule. For MSG, it is only the L form of the amino acid that has potentiating properties. The D form has no activity. Also, the ionic form is very important. The monosodium form (rather than diacidic or dibasic salt) is active while the other ionic forms have little activity.

In the case of the nucleotides, numerous isomers have been examined as possible flavor potentiators. Of the 2′, 3′ or 5′ isomers, only the 5′-nucleotides have been shown to have potentiating activity (Kuninaka *et al.* 1964; Kuninaka 1964). With the 5′ structure, a hydroxy group in the 6 position is required to generate flavor potentiation.

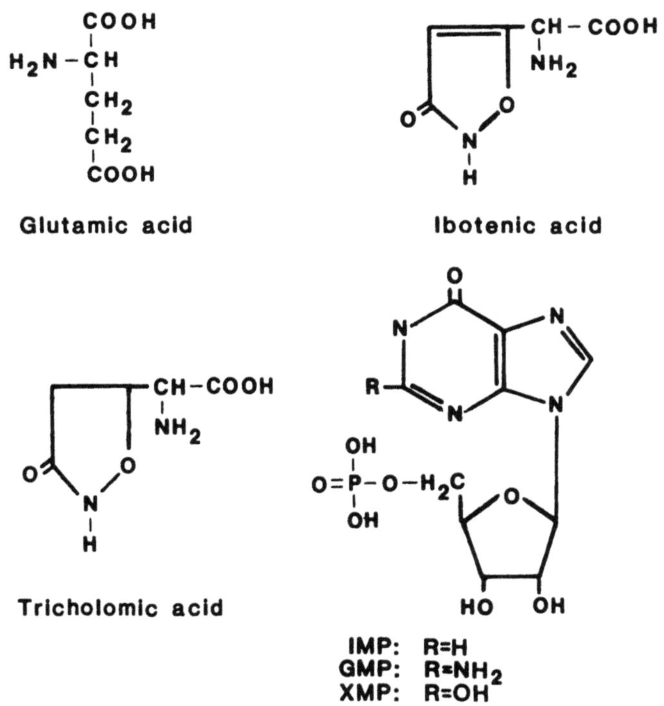

Fig. 9.1. Structure of selected flavor potentiators.

A large number of studies investigating related structures for potentiator activity have been reported (Maga 1983). Other than ibotenic and tricholomic acids, none of the amino acids or amino acid combinations were found to be as effective as MSG. Investigations of related nucleotides for potentiator activity by Imai (1971) demonstrated that sulfur-substituted (2 position) nucleotides were more effective than the original nucleotides. For example, 2-furfurylthio IMP had a flavor potentiating activity approximately 17 times stronger than IMP. Cost of synthesis and necessity of obtaining food additive status has prohibited the sulfur-substituted nucleotides from being commercially available.

2. Stability

As was pointed out earlier, the monosodium salt of MSG is the only form of L-glutamic acid which has flavor potentiating properties. As expected then, pH has a pronounced effect on potentiating properties. Monosodium glutamate is most effective in the pH range of 5.5–8.0. The data of Fagerson (1954) suggest that it is the R_3 form (see Fig. 9.2) of glutamic acid which is responsible for flavor potentiation.

The MSG is liable to destruction during thermal processing. Glutamic acid may lose a molecule of water to form its lactam thereby losing potentiating properties. Glutamic acid may also enter into the Maillard reaction. Free amino acids typically are substantially more reactive than protein bound amino acids. Participation in the Strecker degradation or

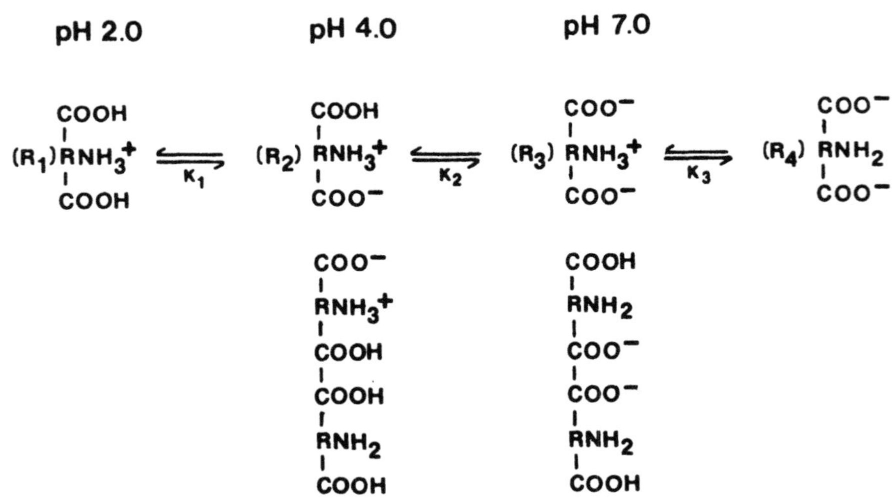

Fig. 9.2. Dissociation form of glutamic acid as a function of pH.
Reprinted with the permission of Fagerson (1954)

any of the numerous browning pathways would result in loss in potentiating activity.

The 5′-nucleotides are most likely to be destroyed by hydrolysis of the ribose linkage to the purine base (Kuninaka 1967). This linkage is weaker than the sugar phosphate bond. While Maga (1983) has reviewed numerous studies on the stability of the 5′-nucleotides, a few of these studies will be summarized here.

Destruction of the 5′-nucleotides during thermal processing shows a slight pH dependency. Product literature (Takeda Chemical Industries, Bull. No. 3) has shown that 40 min of heating (115°C) in solution at pH 3 resulted in 29% loss of IMP and GMP while equivalent heating at pH 6.0 resulted in 23% loss. The IMP and GMP are both quite stable when in the dry form and when exposed to light. Studies by the Ajinomoto Company (Bull. No. 3) have shown that deep fat frying also produces minimal destruction of the 5′-nucleotides.

It appears that canning is the major food process which results in significant losses in the 5′-nucleotides. Hashida *et al.* (1968) have shown that canned short-necked clams lose from 39–48% of their 5′-nucleotides due to canning and 6 months storage at room temperature. Canned boiled lobster was found to lose 35–39% after 3 months of room temperature storage and 39–42% following 8 months storage. Corned beef lost 12% of the 5′-nucleotides when processed at 110°C for 120 min and 22% when processed at 120°C for 60 min. It appears that losses due to canning can be quite significant.

Kinetic studies have shown IMP decomposition to be of first order (Lee 1979). Davidek *et al.* (1972) have provided kinetic data over a temperature range of 60–98°C.

Other means of losing nucleotide activity is through water loss, enzyme activity and irradiation. Since the 5′-nucleotides are water soluble, any loss of moisture from the product (e.g., canning or packaging in water and drip losses following freezing) will result in reduced potentiator activity.

Enzymes can cause either an increase or a decrease in potentiator concentrations in a food. In meats, IMP results from the decomposition of AMP which came from ATP (Terasaki *et al.* 1965). Therefore, IMP levels in the meat will typically increase shortly after death and then gradually decrease as the IMP is enzymatically degraded to inosine and then hypoxanthine. Enzymes may also be present in foods which hydrolyze the phosphate from the 5′-nucleotides (Kuninaka 1967). This would also result in the loss of potentiator effects.

While irradiation has not become a common means of food preservation, studies have shown that the 5′-nucleotides are unstable to irradiation. Terada *et al.* (1968) found room temperature irradiation of 5′-

IMP in distilled water resulted in a 40% loss at pH 7.0 and greater losses at acid pHs. The 5′-IMP is converted to hypoxanthine.

B. Sensory Properties

1. Influence on Taste

The influence of flavor potentiators on the basic sensations of taste is rather controversial. Early studies suggested that MSG intensified the sweetness or saltiness of a food when these tastes were near their optimum level. Sourness and bitterness were found to be suppressed in some food systems by MSG (Cairncross and Sjostrom 1948). However, the literature has become quite contradictory. For example, under the conditions of testing used by Lockhart and Gainer (1950), they found no enhancement of either sweet or salt tastes by MSG. Mosel and Kantrowitz (1952) also found no enhancement of either sweet or salt tastes but substantial reductions in the sensory thresholds of sour (2×) and bitter (30×). Van Cott *et al.* (1954) found the opposite: that sweet and salt thresholds were lower and sour and bitter showed no change. More recently, Yamaguchi and Kimizuka (1979) found no significant influence of MSG on any of the basic tastes.

The literature on the effect of the 5′-nucleotides on the basic tastes is not as confusing. Woskow (1969) found 0.004–0.020% of a 50–50 blend of GMP and IMP suppressed perceived bitterness in model systems. Sour perception was also suppressed by 0.008–0.020% nucleotides. The use of 0.016–0.20% 5′-nucleotides was found to enhance the taste intensity of a 5% sucrose solution. Salt perception appeared to be unaffected at concentrations of nucleotides less than 0.02%. Higher levels were found to lower the threshold to salt.

The flavor potentiators themselves have a taste at high enough concentrations. Maga (1983) noted that MSG has been reported to have a taste threshold ranging from 0.00067 to 0.067%. This range reflects the effects of panel number, age, sex, chemical purity, etc., on threshold determination. A recent study by Yamaguchi and Kimizuka (1979) reported an absolute threshold of 0.010% for MSG. The taste at recognition concentrations has been described as being salty-sweet. The sensory thresholds of IMP and GMP have been reported to range from 0.01 to 0.025% and from 0.0035 to 0.02%, respectively (Kuninaka 1967).

Synergistic effects are quite pronounced with the flavor potentiators. Takeda Industries (Bulletin No. 6) has reported that a 50–50 blend of GMP and IMP have a taste threshold of 0.0063%. When they are used in combination with 0.8% MSG, the taste threshold was reduced to

0.000031% (Maga 1983). Therefore, the flavor potentiators are most commonly used in combination to be cost effective.

2. Influence on Aroma

The influence of flavor potentiators on aroma perception is also controversial. Wagner *et al.* (1963) concluded that IMP could affect the aroma of some foods. They found that 0.01% IMP increased aroma strength of certain notes in beef noodle soup. Similar effects have also been reported by Caul and Raymond (1964) and Kurtzman and Sjostrom (1964). While these studies involved sensory evaluation, Maga and Lorenz (1972) reported on the effects of flavor potentiators on headspace volatiles of beef stock as measured by gas chromatography. They found similar trends for 0.05% MSG and 0.05% 5′-nucleotides with both increasing the volatiles in the headspace. However, the 5′-nucleotides increased overall peak areas by 2.3 times as compared to 1.7 times for the MSG. The mode of enhancing aroma concentration in the headspace is unknown. The low levels of potentiators used are well below the levels needed to produce "salting out" effects.

Yamaguchi (1979) has reported that MSG has no effect on the aroma properties of 16 foods used in her study. This study is quite contrary to much of the other literature.

3. Synergism

Synergism may be defined as the cooperative action of two components of a mixture whose total effect is greater than the sum of their individual effects (Johnson 1978). Flavor potentiators show pronounced synergistic effects. Combinations of the 5′-nucleotides and MSG produce intensity effects as well as taste quality effects.

Synergistic effects are illustrated in Table 9.1. A 1:1 ratio of MSG–

Table 9.1. Taste Intensities of MSG–IMP and MSG–GMP Combinations[a]

Ratio of		Relative Flavor	Ratio of		Relative Flavor
MSG	IMP	Intensity	MSG	GMP	Intensity
1	0	1.0	1	0	1.0
1	1	7.0	1	1	30.0
10	1	5.0	10	1	18.8
20	1	3.5	20	1	12.5
50	1	2.5	50	1	6.4
100	1	2.0	100	1	5.4

[a] Takeda Chemical Industries, Product Bull. No. 3.

IMP resulted in a sevenfold increase in flavor intensity. It was of interest that a 1:1 mixture MSG–GMP produced a thirtyfold increase in flavor intensity. Therefore, it would appear that the most effective flavor potentiator system to use would be a 1:1 mix of MSG and GMP. Due to the relatively higher cost of 5′-nucleotides, they are seldom used at a 1:1 ratio with MSG. Typically a 95:5 ratio of MSG–IMP/GMP (1:1) is used in the industry. This combination yields approximately a sixfold synergistic effect (Maga 1983).

4. Mode of Action

The means by which flavor potentiators enhance flavor is poorly understood. Unfortunately the basic mechanisms responsible for taste and odor perception are not adequately characterized yet to allow an understanding of the effects of flavor potentiators. However some indirect evidence has been gathered which supports some basic theories. These theories will be discussed briefly.

Strong (1968) suggested that flavor potentiators may function by either increasing the amount of flavor compounds actually arriving at the gustatory pores or olfactory cells or by increasing the magnitude of the signal generated by the flavor compound. Schinneller *et al.* (1972) supported the first theory by suggesting that the 5′-nucleotides "unmasked" certain flavor receptor sites. Therefore more receptor sites were available for reaction with the stimulus.

Saint-Hilaire and Solms (1973) proposed that flavor potentiators interacted with collagenlike proteins at the receptor sites. This would create an improved environment for taste receptor stimulation.

The most recent work in the area of taste perception has focused on stimulus–receptor interactions (Torri and Cagan 1980; Cagan *et al.* 1979). It appears that the first step in taste perception involves a weak reversible binding at the surface of the receptor cell (Kruger and Cagan 1976; Beidler 1962). This is followed by transduction processes within the receptor cells, peripheral synaptic transmission and neural transmission with central processing. Data have been obtained which demonstrate that L-glutamate binds preferentially to taste receptor tissues. Also, the 5′-nucleotides have been found to significantly increase the binding of L-glutamate. Therefore it appears that the flavor potentiators operate on the taste receptor cell to enhance flavor perception. It is not clear yet which stage of the perception pathway is actually enhanced by the flavor potentiators.

C. Flavor Potentiators in Foods

1. Naturally Occurring

Maga (1983) has provided a tabulation of the MSG and 5′-nucleotide content of numerous natural foods. There is no question that many foods contain sufficient levels of the flavor potentiators to make a significant contribution to flavor perception. A sampling of Maga's data is presented in Table 9.2.

2. Added to Foods

The application of flavor potentiators in the food industry is determined by the enhancing effects of the potentiator. While there is little agreement on what sensory properties are intensified or masked by the flavor potentiators, there is no question that the flavor potentiators do modify the flavor perception of some food systems. The influence of IMP on different flavor notes is presented in Table 9.3. This data again show that different flavor notes are affected to different degrees by IMP.

Flavor potentiators find wide usage in food products. Vegetables, sauces, soups, meats and other savory foods constitute the major food application. Typical food products and usage levels are presented in Table 9.4. The ability of the flavor potentiators to impart viscosity, drying and fullness is a useful property in soups, gravies, sauces and juices.

Table 9.2. Flavor Potentiators in Natural Food Products[a]

	Amount (%)			
Food	MSG	IMP	GMP	AMP
Beef	0.013–0.088	0.163	0.002	0.006
Cheese, Gruyère	1.333	NR[b]	NR	NR
Cod	0.011	0.043	NR	0.023
Scallop	0.012–0.024	0.000	0.000	0.116
Orange juice	0.015	NR	0.00002	0.00005
Nectarine	1.219	NR	NR	NR
Corn	0.102	NR	NR	0.006
Mushroom, dry shiitake	0.177–0.635	NR	0.56	0.131
Potato, fresh	0.254	NR	0.001	NR
Tomato	0.177	NR	0.001	NR
Tea, green	0.264–0.640	NR	NR	0.012

[a] Maga 1983.
[b] Not reported.

Table 9.3. Influence of IMP on the Flavor, Aroma, and Texture Properties of Foods[a]

Flavor/Texture Characteristic	Aroma	Flavor
Sweet	—	Increase
Salt	—	Decrease
Sour	No change	Decrease
Bitter	—	Decrease
Meaty	—	Increase
Brothy	No change	Increase
Hyrolyzed vegetable protein	Decrease	Decrease
Fatty	No change or decrease	No change or decrease
Buttery	No change or decrease	No change or decrease
Sulfury	Decrease	Decrease
Burnt	No change	No change or decrease
Starch	No change or decrease	No change or decrease
Herb/spice	No change	No change or decrease
MSG effect	—	Increase
Viscosity	—	Increase
Drying	—	Increase
Fullness	—	Increase

[a] Ajinomoto Co., Technical Bull. Part 2.

Table 9.4. Processed Foods' Flavor Potentiators and Usage Levels[a]

Food	Usage Levels: MSG (%)	Usage Levels: 5'-Nucleotides (50/50 IMP and GMP) (%)
Dehydrated soups	5–8	0.10–0.20
Canned soups	0.12–0.18	0.0022–0.0033
Soup powder for instant noodles	10–17	0.30–0.60
Canned asparagas	0.08–0.16	0.003–0.004
Canned crab	0.07–0.10	0.001–0.002
Canned fish	0.10–0.30	0.003–0.006
Canned poultry, sausages, ham	0.10–0.20	0.006–0.010
Sauces	1.0–1.2	0.010–0.030
Dressings	0.30–0.40	0.010–0.150
Ketchup	0.15–0.30	0.010–0.020
Mayonnaise	0.40–0.60	0.012–0.018
Sausages	0.30–0.50	0.002–0.014
Snacks	0.10–0.50	0.003–0.007
Soy sauce	0.30–0.60	0.030–0.050
Vegetable juice	0.10–0.15	0.005–0.010
Processed cheese	0.40–0.50	0.005–0.010

[a] Maga 1983.

3. Source of Commercial Potentiators

Monosodium glutamate was originally a by-product of the sugar industry. However, today the majority of MSG is produced via fermentation (Ogata *et al.* 1976). Fermentation offers the advantage that only the L-isomer, which has the flavor potentiating effect, can be produced. The organism used for MSG manufacture is *Corynebacterium glutamicum* (alternatively named *Micrococcus glutamicus*; Margalith 1981). Under optimum conditions, this organism will produce 30–50 g of L-glutamic acid per liter of ferment. The productivity of this organism, optical purity of the product and ease of isolation from the medium all have made fermentation the production method of choice.

The 5′-nucleotides are obtained from several sources. Maga (1983) listed the following as commercial sources of the 5′-nucleotides: (1) fresh muscle of marine fish; (2) degradation of RNA by 5′-phospho-diesterase; (3) fermentation to produce nucleotides followed by chemical phosphorylation to produce the 5′-nucleotides; (4) direct fermentation to yield 5′-nucleotides; (5) chemical decomposition of RNA to produce nucleotides followed by phosphorylation to yield 5′-nucleotides; and (6) combination of procedures.

Originally, fresh fish muscle served as a major source of the 5′-nucleotides. However problems in maintaining the fresh muscle (enzyme activity) and not being able to economically recover 5′-GMP from the fish muscle have made this process less important. Today the majority of 5′-nucleotides come from the enzymatic hydrolysis of RNA or the fermentation to nucleotides followed by chemical phosphorylation.

D. Toxicity

1. Monosodium Glutamate

The MSG has low toxicity from an acute standpoint. An LD_{50} of 19.9 g/kg body weight in mice has been reported by Ebert (1979a). Long-term feeding (4% of diet) studies have failed to show any significant effects of MSG on several species of animals (Ebert 1979b; Wen *et al.* 1973; Huang *et al.* 1976). If any effects were seen, they were attributed to the large amount of sodium associated with MSG.

Injection of MSG has been shown to induce rapid degeneration of neurons in the inner layers of the retina (Lucas and Newhouse 1957). Degeneration of nerve cells in the brain due to subcutaneous injections of large amounts of MSG have also been observed in various species (Takasaki 1978). Even though ingestion of reasonable amounts of MSG has not been shown to produce any significant toxicity, there remains some concern for toxicity due to the studies using direct injection techniques.

Monosodium glutamate has also been associated with the Chinese Restaurant Syndrome (CRS). Symptoms often associated with CRS are a burning sensation in the back of the neck, facial pressure, chest pain and in some cases, headaches (Schaumburg *et al.* 1969). Additional symptoms provided by Ghadimi *et al.* (1971) included sweating, nausea, weakness and thirst. In fact there appears to be little agreement on the symptoms of CRS or even its existence in other than perhaps a very few, very sensitive individuals. A sampling of the literature reviewed by Maga (1983) on this subject includes the work of Rosenblum *et al.* (1971). They found that typical CRS symptoms could not be induced in 99 healthy males given 5 g of MSG. The existence of CRS remains controversial.

2. 5′-Nucleotides

The acute toxicities of the 5′-nucleotides have been found to range from 2.7 to 14.0 g/kg body weight depending upon the species of animal and the means of administering the 5′-nucleotides (Kojima 1974). While administration of near lethal doses of the 5′-nucleotides will cause temporary depression, clonic convulsion and dyspnea, dietary intakes of 2% (of total diet) or less 5′-nucleotides have not been found to cause observable effects in any species of animals in long-term feeding studies (Kojima 1974; Worden *et al.* 1975). Therefore even though no human toxicity studies have been conducted on the 5′-nucleotides, the lack of observable effects in several species of animals at dietary levels approximately 1000 times typical food applications strongly suggests that there are no significant toxicity concerns for 5′-nucleotides in the diet.

E. Other Potentiators

While MSG and the 5′-nucleotides are the only compounds which have been found to exhibit true potentiating effects in foods, a few other compounds are often included in a discussion of flavor potentiators. Maltol and ethyl maltol are most commonly included in this category. Both of these compounds are used to enhance sweet products. The addition of 5–75 ppm maltol may permit a 15% reduction of sugar in some foods (Sjostrom 1975). It is common to see maltol and/or ethyl maltol used in jams, jellies, fruit or chocolate syrups, baked goods, sweet beverages and other sweet foods. It is of interest that maltol has been found in nature but ethyl maltol has not.

Other compounds included as flavor potentiators are dioctyl sodium sulfocuccinate (adds freshness), *N, N'*-di-*O*-tolylethylenediamine and cyclamic acid (Maga 1983). These compounds have found application in certain dairy products.

References

AJINOMOTO COMPANY. Nucleotides. Part 2. Flavor properties. Ajinomoto Company Inc. Tech. Bull., Tokyo.

AJINOMOTO COMPANY. Nucleotides. Part 3. Physical properties. Ajinomoto Company Inc. Tech. Bull., Tokyo.

BEIDLER, L. M. 1962. Taste receptor stimulation. Prog. Biophys. Biochem. Chem. *12,* 107.

CAGAN, R. H., TORII, K. and KARE, M. R. 1979. Biochemical studies of glutamate taste receptors: The synergistic taste effect of L-glutamate and 5'-nucleotides. *In* Glutamic Acid: Advances in Biochemistry and Physiology, L. J. Filer, eds., p. 1. Raven Press, New York.

CAIRNCROSS, S. E. and SJOSTROM. L. B. 1948. What glutamate does in food. Food Ind. 20, 982.

CAUL, J. F. and RAYMOND, S. A. 1964. Home-use test by consumer of the flavor effects of disodium inosinate in dried soup. Food Technol. *18,* 353.

DAVIDEK, J., VELISEK, J. and JANICEK, G. 1972. Stability of inosinic acid, inosine and hypoxanthine in aqueous solutions. J. Food Sci. *37,* 789.

EBERT, A. G. 1978. Glutamates. *In* Encyclopedia of Food Science. M. S. Peterson and A. H. Johnson, eds., p. 348. AVI Publishing, Westport, CT.

EBERT, A. G. 1979a. Dietary administration of monosodium glutamate or glutamic acid to C-57 black mice for two years. Toxicol. Lett. *3,* 65.

EBERT, A. G. 1979b. Dietary administration of L-monosodium glutamate and L-glutamic acid to rats. Toxicol. Lett. *3,* 71.

FAGERSON, I. S. 1954. Possible relationship between the ionic species of glutamate and flavor. J. Agric. Food Chem. *2,* 474.

GHADIMI, H., KUMAR, S. and ABACI, F. 1971. Studies on monosodium glutamate ingestion: I. Biochemical explanation of Chinese-restaurant syndrome. Biochem. Med. *5,* 447.

HASHIDA, W., MOURI, T. and SHIGA, I. 1968. Application of 5'-nucleotides to canned seafoods. Food Technol. *22,* 1436.

HUANG, P., LEE, N., WU, T., YU, S. and TUNG, T. 1976. Effect of monosodium glutamate supplementation to low protein diets on rats. Nutr. Reports Int. *13,* 477.

IKEDA, K. 1909. On a new seasoning. J. Tokyo Chem. Soc. *30,* 820.

IMAI, K. 1971. Synthesis of compounds related to inosine 5'-phosphate and their flavor enhancing activity: IV. 2-Substituted inosine 5'-phosphates. Chem. Pharm. Bull. *19,* 576.

JOHNSON, K. R. 1978. Synergism. *In* Encyclopedia of Food Science. M. S. Peterson and A. H. Johnson, eds., p. 722, AVI Publishing, Westport, CT.

KODAMA, S. 1913. On a procedure for separating inosinic acid. J. Tokyo Chem. Soc. *34,* 751.

KOJIMA, K. 1974. Safety evaluation of disodium 5′-inosinate, disodium 5′-guanlyate and disodium 5′-ribonucleotide. Toxicology 2, 185.

KRUGER, J. M. and CAGAN, R. H. 1976. Biochemical studies of taste sensation: IV. Binding of L-[H] alanine to a sedimentable fraction from catfish barbel epithelium. J. Biol. Chem. *251*, 88.

KUNINAKA, A. 1964. Symposium on Flavor Potentiation, p. 4. Arthur D. Little, Cambridge, MA.

KUNINAKA, A. 1967. Flavor potentiators. *In* The Chemistry and Physiology of Flavors, E. A. Day and L. M. Libbey, eds., p. 515. AVI Publishing, Westport, CT.

KUNINAKA, A. 1978. Flavor potentiators. *In* Encyclopedia of Food Science, M. S. Peterson and A. H. Johnson, eds., p. 279. AVI Publishing, Westport, CT.

KUNINAKA, A., KIBI, M. and SAKAGUCHI, K. 1964. History and development of flavor nucleotides. Food Technol. *18*, 287.

KURTZMAN, C. H. and SJOSTROM, L. B. 1964. The flavor-modifying properties of disodium inosinate. Food Technol. *18*, 221.

LEE, J. H. 1979. Studies on kinetics of inosine 5′-monophosphate instability. J. Food Sci. *44*, 946.

LOCKHART, E. E. and GAINER, J. M. 1950. Effect of monosodium glutamate on taste of pure sucrose and sodium chloride. Food Res. *15*, 459.

LUCAS, D. R. and NEWHOUSE, J. P. 1957. The toxic effect of sodium L-glutamate on the inner layers of the retina. AMA Arch. Opthalmol. *58*, 193.

MAGA, J. A. 1983. Flavor Potentiators, CRC Crit. Rev. Food Sci.Nutr. *18*(3), 231.

MAGA, J. A. and LORENZ, K. 1972. The effect of flavor enhancers on direct headspace gas–liquid chromatography profiles of beef broth. J. Food Sci. *37*, 963.

MARGALITH, P. Z. 1981. Production of flavor and flavor enhancing compounds by microorganisms. *In* Flavor Microbiology, p. 256. Thomas, Springfield, IL.

MOSEL, J. N. and KANTROWITZ, G. 1952. The effect of monosodium glutamate on acuity to the primary tastes. Am. J. Psychol. *65*, 573.

OGATA, K., KINOSHITA, S., TSUNODA, T. and AIDA, K. 1976. Microbial production of nucleic acid-related substances. Kodansha Ltd., Tokyo.

ROSENBLUM, I., BRADLEY, J. D. and COULSTON, F. 1971. Single and double blind studies with oral monosodium glutamate in man. Toxicol. Appl. Pharmacol. *18*, 367.

SAINT-HILAIRE, P. and SOLMS, J. 1973. Interaction of 5′-nucleotides with gelatin—a model system for flavor potentiation. J. Agric. Food Chem. *21*, 1126.

SCHAUMBURG, H. H., BYCK, R., GERSTL, R. and MASHMAN, J. H. 1969. Monosodium-L-glutamate: Its pharmacology and role in the Chinese-restaurant syndrome. Science *163*, 826.

SCHINNELLER, D. J., DOUGHERTY, R. H. and BIGGS, R. H. 1972. Influence of selected 5′-nucleotides on flavor threshold of octanal. J. Food Sci. *37*, 935.

SJOSTROM, L. B. 1975. Flavor potentiators. *In* Handbook of Food Additives, 2nd edition. T. E. Furia, ed., p. 513. CRC Press, Boca Raton, FL.

STRONG, A. M. 1968. Flavours—their uses and abuses. Flavour enhancers. Food Technol. Aust. *20*, 574.

TAKASAKI, Y. 1978. Studies in brain lesion by administration of monosodium-L-glutamate to mice: I. Brain lesions in infant mice caused by administration of monosodium-L-glutamate. Toxicology *9*, 293.

TAKEDA CHEMICAL INDUSTRIES. The synergistic effect between disodium 5′-inosinate, disodium 5′-guanylate and monosodium glutamate. Product Bull. No. 3, Tokyo.

TAKEDA CHEMICAL INDUSTRIES LTD., Stability of ribotide against heating. Takeda Chemical Industries Ltd., Product Bull. No. 6, Tokyo.

TAKEMOTO, T. and NAKAJIMA, T. 1964A. Studies on the constituents of indigenous fungi: 1. Isolation of the flycidal constituent from *Tricholoma muscarium*. J. Pharm Soc. Japan (Yakugaku Zasshi) *84*, 1183.

TAKEMOTO, T. and NAKAJIMA, T. 1964b. Structure of tricholomic acid. J. Pharm. Soc. Japan (Yakugaku Zasshi) *84*, 1230.

TAKEMOTO, T., NAKAJIMA, T., and SUKUMA, R. 1964a. Isolation of a flycidal constituent "ibotenic acid" from *Aminita muscaria* and *A. pantherina*. J. Pharm Soc. Japan (Yakugaku Zasshi) *84*, 1233.

TAKEMOTO, T., NAKAJIMA, T., and YOKOBE, T. 1964b. Structure of ibotenic acid. J. Pharm. Soc. Japan (Yakugaku Zasshi) *84*, 1232.

TAKEMOTO, T., YOKOBE, T. and NAKAJIMA, T. 1964C. Studies on the constituents of indigenous fungi: 2. Isolation of the flycidal constituent from *Aminita strobiliformis*. J. Pharm Soc. Japan (Yakugaku Zasshi) *84*, 1186.

TERADA, J., HATA, K., MOURI, T., HASHIDA, W. and SHIGA, I. 1968. Effects of γ-irradiation on the stability of 5'-nucleotides in solution. Shokuhin-Shosha *3*, 160.

TERASAKI, M., KAJIKAWA, M., FUJITA, E., and ISHII, K. 1965. Studies on the flavor of meats: I. Formation and degradation of inosinic acid in meats. Agric. Biol. Chem. *29*, 208.

TORRI, K. and CAGAN, R. H. 1980. Biochemical studies of taste sensation: IX. Enhancement of L-[H] glutamate binding to bovine taste papillae by 5'-nucleotides. Biochim. Biophys. Acta. *627*, 313.

VAN COTT, H., HAMILTON, C. E. and LITTELL, A. 1954. The effects of subthreshold concentrations of monosodium glutamate on absolute thresholds. Paper presented at 75th Annu. Meet. of the Eastern Psychological Assoc., New York, April 10.

WAGNER, J. R., TITUS, D. S. and SCHADE, J. E. 1963. New opportunities for flavor modification. Food Technol. *17*, 730.

WEN, C., HAYES, K. C. and GERSHOFF, S. N. 1973. Effect of dietary supplementation of monosodium glutamate on infant monkeys, weanling rats, and suckling mice. Am. J. Clin. Nutr. *26*, 803.

WORDEN, A. N., RIVETT, K. F., EDWARDS, D. B., STREET, A. E. and NEWMAN, A. J. 1975. Long-term feeding study on disodium 5'-ribonucleotide on reproductive function over three generations in the rat. Toxicology *3*, 349.

WOSKOW, M. H. 1969. Selectivity in flavor modification by 5'-nucleotides. Food Technol. *23*, 1364.

YAMAGUCHI, S. 1979. The unami taste. *In* Food Taste Chemistry, J. C. Boudreau, ed., p. 33. Amer. Chem. Soc., Washington DC.

YAMAGUCHI, S. and KIMIZUKA, A. 1979. Psychometric studies on the taste of monosodium glutamate. *In* Glutamic Acid: Advances in Biochemistry and Physiology, L. J. Filer, ed., p. 35. Raven Press, New York.

10

Flavorists and Flavor Creation

One cannot consider the subject of flavor creation (imitation would perhaps be a better word as the term creation more nearly applies to abstract perfumes than to flavors) in the laboratory without first examining the type of persons who carry out this work (i.e., in the flavor industry—the flavorist; in food and related industries—the flavor technologist), the environment of a flavor laboratory and the functional interfaces which influence the development of a successful flavoring product. This discussion will be centered on the flavor industry but much can be applied to technologists involved with the flavor of end products.

A. The Flavorist

It is recognized that flavorists have to be multitalented and have a wide range of interests and expertise, not only within the flavor industry, but in the technological industries which it serves and the complementary areas of organic chemistry and process engineering which are fundamental to these industries. Not only must flavorists be scientists but they should have an artistic talent and a creative flair; they should not be bench bound but have a direct involvement in the commercial marketplace; they must be equally at home in the midst of research chemists and sophisticated instruments as with process technologists and the production plant as with potential customers; they must be informed about biochemistry, enzyme systems, sensory assessment and a host of other related topics; and they should not be parochial but have a worldwide overview, particularly with regard to legislation, social responsibilities to consumers, and safety of their products in foods.

The core of the activities of a flavorist fall within three broad areas: (a) the creation of flavorings which meet the requirements of present or potential end users of such products; (b) the participation in the marketing process for such flavorings, including technical presentation and collaborative work on specific applications; and (c) the advising on the legislative aspects of flavor usage worldwide.

Within this framework, the flavorist is called upon to collaborate with colleagues to ensure that his/her activities are not being carried out without regard to the successful commercial exploitation of the product.

1. Selection of Flavorists

The selection of young men or women for training for this demanding vocation requires careful consideration of those attributes which long experience has shown to be necessary in a successful flavorist; although it should be remembered that after selection it will take at least 10 years of study and practical bench experience in the science and art of flavor creation and application to develop these aptitudes before the trainee is ready for acceptance as a flavorist. There are no recognized university or college courses in this specialized field, indeed most food technology courses allocate a minimum of time to flavor topics. Hence, those who choose to become flavorists must be accepted by and trained within the flavor industry itself. This system has been operating successfully for very many years although the current economic climate, in forcing a reduction in staffing levels, is now placing such long-term commitments in jeopardy.

Apart from those general character attributes, which apply equally to other professions (e.g., self-confidence, imagination, articulateness, enthusiasm), the would-be flavorist must have (a) an artistic flair for this type of work which calls for a lively imagination, intuitive creativity tempered by patience, perseverence and a desire for perfection; (b) a willingness to learn and apply knowledge; (c) normal sensitivity to sensory stimuli; (d) a retentive memory, particularly of sensory experiences; (e) an orderly approach to problem solving and a tidiness in recording and presenting experimental data; (f) an ability to communicate verbally and in writing and (g) an insatiable curiosity about food and flavors.

Careful screening of candidates, by interview and simple practical tests (Heath 1981) is necessary at the outset if valuable time and costly training are not to be wasted. The preceding applies mainly to trainees within the flavor industry, but as food technologists are increasingly called upon to carry out a degree of flavor creation as part of food product development, such technologists should consider the attributes just listed and honestly assess their own potential and aptitudes before accepting innovative commitments. Some of the problems associated with new food product development and a low rate of acceptance stem from the involvement of a staff that does not have the right attitude and aptitude for this creative work.

2. Training of Flavorists

Trainee flavorists are now at a disadvantage with their older colleagues. Until comparatively recently, training was carried out on a one-for-one basis under the direct personal day-to-day guidance of an experienced flavorist. Although very active, the course was leisurely stretched over very many years during which the practical experience of the senior flavorist was transferred to the trainee who was constantly encouraged to try out new flavor experiences and to build up a personal record of them for future reference. Today the pace is a great deal faster, the need to learn so much more urgent and, in consequence, the acquired knowledge is more superficial and less wide ranging, and there is a greater reliance on memory than on written records. In most of the major flavor houses trainees are required to undertake an organized course of studies to augment their day-to-day bench work. They are encouraged to take an active part in their own and related professional societies and attend seminars and symposia related to their work. So long as the trainee's bench work is subjected to critical appraisal this system works quite well. Of course, not all flavor manufacturers can afford this degree of commitment of resources and any training becomes the responsibility of the senior flavorist who may or may not have the time to spare. More frequently it falls to the trainee to carry out a course of self-education and, as a guide, Heath (1981) and Jellinek (1985) have set out a suggested syllabus of reading which must be augmented by an organized practical study of flavoring materials, their characteristic profiles and relative strengths and the application of flavorings in end products. Training by reading is a very poor substitute for real practical guidance by an experienced flavorist. Fischetti (1984) discusses many aspects of training candidates to be flavorists.

B. Working Environment

Within the flavor industry laboratory facilities range from excellent to barely adequate; from purpose-designed and superbly equipped laboratory complexes to small all-purpose units providing only the minimum of necessities. In the majority of food factories the facilities for work on flavorings are generally part of the quality control setup although most majors do have very adequate, purpose-designed, development kitchens and associated product testing rooms, etc. Although not essential, such planned facilities do make the work of the flavorist/flavor technologist that much easier and more reliable.

A laboratory suitable for flavor creation should include the following minimum fitments: (a) a clear run of benching, preferably with an impervious surface for easy cleaning; (b) a bank of narrow shelving to carry sample bottles of flavorants and other raw materials, laid out systematically and within hand-reach of the flavorist; (c) adequate cupboard storage space; (d) natural or indirect overhead lighting—where direct sunlight cannot be avoided blinds should be fitted; (3) a stainless steel sink unit fitted with hot and cold water and with adequate drainage; (f) air conditioning—desirable but not essential and (g) if possible, a separate area for use as an office by the flavorist.

The flavor laboratory should not be used for application work for which a separate laboratory/experimental kitchen should be provided. It is often convenient for a sensory evaluation/discussion area to be sited between the two laboratories to enable joint use. A typical layout is given in Fig. 10.1.

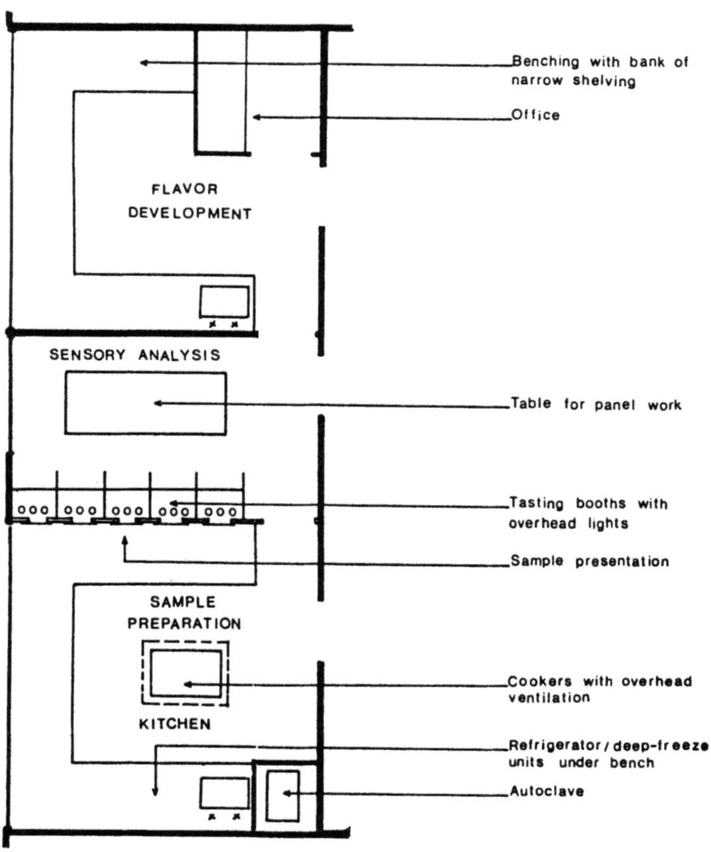

Fig. 10.1. Suggested layout for flavor development laboratories.

The flavorist must be equipped with the following: (a) a sensitive and accurate direct reading balance fitted with a tare control; (b) an adequate supply of glassware, particularly beakers and pipettes; (c) an electric hot plate/steam bath—open gas heaters should be avoided as they pose a fire hazard and (d) a small refrigerator.

The flavor application laboratory should have, in addition, refrigeration and deep-freeze units, an electric oven preferably with overhead ventilation to reduce cooking smells, a blender and/or food processor together with all the usual kitchen ancillaries. Preferably, such laboratories should be able to reproduce on a small scale the processing parameters encountered in the equivalent manufacturing department, as an ability to produce end products under the correct processing conditions offers the best opportunity for establishing reliable flavor usage data.

The aim throughout should be to create a working environment for the flavorist which provides adequate room for several flavor projects to proceed concurrently without danger of cross-contamination, a calm atmosphere as free from distractions as possible, suitable facilities for the sensory evaluation of flavorings and products with access to product application facilities.

C. Flavor Creation

1. Incentives

The food, soft-drink, confectionery and other related industries,whose products depend upon their flavor, traditionally look to the flavor industry to provide them with flavoring materials that are not only technologically compatible with their particular products but give to them a pleasingly attractive character capable of promoting optimum acceptance by consumers. It is the flavor industry's concern to anticipate its end users' requirements and, within the known constraints, provide new flavorings which will fully satisfy these needs. This external stimulus to flavor creation is not the only incentive. Through imaginative application of current research data and advances in technology, the flavor industry responds to competitive pressures and creates new flavorings which are demonstrably ahead of the competition. A third cause for flavor creation is the need to compete for existing business by the matching of competitive flavoring products—though most flavor manufacturers do not rate this activity very highly and avoid matching competitive products as much as possible. Whatever the incentive, it is necessary for the flavorist who will be carrying out the bench work to have a precise target profile which is to be imitated either in the form of a detailed description or, preferably, as a stable material sample. The flavorist must establish what flavorants best represent the different attributes of the target flavor and what the anticipated end use of the resultant flavoring will be. Once these initial parameters have been finalized, the flavorist is free to proceed with the project and apply his/her interpretation to the development process.

2. Information

The imitation of natural aromatic profiles calls for a knowledge of the biochemistry concerned, an ability to interpret available research data and an application of all relative information in order to determine what permitted flavoring chemicals may best be used. From this mass of data the flavorist must assemble the prime ingredients, assess what interactions may take place when they are blended in the presence of a solvent, determine what harmony or discords may result in the odor and flavor profiles and see how these are likely to be affected by conditions encountered during processing. To assist in the setting up of the project, the flavorist has recourse to consult the following sources of information:

a. Existing well-proven formulations on file within the flavor company. These, generally, represent many years of experience and, although once totally acceptable, may now leave something to be desired in terms of a match for the target profile but form an excellent start point.

b. Research reports on the chemistry of flavoring components in the target material. The past two decades have seen an enormous increase in characterized aromatic components of most fruits and vegetables but the relative importance of these to the profile has yet to be established and not all are permitted for use nor are available commercially.

c. Technical literature, particularly the published proceedings of symposia covering appropriate flavoring topics (Land and Nursten, 1979; Teranishi *et al.* 1981).

d. Personal records and practical experience of the odor and flavor profiles of permitted flavorants which is necessary to enable reported data to be interpreted in terms of practical aromatic effects.

During these initial stages of the project, it is necessary for the flavorist to determine the processing parameters likely to apply to the new flavoring product, to ensure that components comply with current legislation and do not impose an impossible cost burden, availability problems or manufacturing constraints which would endanger the ultimate commercial success of the product.

3. Functional Interfaces

The flavorist cannot and should not endeavor to work in a vacuum as it is all too easy to become too close to the problem, to be oversensitive to counteropinion and to have a closed mind to the suggestions of others. Success depends on the quality of interdisciplinary communications and collaboration with colleagues and others outside the company. These functional interfaces are particularly necessary with (a) the research department of the company or, failing this, with an appropriate industrial/ university research laboratory involved in a related field, in order to discuss possible new approaches to achieving the desired aromatic profile; (b) other flavorists within the company, in order to reduce bias and a too narrow view of the problem—these contacts would be normal but gaining ideas from flavorists working for other companies is not easy and may call for considerable diplomacy; (c) application technologists, in order determine the best form for the projected flavoring; (d) the purchasing department, to ensure commercial availability of proposed ingredients; (e) the quality controller, to agree on standards and methods of quality assurance as they apply to the new product and (f) the marketing/sales department to establish cost constraints which affect the commercial viability of the proposed flavoring product.

These interfaces have been critically examined by Heath (1981) and behavioral and commercial aspects of flavors are reviewed by Apt (1977).

4. New Products—The Creative Process

The process of flavor creation and new product development depends on which of the following describes the product's type: (a) the creation of a new flavoring formulation in imitation of an existing flavor (e.g., ripe mountain strawberry; bourbon vanilla, a competitive flavoring); (b) the blending of herbs, spices and other ingredients to give seasonings (e.g., curry powder, salami seasoning) and (c) the development of an overall flavor profile in an end product involving numerous intrinsic and additive flavor sources as well as those produced during processing (e.g., apple pie, pizza).

Each of these calls for a quite different approach to the creative process and will be considered separately.

a. Imitation Flavorings

The two basic methods used in the creation or reconstruction of flavors are as follows:

a. Traditional—based on the building up of the formulation by the use of known major constituents of the target flavor and the introduction of aromatic nuances by the use of chemicals having profiles reminiscent of the various attributes or notes detected in the target profile. One might think that this method is hit and miss but, in reality, it is the expertise of the flavorist in insight problem solving, involving a flexible approach and an ability to apply different but associated sensory relationships, that has made this method so effective over very many years.

b. Analytical—relying on the analysis of the target flavor by instrumental techniques, mainly GC/MS, and the use of the results to determine the composition of the imitation flavor. The quantitatively significant components of a flavor can often readily be identified but experience has shown that frequently this knowledge is insufficient to compound a satisfactory flavoring which imitates the original. The flavor attribute which typifies a particular material is often due to compounds which are present in trace amounts. The threshold values of these compounds are usually very low so that their effective contribution to the odor and flavor profiles may be disproportionately higher than their quantitative presence would suggest.

In consequence, neither method is fully satisfactory—the traditional is too dependent on the subjective experience of the flavorist and is time consuming, the purely analytical results in an incomplete or unbalanced profile due generally to volatile losses, breakdown of unstable components and the formation of artifacts during the analysis or to the noncharacterization or nonavailability of trace components which may have a significant effect on the profile. The most satisfactory approach is a combination of each using all available analytical data and technological know how combined with an artistic interpretation of the profile. This calls for the subjective judgment of the flavorist based on knowledge, practical experience of flavorants and creative flair.

The creation of a new flavoring consists of seven stages, which must be carried out systematically and each completed to the satisfaction of the flavorist before proceeding to the next stage. These are as follows:

1. Establishment of the target flavor. The target material may be either a natural fruit, vegetable, or other flavorful product or may be an existing flavoring. Whatever its source, it must be critically examined in order to define its odor and flavor profiles and the relative impact of the several attributes present. This initial sensory assessment may be carried out by the flavorist or by a team of trained assessors which should include the flavorist.

Using the target profile as a guide, the flavorist critically reexamines the material, confirms each designated attribute and associates each with known constituents or the known profiles of other flavorants not necessarily present in the target material. This reappraisal is very subjective to the flavorist who generally has his/her own system of equating odor and flavor categories with specific chemical entities. The result of this examination is a profile restated as key notes and nuances in terms of the chemical(s) directly associated with each (Seidman 1979).

2. The assembling of data on the target material and likely chemical components, essential oils, etc.

3. Preparation of a trial blend. The flavorist prepares a first blend of ingredients. This is most unlikely to be sufficiently near to the target profile and may indeed be something of a caricature as the main aromatic features are likely to be exaggerated and the minor nuances swamped or not even present. This first rough model must then be continuously modified by trial and error until a reasonably close match is achieved. Before the evaluation of any trial mixing it is necessary to allow the mixture to stand for a period in order to let the components blend together. If the assessment is carried out too soon wrong judgments may result.

4. Subjection of the created flavor to appraisal by other flavorists and GLC examination against the target material.

5. The preceding stages are repeated as often as is necessary to achieve a totally acceptable composition. The gas-chromatogram of the final blend is likely to be very similar to that of the target but not necessarily identical.

6. Establishment of application data. After an adequate period to enable the blend to mature, it is submitted for evaluation in either a nominated product or range of end products. This stage will prove the efficacy of the flavoring in various media and draw attention to any changes which result from processing.

7. Commercial exploitation of the product. On the completion of the creative process, the product has to be offered for sale either to a specific customer or to food manufacturers generally. The flavorist may be called upon to make a presentation of his/her new product and to demonstrate its value in appropriate end products.

The flavorist who created the new flavoring is also most likely to be involved in establishing quality norms and to be assisting in establishing its production (Heath 1981).

b. Blending of Seasonings

The blending of herbs, spices and other food ingredients (e.g., salt, MSG) into seasonings calls for quite a different expertise on the part of the flavorist. Within the industry, such an activity is not normally carried out by highly trained flavorists but by flavor technologists more closely concerned with flavor applications in food and often working in direct collaboration with technologists in the user industry. The methods used in building up an acceptable seasoning are roughly the same as those for the creation of a flavoring but the number and nature of ingredients is quite different.

Generally, ground herbs and spices and/or dispersed or encapsulated spice extractives are blended together to suit some specific application dictated by the end-product formulation and its method of manufacture. Rarely does the flavoring effect of any one herb or spice suffice to make a well-rounded and attractive flavor in the end product, but, traditionally, seasonings are relatively simple compositions which may include other nonspice ingredients (Table 10.1). Skill in blending seasonings is only acquired by practical experience of the raw materials available, their relative flavoring strength and qualities as well as their interrelationships with the profiles of the main dietary items in the end product. This aspect of seasoning formulation is discussed in greater detail in Chapter 6.

The blending process parallels that used in the creation of flavorings but, unlike chemical flavorants, herbs and spices do not have odor and flavor characters capable of concise description but are themselves complex. They have profiles comprising numerous attributes which have different sensory impacts perceived in different sequences (Heath 1978).

The stages of building up a seasoning depend upon whether one is attempting to match an existing seasoning or formulating one for a designated end use. In the latter case, the process comprises the following:

1. Establishing the target profile. This implies a knowledge of the composition and processing parameters of the product in which the

Table 10.1. Ingredients Used in Seasonings

Ground herbs and spices
Herb and spice essential oils, usually plated or dispersed on an edible carrier
Encapsulated herb and spice extractives
Flavor enhancers
Preservatives (as permitted)
Food colors (as permitted)
Food acids (e.g., citric acid)
Imitation flavorings (e.g., smoke) (as permitted)
Salt, dextrose, sucrose, etc.
Artificial sweeteners (as permitted)
Milk whey, sodium caseinate
Hydrolized vegetable protein, yeast extract, etc.

seasoning will be used. The flavor profile of the basic unseasoned product after cooking significantly determines the type of seasoning required. In general, the weaker the flavor of the basic product, the lower the level of added seasoning required (e.g., a fish product requires only a lightly balanced herbal seasoning; a product based on beef requires a heavy, spicy seasoning with a distinct pungency).

2. Deciding the balance required between the flavoring attributes of the seasoning ingredients. This subject has already been discussed but ingredients should be assembled in the following flavor groups: (i) light, sweetly herbaceous, (ii) medium aromatic, (iii) heavy, full-bodied, spicy and (iv) pungent. These should be supplemented, as desirable, with nonspice materials such as MSG, hydrolyzed vegetable protein, yeast extract, salt, and dextrose.

3. Preparation of the preliminary seasoning mix. This involves the creative abilities of the flavorist who decides on the most acceptable balance between the various flavoring categories.

4. Evaluation of the mix, in a hot white sauce or neutral soup base (Heath 1978) for its odor and flavor profiles.

5. Repetition of these stages until a well-balanced seasoning is achieved.

6. Evaluation of the seasoning in an end product processed under normal conditions. Seasoning effects are significantly altered by cooking and the presence of other materials, particularly meat. If the end product is to be deep frozen then this test must include a period of deep freeze of at least 24 h.

The matching of an existing seasoning involves first an agreement on the target sample and its profile. This might sound obvious, but whereas liquid flavorings are very stable and show little change over long periods, blended seasonings have only a limited shelf life and lose their flavoring power. If the sample is old or has been incorrectly stored the development of a matching product will be a waste of time. Once this is agreed it is necessary to carry out some preliminary analysis to determine salt content, coloring matter, pungency level, MSG, etc. Thereafter the building up of the matching seasoning follows the usual lines of blending appropriate ingredients and repeated testing until a matching product is achieved. The nearness of trials to the target material is assessed in a white sauce or neutral soup, using a triangle test procedure, adjustments being made until an indistinguishable product is obtained.

Again, the flavorist will be involved in the marketing process as well as in quality assurance and manufacturing procedures applying to the new product.

D. Flavor Development in End Products

The development of a well-balanced and attractive flavor in comestibles is complicated not only by the diversity of the raw materials available but by the need to recognize that the acceptability of the end product rests with the consumer. In this respect, the opinion of the flavor technologist or product developer may be at variance with public taste so that evaluation of his/her product rests, ultimately, not with an individual or a small trained panel but with a very large untrained body of consumers. The whole subject of food product development lies outside the scope of this book but certain aspects directly associated with flavor are worthy of comment.

1. Nature of Flavor

Flavors fall roughly into one of the following four categories: (a) natural—fresh fruits, raw vegetables, herbs and spices and products made from them; (b) processed—arising in end products as a result of cooking, enzymatic action, fermentation, etc.; (c) compounded—formulated to produce a desired flavoring effect using permitted chemicals and other specifically identified flavoring materials of natural or synthetic origin and (d) degraded—off-flavors produced by oxidation, etc.

Flavor cannot be considered as a static attribute as it is in dynamic equilibrium and everyone recognizes how it may change. In raw materials, it changes during growth and maturation in both animal and plant sources. The flavors of unripe, fully ripe and overripe fruits display wide differences; young and old meat is markedly different in both flavor quality and strength; cooked flavors depend on the degree of cooking and the temperature involved, the resulting flavors being very different from the raw start material, and the flavor of a freshly cooked product may become quite different on storage. Flavor changes may be due to transformation or losses arising from physical, chemical or biological conditions. Flavor substances, being mainly volatile, may gradually evaporate from food due to heat or exposure. As aromatic compounds volatilize at different rates and are perceived at different times, some being fleeting, others persistent, changes in their relative concentration lead to an upset of flavor balance and often to the production of unpleasant and even objectionable off-notes. Losses in flavor may also occur due to absorption onto solid surfaces within the product and to differential partition between aqueous and lipid phases by which some flavor elements may be removed from the sensory impact of the system. It is for such reasons that the formulation and production of foods is far from a precise science. It is this variability which calls for considerable effort on the part of the food processor to maintain consistency of product, particularly its flavor, in the interests of consumer satisfaction.

Figure 10.2, originally proposed by Jaubert (1980), sets out a schematic representation of the various flavor changes which may occur in a food product.

2. Flavor Control

The development of the flavor profile of an end product is really a process of profile alteration. Any food product, at the outset, has a basic profile comprising, in the main, items of low flavor intensity included

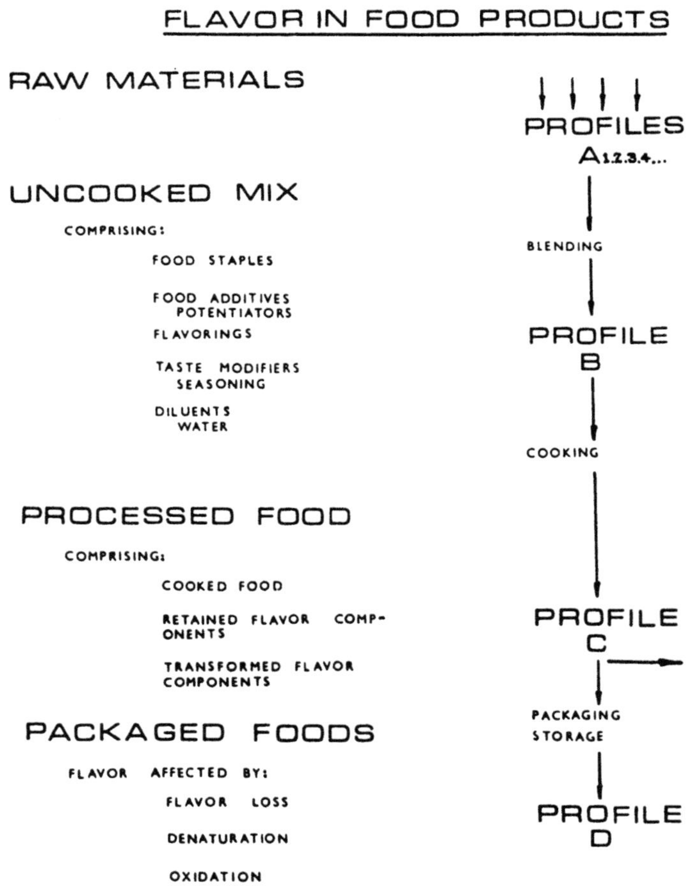

Fig. 10.2. Schematic representation of various flavor changes which may occur in food products.
From Heath (1981a) and Jaubert (1980)

primarily for their nutritional or textural contribution but also containing flavorless precursors which are only activated and produce flavor during heat processing or cooking. Other components may be added at this stage to modify the flavor profile, to change its intensity or even to impose a totally new profile. The culinary art involved in food product development is concerned with creating this initial balance so as to achieve the desired flavor profile at the end of the processing/cooking cycle. This involves the following four quite different sets of circumstances: (a) the selection and balancing of existing or potential flavors, which must be done within the constraints of nutritional necessity and the concept of the end product; (b) the adjustment of the resulting flavor profile to suit particular palates or consumer anticipations; (c) correction to overcome some preexisting flavor defect(s) and (d) the imposition of an entirely new flavor to the end product, where basic materials are either bland or flavorless.

The techniques employed are numerous but may involve either of the following:

a. Flavor intensification—comprising the following: (i) the concentration or addition of the same flavor (e.g., tomato puree added to canned tomatoes); (ii) the incorporation of ingredients of similar or supportive flavor profile (e.g., use of low levels of garlic to enhance the flavor of onions); (iii) the addition of an imitation flavoring, such as a topping note, to enhance the profile of a weak constituent (e.g., the addition of imitation cherry flavor to a cherry pie filling) and (iv) the judicious use of flavor potentiators or enhancers (e.g., the addition of MSG or ribonucleotides to canned meats).

b. Flavor suppression—involving the following: (i) removal of unwanted sensory characters, usually by heat treatment or the incorporation of some absorbent material such as starch; (ii) the neutralization or conversion of flavor by adjustment of the salt/sweet/acid balance or by marinading; (iii) by dilution and (iv) by masking or disguise with a much stronger flavoring.

The whole field of flavor control is one of subjective judgment and it is for this reason that the food industry relies heavily on consensus opinion derived from large consumer panels (Franzen and Kinsella 1975).

The creation and preparation of most food products, beverages, confectionery, etc., is far from simple, particularly in achieving optimum flavor acceptance. Increasingly, product development, whether it be the creation of a new flavoring, the blending of a new spice seasoning or an entirely new food product calls less for individual flair and panache and more for teamwork of the highest order and the coordination of all the multidisciplined expertise available, assisted by computerized data processing (McLellan and Cash 1983).

E. Sensory Assessment

In any flavor development project it is necessary to carry out a continuous sensory assessment of trial samples in order to establish the nature and extent of differences between the trial and the target flavor profile. The purpose of this section is to draw attention to the preparation of the material for examination, to the techniques of odor and flavor assessment and to certain precautions which must be taken in the testing of high flavorful materials. The methodology of sensory analysis is described in Chapter 14, as part of quality assurance.

1. Preparation of Sample

The degree of preparation necessary is determined by the nature of the material under examination: (i) unmilled spices and broken herbs—reduce to a fine powder having a particle size of not greater than 250 μm; (ii) spice oleoresins and essential oils—disperse on a suitable carrier prior to tasting but evaluate directly for odor; (iii) concentrated flavorings—dilute to an acceptable level prior to tasting but evaluate directly for odor; (iv) aromatic chemicals—dissolve in ethanol and taste at a suitable dilution as just stated; and (v) end products—prepare directly for consumption as appropriate.

In the case of concentrated preparations, it is necessary first to dilute these to an acceptable level to permit direct evaluation for tasting. This is achieved in one of the following three ways: (a) admixture with a simple chemical substance (e.g., dextrose, sucrose, lactose or salt) as appropriate; (b) dilution with an acceptable solvent (e.g., ethanol) or (c) incorporation into a neutral food product or carrier at the correct dosage rate. The carriers recommended for this purpose include the following: (i) Neutral soup comprising 60% corn starch, 22% caster sugar and 18% salt, used at 4.5% with boiling water and thickened by cooking for 1 min. The sample to be evaluated is added to the prepared soup. This medium is particularly suitable for spices, culinary herbs and blended seasonings. (ii) White sauce, made with either butter or margerine, plain flour and milk. (iii) Sugar syrup—10% sucrose in potable water. This medium is the standard one used for the evaluation of flavorings, essential oils, and sweet spices (e.g., ginger, cinnamon, cassia). (iv) Reconstituted mashed potato, made in accordance with the manufacturer's instructions but omitting any butter. This is used in the evaluation of onion, garlic, paprika and blended seasonings. (v) Fondant, using an instant fondant base reconstituted as required. The flavored base is formed in cornstarch molds. (vi) Sugar boilings, using the following base: sucrose 120 g, water 40 ml, glucose syrup 40 g. The glucose and water are boiled and mixed until dissolved, the sugar is added and boiled rapidly at 152°C (300°F), the product then being removed from the heat and any flavor, color or citric acid stirred in. The mix is then poured onto a greased slab, allowed to cool and passed through drop rollers to form the product. (vii) Pectin jellies, using the following base: sucrose, 100g; water, 112.5 ml; glucose (43 DE), 100 g; citrus pectin (slow set), 6.25 g; citric acid solution, 50%, 2 ml. The pectin and sugar are mixed together and slowly added to the water at 77°C (170°F) with continuous stirring. The temperature is raised to 106°C (220°F) and the product removed from the heat; the flavoring, any color and the acid solution are quickly stirred in and the mass poured into cornstarch molds. The poured product is allowed to stand for 12 h before removing from the molds. (viii) Carbonated soft-drink base, using the following general formulation which may be modified to suit particular products: sodium benzoate solution, 10%, 0.25 ml; citric acid solution, 50%, 1 ml; 8-lb sugar syrup to make 160 ml. The flavor is stirred into the base at the appropriate dose level and the concentrate diluted 1 plus 5 with potable water prior to bottling and carbonation. (ix) Milk, a useful carrier for the preliminary evaluation of flavorings for ice cream and other dairy products. The milk may be sweetened with 8% sucrose if desired.

The base most appropriate to the material under examination is really one of preference, but the American Society for Testing and Materials (ASTM 1968) lays down the following guidelines for the preparation of samples for flavor evaluation: (a) The method of preparation shall impart no foreign tastes or odors to the sample. (b) All samples shall be prepared in a similar manner. (c) For difference testing, the method used shall be that most likely to permit the detection of a difference, keeping the preparation as simple as possible. (d) Avoid preparations which may add flavor to the sample (e.g., frying). (e) For preference testing, use a method suitable to the normal use of the product. (f) Use any supplementary food carriers as may be judged necessary (e.g., ice cream for a topping syrup evaluation).

The aim is to present the assessor with the sample in the best form and under the best conditions to enable a reliable judgment to be made.

2. Odor Evaluation

Although obvious, the conditions for odor evaluation are most important and particular care is necessary to ensure that no extraneous odors interfere with the judgment of the assessor. Unfortunately, all too often scant regard is paid to this simple precaution. Complete concentration on the part of the assessor is essential and even minor distractions can significantly reduce the efficacy of the judgment.

The prime requirement for all smelling tests is the presentation of the sample to the nasal receptors in a consistent and preferably standardized manner. The temperature and relative humidity of the environment are important and, if possible, these should be maintained in a steady state throughout the testing period. The actual smelling process is one of personal preference but may be either natural breathing using both nostrils, forced sniffing using one nostril, or "blasting" from an automatic dispenser.

Presentation of the sample to the assessor is generally carried out in one of the following two ways:

a. Liquid materials—using a standard smelling strip (usually 140–150 mm long by 5–6 mm wide), cut from suitable absorbent paper free from any extraneous odor. These are dipped into the sample to a fixed level (1–2 cm) and the vapor inhaled through the nostril(s) from a distance of about 1 in. from the nose for a period of 2–3 sec at a time. The sequence of smelling should be rhythmic. It is bad practice to compare two odors by using first the left hand-nostril and then the right-hand nostril. Similarly it is bad practice to smell dipped strips mounted together on a spring clip. Such practices are frequent but do not result in reliable judgments.

b. Odorants in sprays—using standard smelling cards which are cut from heavier absorbent paper; otherwise used as in item a. The spray must be applied from a standard distance for a standard time to ensure uniformity of sample dose.

Odor evaluation is most reliable when confined to the comparison of one sample directly with another of similar identity and strength. It is least reliable when either multiple comparisons involving more than three samples are made or no comparison is possible, the assessor having to rely on remembering what the odor should be.

The odor pattern over a period of time is often most informative and assessment should be carried out at the following intervals: immediately after dipping; after 1 h, 2 h and 6 h; after standing overnight or for a period of not less than 18 h. During the intervals the strips should be mounted in suitable clips, well separated from each other, at room temperature and away from any high concentration of extraneous odors.

An alternative to smelling strips is the presentation of the sample in a closed vessel in which the vapor is allowed to reach equilibrium prior to smelling. In the case of liquid materials, oleoresins, etc., use 5 ml/g in a 100-ml stoppered container. For dry materials, place a representative sample in the testing vessel so that the level is about one quarter the total height. The prepared container is allowed to stand at a steady temperature (20–25°C) for at least 1 h before evaluation is carried out. This must be by controlled sniffing so that only the vapor phase is inhaled. The vapor must be allowed to equilibrate between repeated smellings and an interval of 5 min is usually sufficient for this.

The odor of encapsulated flavorings cannot be assessed directly. Such products are evaluated by taking 5 g of the material and adding 25 ml of boiling water, swirling the contents in a loosely covered vessel and allowing it to stand to attain equilibrium. After about 20 min the temperature will have fallen to about 50°C and the odor can then be assessed.

3. Flavor Evaluation

If anything, the conditions for the evaluation of flavor are more critical than those necessary for reliable odor assessment. Wherever possible, such tests should be carried out in a room set aside for the purpose but all too often taste tests are conducted in the open laboratory alongside other activities. The absence of distraction is all important to the achievement of meaningful results and experience shows that the sensitivity of tasters is significantly lowered when other tasters are present, as in a panel session, and when the evaluation is carried out under noisy or hurried conditions. Even if purpose-designed tasting booths are not possible, the provision of individual tasting bays is highly desirable. These can readily be made from interlocking hardboard screens which can be mounted temporarily on a table or bench as required and stored away when not in use (Jellinek 1985).

The following additional precautions should be taken in the conduct of flavor evaluation.

a. Foreign odors. Assessors are frequently the source of foreign odors arising from their use of perfumed soap, perfumes, toiletries, etc. Prior to a tasting session, all assessors should wash their hands in nonperfumed soap or detergent and refrain from the use of odorous cosmetics, etc. Smokers are a very bad source of foreign odors particularly detectable by the nonsmoker. It is strongly recommended that smokers be requested to refrain from smoking at least 30 min before the tasting session.

b. Lighting conditions. Assessors must be comfortable (but not too comfortable) during the test period and too bright or too low a level of lighting can affect judgment. The use of abnormally colored lighting may be useful in the elimination of differences in color and/or shade in samples but may have an inhibiting effect upon assessors. The need to eliminate irrelevant characteristics must be carefully controlled. End products are best judged in their normal state even if extraneous characters are present. If necessary, these can be declared to assessors before the test and their relevance or irrelevance determined.

c. Utensils. All utensils must be scrupulously clean and free from contaminating odors or tastes. Preferably, identical containers should be used for sample presentation. Disposables should be carefully examined for foreign odors and rejected after the test.

d. Water. Always use potable and not distilled or demineralized water in the making up of samples. Any flavor detectable in the municipal water supply will be common to all samples and can be disregarded.

e. Test conditions. Samples must be compared at the same time under the same conditions. Assessors should receive sufficient amounts of each sample, at the right temperature, to allow for several taste trials until a judgment is reached. Whether a sample should be swallowed or masticated and rejected is a matter for the test controller to stipulate.

f. Sample coding. All tests must be designed to ensure random presentation and should be identified with a three-digit random number code. Simple letters, such as A, B, C, must be avoided as these can introduce an unwarranted judgment factor.

g. Results. Findings or judgments should never be discussed until after the completion of the test by all assessors. Preferably, results should be written on predistributed pro forma involving minimal writing by the assessor.

4. Palate Clearing

There is considerable disagreement over the need or advisability of palate clearing between samples or taste trials. With end products there is little need for this practice but in the assessment of powerful flavorings it may be desirable to recondition the palate after each tasting so as to reestablish the initial conditions as much as possible. Each assessor should be allowed to develop his/her own technique which will be effective so long as it is applied consistently.

The following materials are the most widely used for this purpose.

a. For most fruit flavors, herbs and mild seasonings: Water (potable or carbonated), salt-free crackers, puffed rice, fresh bread (white crust free), skimmed milk and buttermilk.

b. For flavorings having a bitter, strong or oily character and/or aftertaste: Diluted lime juice, apple slices and apple juice (slightly sweetened).

c. For strong and pungent spices: Natural yogurt, diluted syrup (10% sucrose) and mashed potatoes.

5. Special Precautions

One can readily assess highly aromatic materials or very flavorful end products by obeying the following three prime rules:

a. The material to be examined must be diluted to a suitable level so that the nose and palate are not swamped.
b. Due time must be allowed between samples and between tests for the senses to recover.
c. Repeat tests are carried out in reverse order on a separate occasion to eliminate any carryover effects.

So far as item a is concerned, the degree of dilution will be dictated by the strength of the initial sample and the test medium chosen, but the following are general recommendations for the dilution of a wide range of materials:

Aromatic chemicals: Dissolve 0.1 g. (0.1 ml for liquids) in 5 ml of 95% ethanol; add 0.02 ml of the solution to 100 ml of test medium.
Essential oils: Mix 0.25 ml with 10 ml of 95% ethanol; add 0.05 ml of the solution to 100 ml/g of the test medium.
Oleoresins: Dissolve 0.25 g in 10 ml of 90% ethanol; shake well and allow any insolubles to settle; add 0.5 ml of the clear supernatant solution to 100 ml/g of the test medium. Alternatively, take 0.25 g and disperse this on 100 g of salt, using a suitable mixer; take 0.5 g of the dispersion for tasting in the test medium.
Flavorings: Dissolve 0.02 ml (0.02 g for powder flavorings) in 50 ml/g of the test medium.

Some products have unusual organoleptic properties which may pose problems. The following require particular attention:

a. Pungency

Pungency is detected as an irritation or pain in the oral cavity and/or upper throat. The location of the stimulus is often an indication of its source and at normal dosage levels, spices have the following pungent effects. Pepper is felt along the front edge of the tongue; ginger, along the roots of the tongue; capsicum, deep in the throat; mustard, throughout the whole mouth; horseradish, similar to mustard.

Pungency also creates a difficulty in strength and quality ratings due to a rapid buildup of the stimulus which effectively swamps and tires the sensory area so that later samples have no effect. In consequence, the evaluation of such materials as curry powder, piquant seasonings containing high levels of capsicum, as well as the individual spices and/or spice products concerned, must be carried out so as to reduce this effect to a minimum. This involves (a) evaluation at low concentrations near to the threshold of detection; (b) allowing sufficient time for the sense organs to recover; (c) repeat testing of the samples in reverse order of presentation and (d) limitation of the test to the assessment of a single attribute (e.g., comparative pungency).

b. Bitterness

Spices such as nutmeg and mace are very bitter and this attribute can influence their evaluation. As the palate is very sensitive to bitterness, its relative assessment must be carried out at low concentration.

c. Garlic and Onions

The assessment of any of the alliaceous vegetables, of which garlic and onion have the strongest flavors, is difficult due to the carryover of flavor in the mouth. The sulfur compounds responsible are very persistent and garlic, in particular, may taint the breath for hours after consumption. In addition, the test room(s) may carry a residual odor of garlic or onion unless the air conditioning is effective. If many such tests are carried out, a separate area is advisable.

Samples must be prepared well away from the test area and those involving garlic or onion oils, preferably in a well-ventilated fume-cupboard. Prepared samples must be well covered to minimize atmospheric contamination.

Onion and garlic powders, or products containing high percentages of these materials, should be evaluated in reconstituted mashed potato, using diluted lime juice as a palate clearer. Such materials are best tested in the cold as aromatic differences are much less noticeable in a hot medium.

d. Physiological Effects

Certain spices and essential oils have a marked physiological effect on the nose and palate and this can interfere with their quality assessment (Heath 1978). Clove and allspice contain high levels of the phenol, eugenol, which is a mild anesthetic, and tannins, which are astringent. Oil of peppermint, characterized by a high content of menthol, has a strong cooling effect in the mouth and is somewhat anesthetizing in the nasal cavity. These effects can be minimized by careful dilution and by allowing adequate recovery time between the presentation of samples.

F. Conclusion

Although flavor is very important, many things are necessary to ensure the success of a food product, and articles in an overview on sensory testing in product development and optimization (Schaefer 1981; Rose 1981; Kraft 1981; Pangborn 1984; Gridgeman 1984) draw attention to the main procedures currently used by food manufacturers to establish consumer views as an essential input to their development programs.

References

APT, C. M., ed. 1977. Flavor: Its Chemical, Behavioral and Commercial Aspects. A. D. Little Symp., 1977. Westview Press, Boulder, CO.

ASTM. 1968. Manual on sensory testing methods. ASTM Special Tech. Publ. No. 434. Amer. Soc. Testing Materials. Philadelphia, PA.

FRISCHETTI, F. JR. 1984. Training candidates to be flavorists who can make quality flavors. *In* The Development and Application of Natural and Artificial Flavor Systems, E. J. Merwin, pp. 7–21. Allured Publishing, Wheaton, IL.

FRANZEN, K. L. and KINSELLA, J. E. 1975. Physiological aspects of food flavouring. Chem. Ind. (June 21), 505–509.

GRIDGEMAN, N. T. 1984. Tasting panels: Sensory assessment in quality control. *In* Quality Control in the Food Industry, S. M. Herschdoerfer, ed., pp. 299–350. Academic Press, Orlando, FL.

HEATH, H. B. 1978. Flavor Technology: Profiles, Products, Applications. AVI Publishing, Westport, CT.

HEATH, H. B. 1981. Source Book of Flavors. AVI Publishing, Westport, CT.

HEATH, H. B. 1981a. Flavours in food products—art or science? Royal Soc. Health J. *101*(1), 6–12.

JAUBERT, J. -N. 1980. Arôms et aromatisants. Indust. Aliment. Agricol. (May), 453–456. (In French.)

JELLINEK, G. 1985. Sensory Evaluation of Food: Theory and Practice. Horwood, Chichester, UK.

KRAFT, L. 1981. Focus groups: Letting consumers think about your new product ideas. Food Technol. *35*(11), 70–72.

LAND, D. G. and NURSTEN, H. E., eds. 1979. Progress in Flavour Research. Appl. Sci. Publ., London.

McLELLAN, M. R. and CASH, J. N. 1983. Computerized sensory evaluation: A prototype data collecting system. Food Technol. *37*(1), 97–99.

PANGBORN, R. M. V. 1984. Sensory techniques in food analysis. *In* Food Analysis: Principles and Techniques, D. W. Gruenwedel and J. R. Whitaker, eds., pp. 37–94. Dekker, New York and Basel.

ROSE, S. 1981. The laboratory test market: An approach to estimating product potential. Food Technol. *35*(11), 73–74.

SCHAEFER, E. E. 1981. Sensory discovery, sensory scale-up and sensory cycles. Food Technol. *35*(11), 65–66, 69.

SEIDMAN, M. 1979. Sensory methods in the work of the flavour chemist. *In* Progress in Flavour Research, D. G. Land and H. E. Nursten, eds., pp. 15–26. Appl. Sci. Publ., London.

TERANISHI, R., FLATH, R. A. and SUGISAWA, H., eds. 1981. Flavor Research, Recent Advances. Dekker, New York and Basel.

11

Flavor Production

Flavorings are commonly divided into classes based on the physical states, i.e., liquids, emulsions, pastes or solids. Liquid- and emulsion-type flavors are used in liquid food products. These food products may be aqueous or fat based with the corresponding flavor being either in an aqueous (alcohol or propylene glycol) or fat (benzyl alcohol, triacetin, triethyl citrate or vegetable oil) solvent. Paste flavorings are a mixture of components which typically are either very high in natural flavorings (e.g., botanical extracts) or components which differ in solubility. An example of a paste flavoring would be a cheddar cheese flavor. The top-notes of cheddar are composed of rather light carbonyls and fatty acids. The base-notes are made up of various salts and amino acids. Since these components are mutually insoluble, it is virtually impossible to make a complete cheddar flavor without it being a paste.

Solid flavorings may again be products very high in natural flavorings or flavor chemicals dispersed in a solid fat, dry gum or starch matrix. This chapter will discuss the methods used to produce liquid, emulsion and dry flavoring materials.

A. Liquid Flavorings

The liquid flavor production area is no more than a chemical blending operation. The production area consists of a store of raw materials and weighing, mixing and packaging equipment. While some of the less stable chemicals are held in large drive-in coolers and the very potent flavor chemicals (e.g., mustard oil) are held in well-ventilated rooms, the vast majority of chemicals are held at room temperature. The flavor compounding area also contains a variety of mixing tanks being stirred with air driven motors to minimize the danger of fire or explosion. These tanks will range in size from only 5 gal. up to several hundred gallons. The flavor compounder will take a formula and weigh each ingredient into the mixing tank. Order of mixing is often important since solubility and reactivity may depend on the particular chemicals present and their concentration. Obviously, the solvent(s) are typically weighed in first so that these problems are minimized. Once the formulation has been completed, a sample is taken for quality control. When quality control has given approval of the lot, it is packaged for transport to the customer.

It is of interest that sanitation is of little concern in this area. The products being handled will not support microbial growth and, more than likely, are toxic to microorganisms. The primary reason for cleanliness in a liquid formulation area is to avoid contamination of one flavor with another. Due to the very low flavor thresholds of some flavor constituents, cleanliness is essential.

B. Emulsions

Emulsions typically are used in the flavor industry to carry product flavor (e.g., beverage or bakers' emulsions) or impart turbidity (cloud emulsion) to a product. The emulsions are virtually always an oil (or flavor) in water (continuous phase) emulsion. Emulsions offer a simple means of incorporating an insoluble flavor into an aqueous food system. They also offer cost advantages since water is a very inexpensive solvent. The primary disadvantages of emulsions are that they are physically unstable (either ring at the top of a beverage or form a sludge at the bottom) and require preservation from microbial spoilage.

Emulsions are typically classified on the basis of particle (or droplet) size. A microemulsion is one that has an average particle size $<0.1\ \mu m$. This emulsion is clear in appearance. This very small particle size is typically achieved only with the use of very effective emulsifiers (e.g., Tween 80). The microemulsions are used in rather highly flavored products such as toothpaste or mouthwash since the high level of emulsifier often imparts an off-flavor to less flavorful products.

Beverage and cloud emulsions are usually composed of an oil phase (terpenes or vegetable oil for clouds), water and emulsifier such as gum arabic (Table 11.1). The beverage emulsion is a flavoring used to flavor a beverage while a cloud emulsion is used specifically to impart turbidity to the beverage. Typically a beverage contains too little real juice to impart the turbidity associated with the natural counterpart (e.g., lemonade with fresh lemon juice vs. a lemon oil product). Therefore the beverage manufacturer may use a beverage emulsion to impart flavor and some turbidity and additional cloud to get the remaining turbidity. Beverage and cloud emulsions present the greatest problem with physical stability since particle size must be large enough for the solution to appear turbid. Stokes' law defines the parameters which determine sta-

Table 11.1. Examples of Food Flavors Emulsions

Ingredients	Emulsion Type (%) Beverage	Cloud	Baker's
Xanthan	—	—	0.70
Gum arabic	9.45	25.00	2.00
Citrus oil	4.00	10.00[a]	10.80
Sodium benzoate	—	[b]	0.20
Citric acid	—	[b]	0.39
Brominated vegetable oil	0.50	0.45	—
Glyceryl abietate[c]	3.30	5.00	—
Propylene glycol	9.45	—	—
Water	73.30	59.55	85.91

[a] Vegetable oil or orange terpenes would be used.
[b] Optional ingredient.
[c] Also called "ester gum," trade name of Hercules, Inc., Wilmington, DE.

bility of these emulsions as follows

$$V = \frac{2gr^2(\rho_2 - \rho_1)}{9\eta}$$

where V is velocity of separation; g, acceleration of gravity; r, radius of droplet; ρ_2, density of continuous phase; ρ_1, density of dispersed phase and η, viscosity of continuous phase.

An examination of Stokes' law indicates that the only parameters which can be varied in beverage applications are the radius of the droplets and the density differences. Viscosity is of significance to the cloud or beverage emulsions since these products are quite viscous, however, the finished beverage cannot be viscous. Radius of the dispersed droplets should be as close to 1 μm as possible. This is not easily accomplished with only gum arabic or a modified starch as the emulsifier (King *et al.* 1976). Therefore, despite the use of efficient homogenizers (two stage), particle size most often ranges from 0.5 to 7 μm in diameter which is too large to promote stability in itself (Fig. 11.1). The other parameter, density, may be manipulated via the use of weighing agents. The commonly used weighing agents and legal limits are presented in Table 11.2. A flavoring such as orange oil or orange terpenes has a specific gravity of 0.84 while a vegetable oil will have a specific gravity of about 0.94. Since a carbonated beverage will have from 0 to 13% sugar depending on the type of beverage and whether it is artificially sweetened, the specific gravity of the finished beverage will range from about 1.00 to 1.06. Prior to 1970, brominated vegetable oil (BVO) was used to increase

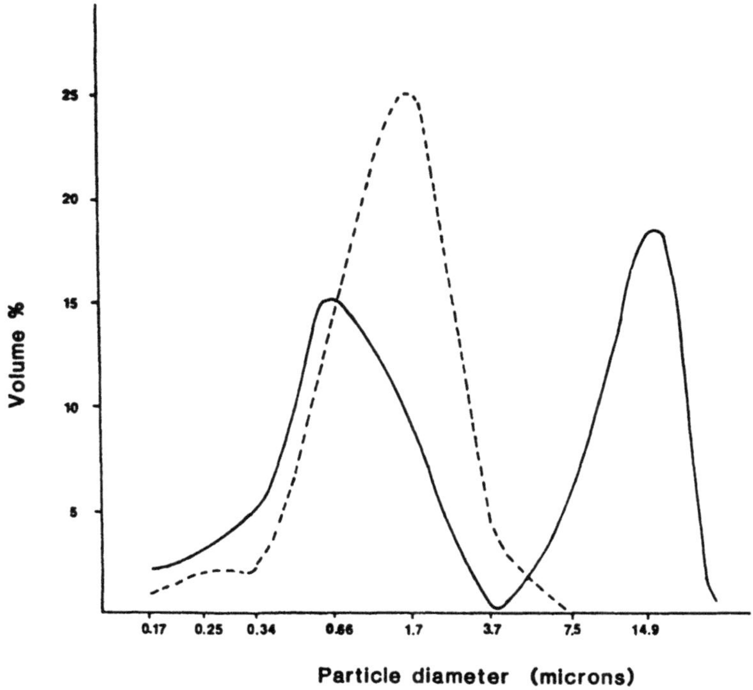

Fig. 11.1 Particle size distributions of two commercial cloud emulsions: —, emulsion A; - - -, emulsion B.

the specific gravity of the oil phase. However, in 1970, BVO was legally limited to 15 ppm in the finished product. This meant that emulsion manufacturers could no longer use enough BVO to adjust the density of the dispersed phase to that of the continuous phase. Manufacturers had to search for substitutes for BVO and came up with glyceryl abietate,

Table 11.2. Weighing Agents for Flavor Oils Used in Emulsions

Ingredient	Specific Gravity	Legal Status[a]
Brominated vegetable oil (BVO)	1.33	15 ppm
Glyceryl abietate (ester gum)	1.10	100 ppm
Sucrose acetate isobutyrate	1.15	10–50 ppm[b]
Glycerol tribenzoate	1.14–1.20	GMP[c]
Damar gum	1.05	[b]
Wood rosin	1.06	[b]

[a] The ppm in finished product.
[b] Not permitted in the United States, but permitted in some other countries.
[c] Total amount of this and other weighing agents may not exceed 0.5%.

sucrose acetate isobutyrate (SAIB) and benzyl esters. As can be seen in Table 11.2, none of these substitutes has as high a density as BVO. Therefore, larger proportions of the weighing agents have to be used. In fact, not enough of these weighing agents can be used due to solubility and legal limits to bring the density up to the density of the beverage. The manufacturer can generally add only enough weighing agents to bring the density of the oil phase up to about 1.02. While this is adequate for stability in artificially sweetened beverages, it is too low to obtain good stability in naturally sweetened products.

Baker's emulsions are very viscous macroemulsions. They typically are composed of the flavoring, gum arabic, gum tragacanth, propylene glycol and water. The amount and proportion of gums are chosen to provide an emulsion which is stable primarily due to its high viscosity. The viscosity also facilitates incorporation of the flavor into a viscous filling or dough. Propylene glycol is added as a microbial preservative. A typical formula for a baker's emulsion is presented in Table 11.1. There are no significant problems with stability of these emulsions because of their final applications.

C. Dry Flavorings

Most flavor chemicals are liquids, however, many products such as gelatins, dry beverage mixes, cake mixes, and frosting mixes, require that the flavor be a free-flowing solid. This requires that the liquid flavor chemicals be either adsorbed on a dry carrier or be encapsulated in an inert edible polymer such as gum arabic or a starch-based product.

1. Extended or Plated Flavors

The simplest process of producing a dry flavoring is to "extend" the liquid flavor on a flavor base. This base may be salt, sugar, silicates or cyclodextrins, depending upon use. It is not uncommon for a sausage manufacturer to disperse essential oils and oleoresins over salt. The salt/flavoring is then blended into the sausage at the time of manufacture. This method of producing a dry flavoring has a major advantage of being low in cost. Flavor dispersion by this technique also insures uniform distribution of the relatively potent insoluble flavoring throughout the food material and weighing is facilitated since the flavor oil is diluted by the carrier. A significant disadvantage is that the flavor is afforded little or no protection by the carrier. The flavor may evaporate to the storage area or become oxidized due to air contact.

The use of flavor carriers which form inclusion complexes with the flavor do afford some protection to the flavoring material. However, the formation of inclusion complexes can be slow, incomplete and weak depending on the carrier and the flavor compound.

Decanal has been shown to form an inclusion complex with gelatinized potato starch in less than a minute. Compounds such as limonene and menthone will require several days to accomplish any appreciable complexing at room temperature (Kitson and Sugisawa 1974). Encapsulation of acetaldehyde on starch hydrolysate has been shown to be possible by inclusion or adsorption (Pitchon *et al.* 1980). This process involves spraying an aqueous solution of acetaldehyde solution on a starch hydrolysate (DE 10–13) at 5°C under constant agitation. A contact time of 15 min results in a product having 4.1% acetaldehyde. This product does not lose acetaldehyde when stored at 40°C for one week.

Flavors may be encapulsated via the use of β-cyclodextrins. Cyclodextrins are produced by the action of a glycosyl transglycolase enzyme on starch (Szejtli 1982a). Cyclodextrins consisting of 6, 7 and 8 glucose units (α, β, and γ, respectively) result from this enzymatic action. Of these common cyclodextrins, the β-form has been found to be most suitable for flavor encapsulation via the formation of inclusion complexes (Szejtli 1982a).

The preparation of inclusion complexes with β-cyclodextrin is rather simple. An alcoholic solution of the flavor is added to an aqueous–alcoholic solution (2:1) of β-cyclodextrin at about 50°–55°C. The solution is cooled slowly to room temperature and the insoluble β-cyclodextrin/flavor precipitate is removed by filtration. The product is then air dried. The active core material load attained is generally from 6 to 15% (w/w).

The stability of volatile and thermally or chemically labile materials encapsulated in cyclodextrins is exceptional. Szejtli (1982b) reported the uptake of oxygen by a β-cyclodextrin/benzaldehyde inclusion complex to be about 4% of the uptake of oxygen by the pure benzaldehyde (2 μl O_2/mg versus 55 μl O_2/mg, respectively) at 37°C in 10 h using pure oxygen as the vapor phase in a Warburg apparatus. Venczel (1982) reported similar data for a lemon oil system. A vanilla/β-cyclodextrin complex was shown to lose 19% of the initial vanilla in 180 days when stored in open Petri plates at ambient temperatures (Szejtli 1982b). No further losses were observed after 2 years of storage. A vanilla/glucose control sample lost nearly 100% of the initial material in 240 days. Placing a cyclodextrin/spice oil complex in a vacum at 150°C for 24 h resulted in a 25–30% loss of the spice oil. Nearly 100% of the flavoring was lost if the whole spice was given the same treatment.

At the present, cyclodextrins are not approved for food use in the United States or Western Europe. They are approved, however, for use in Japan and Eastern Europe.

2. Phase Separation/Coacervation Processes

This technique employs a conventional three-phase system: the manufacturing vehicle (solvent), the flavor carrier (wall material), and the flavor (core material). The process is further classified into aqueous and nonaqueous phase separation, depending upon whether water is used as the manufacturing vehicle or not. While organic solvents could be used as the manufacturing vehicle, aqueous phase separation is most common since flavor components are hydrophobic in nature. Also, the shell material must be soluble in the manufacturing vehicle and the finished food product which is most commonly an aqueous system. The basic principle involved in this method is the formation of an emulsion and subsequent precipitation of the continuous phase around the droplets of the discontinuous phase. The mechanism of capsule formation by this process has been reviewed in the literature (Bakan 1973). There are several patented processes available for flavor encapsulation employing phase separation. The essential differences between the various processes arise due to the differences in the method of inducing phase separation.

The details and importance of each of the stages in coacervation are discussed in the following.

a. Addition and Mixing

The flavor carrier, or mixture of carriers, is initially dissolved in water. In order to facilitate dissolution of the carrier, the solution is generally heated and gently agitated. Vigorous agitation is avoided since high shear rates at elevated temperatures often result in structural changes of the carrier material. These structural changes may significantly alter the molecule and thereby change physical stability to precipitation. After the carrier solution is cooled (generally to ambient temperature), the flavor system is added.

b. Emulsification

Emulsification is a high shear operation intended to form a stable emulsion of the flavor system in the solubilized carrier. The immiscible flavor and the carrier are brought into intimate contact and the capsule wall forms around the flavor droplets. This processing step determines the ultimate particle size of the capsule (5–5000 μm). When the flavor carrier does not have the required emulsifying properties, emulsifiers can be added to the system. However, it is reported that the addition of an emulsifier does not provide the same intimate contact as a carrier having inherent emulsifying and lipophilic properties (King *et al.* 1976). A colloid mill or homogenizer is employed to effect emulsion formation.

c. Solidification and Hardening

The solidification of the continuous carrier film around the flavor can be brought about by physical and/or chemical means. Phase separation via a change in the system temperature is one of the simplest and most effective methods. Solubilities of various chemical compounds are highly temperature dependent. Lemon oil encapsulation in dextran has been accomplished using this principle. Lemon oil is dispersed in a dilute solution of methyl cellulose and subsequently a dextran solution is added. Upon emulsification and heating the system to 60°C, microcapsules separate. Methyl cellulose has decreased solubility in water at higher temperatures and thus induces phase separation.

Mutual incompatibilities and coprecipitations are another means of inducing phase separation. Addition of one polymer can cause another polymer to precipitate by decreasing its solubility. Solubility of methyl cellulose decreases substantially upon the addition of a dilute solution of carboxymethyl cellulose. Encapsulation of lemon oil and menthol has been accomplished using this method of phase separation (National Cash Register 1967). Another example of such solubility changes is coalescence of gelatin upon the addition of gum arabic solution.

Several polymers used in flavor encapsulation are charged when dissolved in water. Some polymers in this category are proteins, peptides, and carrageenan. Alteration of the pH of an emulsion may induce phase separation when a charged polymer is involved. When a single charged polymer is present in the system, the precipitation occurs at its isoelectric point. However, precipitation can be induced above or below the isoelectric pH if mixtures of proteins, peptides and other polymers (e.g., gum arabic, gum tragacanth) are used (Gibbs *et al.* 1968; Hiroharu *et al.* 1972; Saeki *et al.* 1976).

Sanenari and Shozo (1975a,b) have developed a process for encapsulating lemon oil in a mixture of gum arabic, gelatin, sodium alginate and sodium carboxymethyl cellulose. The underlying principle of inducing phase separation is selective complex formation of the various polymers with bivalent alkaline earth metal ions (e.g., Ca^{2+}) or trivalent ions (e.g., Al^{3+}). Upon complex formation, the microcapsules fall out of solution. The process can be completed without any heat treatment. This may be an added advantage since higher temperatures tend to alter the flavor components. However caution should be exercised in the choice of metal ions since they can catalyze oxidative degradation (e.g., Fe^{3+} and Cu^{2+}) of the flavor. There is also the potential for human toxicity (e.g., Cu^{2+}).

While encapsulation occurs simultaneously with phase separation, the walls of these microcapsules are soft and the product is not very amenable to further processing and handling. Thus, the capsule walls need to be hardened sufficiently to permit further processing. One of the general methods is to treat the capsules with an aldehyde. This is particularly common when a protein or peptide is part of the wall material. Glutaraldehyde is the commonly used hardener in flavor encapsulation. When an aldehyde is used for hardening the capsules, residual odor and impared solubility could pose practical problems. Another approach is to heat the capsules if the wall material has decreased solubility at higher temperatures.

d. Separation

The capsules are generally separated from the manufacturing vehicle via centrifugation or filtration. While centrifugation may be more applicable to continuous processes, the capsules can rupture under the centrifugal forces in the centrifuge.

e. Washing

Washing the capsules with water or a suitable solvent (ethyl alcohol) is essential. The surface of the capsules is likely to contain unencapsulated flavor (surface oil) and residual salts or aldehydes depending on the process used. Removal of these residues is important for obtaining a shelf-stable product free from residual off-flavors.

f. Drying

Any of the common dehydration techniques can be employed to dry the microcapsules. Fluidized bed, tray, spray, vacuum tray or freeze drying are all suitable for the final dehydration.

Dehydration of the capsule may be an optional step in this process. For several applications (e.g., liquid compositions and pastes), wet capsules are adequate and appropriate. The wet capsules have a more continuous film of the carrier and offer more protection to the core material (Moxey 1964). It should also be pointed out that upon dehydration, the capsule wall may develop cracks and become permeable.

3. Dehydration Processes

Dehydration processes offer a more economical, simple and flexible alternative to the coacervation method for flavor encapsulation. Dehydration methods yield a particulate powder which contains active flavor ingredients uniformly distributed in the carrier matrix. While any method of dehydration could be used to provide an encapsulated flavor, spray drying and freeze drying are the methods of choice. These two methods are reasonably economical and yet produce a dry flavor of good quality.

A product of satisfactory quality can be produced only when the drying method results in selective loss of water rather than flavor from the drying mass (Bangs 1980; Reineccius *et al.* 1982; Takayoshi 1978). Spray drying meets this requirement due to a shell of high solids (carrier material) forming around the drying droplet while in the spray dryer. This high-solids shell is permeable to the relatively small soluble water molecules but impermeable to the larger insoluble flavor molecules. This semipermeable shell forms early during drying of the droplet and, therefore, little flavor is lost during the drying operation yet the water is nearly completely lost. Temperatures of the drying droplets of atomized flavor emulsion never exceed wet bulb temperature (or exit air temperature) and residence time in the spray dryer is generally less than 30 sec. Therefore, there is minimal opportunity for thermally induced flavor changes to occur.

Freeze drying is a slower more costly process of dehydration than spray drying. The finished product must be ground to achieve a particulate nature. Grinding results in exposure of the otherwise encapsulated flavoring to air. This results in evaporative losses and increased potential for oxidation of the flavor during storage. The cost factors, availability of processing equipment, grinding requirement and increased susceptibility to oxidative spoilage during storage result in freeze drying being rather uncommon for flavor encapsulation. However, recent market needs have renewed interest in the freeze drying process. The concept of "flavor crystals" being visible in the product has received favorable market response. Also the loss of flavor to the package headspace due to evaporation may be desirable. For example, when the consumer opens a package of green apple dry beverage and is struck by the beautiful apple aroma, it makes a positive impression on the consumer. Therefore, aroma impact of the dry flavor and visible flavor crystals are attributes of freeze-dried flavors which may result in increased use of freeze drying for the production of dry flavorings.

Dehydration processes involve only two major unit operations: emulsification and dehydration.

a. Emulsification

The emulsification process is similar to that described under the coacervation method. However, it should be noted that unlike the coacervation method, the finished capsule size is not determined by emulsion size. The dehydration process and/or further processing (e.g., grinding or agglomeration) is the step which determines the ultimate capsule size.

Physical stability of the emulsion through a range of temperatures is desirable. When the carrier has inherent emulsifying properties (e.g., gum arabic and Capsul), emulsion stability is easily accomplished. However when the carrier lacks any emulsifying ability (e.g., most starches and hydrolyzed starches), an emulsifier may be used to facilitate emulsion stability. Ethoxylated sorbitan tristearate or Tween, decaglycerol distearate (Noznick and Tatter 1973) and decaglycerol tristearate (Noznick and Tatter 1971) are some emulsifiers which are used to aid in emulsion formation.

The viscosity of the emulsion is a second factor which influences emulsion stability. If an emulsion is sufficiently viscous, it will not break during processing. Therefore inherent emulsifying properties of the carrier material are not overly important during processing if a sufficiently viscous emulsion is prepared for the following dehydration step. Viscosity does have an upper limit depending upon the dehydration process to be used. Solution viscosity must be low enough to permit pumping, homogenization and subsequent atomization in the spray dryer. If freeze drying is to be used, the emulsion must only be low enough in viscosity to be pumped and homogenized.

b. Dehydration

As mentioned earlier, the two methods of choice are spray drying and freeze drying. The choice of method is determined by economics, nature of the flavor to be encapsulated, equipment availability and product characteristics desired.

i. Spray drying Spray drying is the most widely used process for flavor encapsulation (Fig. 11.2). The flexibility, cost, availability of equipment and technology as well as good product quality have made this an extremely desirable process for flavor encapsulation.

There are several processing parameters which influence product quality during spray drying (Reineccius *et al.* 1982). Briefly, flavor retention is primarily influenced by infeed solids content, dryer air temperatures and perhaps particle size distribution. Infeed solids content is the most important parameter determining flavor retention during spray drying (Leahy *et al.* 1985; Reineccius and Bangs 1985). Studies have shown that flavor retention may be nearly 100% if the infeed solids level is 50% or greater. Solids levels of 10–20% going into the dryer typically result in only 30–40% retention of volatile flavoring materials in the finished powder.

Theoretically, the best retention of flavor occurs when one uses the lowest inlet (155°C) and highest exit air temperatures (100°C). In practice,

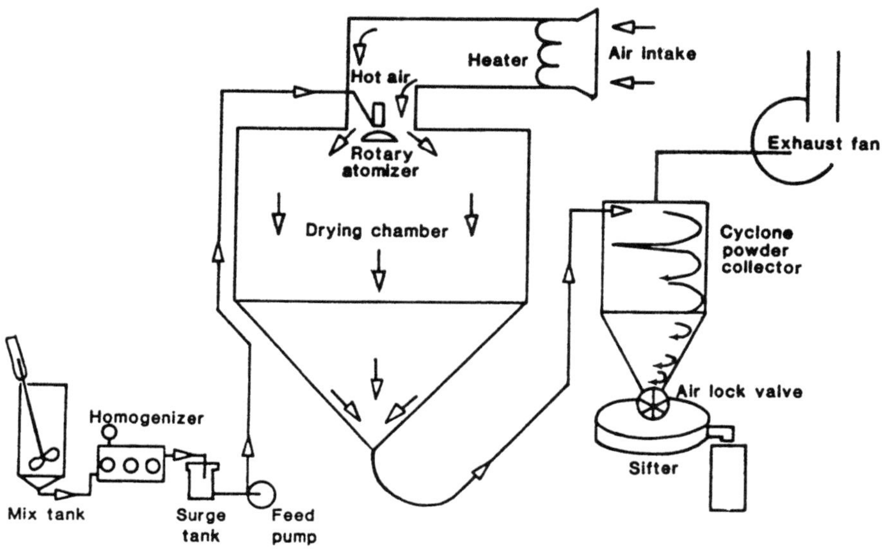

Fig. 11.2. Spray dryer for flavor encapsulation.

however, inlet and exit temperatures are generally maintained in the range of 200°–300°C and 80°–100°C, respectively, in order to maintain production capacity. Maintaining as low an outlet temperature as possible and minimizing the residence time in the dryer are beneficial if the flavor is chemically unstable. These conditions tend to minimize oxidative degradation of the flavor compounds. Cinnamaldehyde in cinnamon bark oil and unsaturated monoterpenes (*d*-limonene, γ-terpinene, terpinolene, etc.) in citrus peel oils can undergo extensive oxidative degradation at higher drying temperatures and longer residence times.

Particle size and distribution can be controlled during spray drying. The two common types of atomizers used in spray drying are the pressure spray nozzle and centrifugal wheel. The diameter of the nozzle orifice and the speed of rotation are so chosen to achieve the desired particle size in each of the respective methods. Particle size obtained by either method of atomization seldom exceeds 100 μm. This small particle size has a disadvantage of being difficult to disperse in beverage applications. Therefore, some spray-dried flavors are agglomerated following spray drying. The agglomeration process involves rewetting the dry powder, often with live steam, sufficiently to produce a sticky particle surface. Particles will stick together to form larger, more dispersible agglomerates. The agglomerates are then redried to the desired final moisture content. Two-stage drying is starting to find application in the flavor area. Two-stage drying is a process whereby the powder comes out of the initial spray dryer at about 10–12% moisture. This somewhat sticky powder then goes into a fluidized bed second dryer which lowers the moisture to 3–5%. Agglomeration and drying takes place in this second stage of drying.

Several water immiscible natural flavors have been satisfactorily encapsulated via spray drying. Table 11.3 presents some typical applications described in the patent literature for the preparation of encapsulated citrus flavors. A wide variation exists in choosing the carrier, composition of the feed and drying temperatures even when very similar flavors are encapsulated.

Spray drying can be used for encapsulation of water soluble flavorings as well. Shaffer and Burke (1974) describe a process for encapsulating a water-soluble imitation butter flavor. Satisfactory encapsulation was accomplished by spray drying a feed with 1:4 flavor-to-carrier (dextrin) ratio.

Modifications of the conventional spray drying process have also been used for encapsulation. Modified methods include the use of a dehydrating solvent as the drying medium rather than hot air as in the conventional process. The feed is sprayed into a suitable dehydrating solvent such as ethyl alcohol or polyglycols. Coffee aroma in gum arabic (Balassa 1970) and lemon oil in dextrin (Balassa 1973) have been encapsulated using ethyl alcohol and polyglycols as the dehydrating medium.

ii. Freeze drying Freeze drying has found only very limited application in flavor encapsulation. One of the major deterrents in ex-

Table 11.3. Examples of Citrus Oil Encapsulation via Spray Drying

Flavor	Carrier	Water (g)	Flavor/Solids Ratio	Dryer Temperatures[a] (°F)	Reference
Lemon oil (50 g)	Succinylated gelatin (15 g) Dextran (37.5 g)	247.5	1:1.1	—	Theodore *et al.* 1974
Lemon oil (1811 g)	Lactose (1613 g) K-Dex (1177 g)	1794	1:1.5	356–212	Marotta *et al.* 1969
Lemon oil (543 g)	Maltrin M100 (4047 g)	1443	1:7.5	380–210	Merritt 1981
Orange oil (1500 g)	Sucrose (1800 g) K-Dex (1200 g)	5700	1:2	300–180	Giel 1977
Orange oil (1811 g)	Lactose (1766 g) K-Dex (1177 g)	7200	1:1.6	300–180	Giel 1981

[a] Inlet air temperature–exit air temperature.

ploiting this mild dehydration method has been the higher operating cost. Freeze drying is simple, involving freezing the flavor emulsion and removal of water from the frozen sample by sublimation of the ice crystal at low pressures (<0.01 mm Hg). This is a rather slow process in comparison to spray drying. However, encapsulation of very low boiling components can be readily achieved by this process due to the milder temperatures employed. Encapsulation of flavors via freeze drying remained in the realms of academic interest until only recently. Ono and Saigusa (1980) have investigated the encapsulation of eugenol, limonene, linalool and benzyl alcohol in a mixture of corn syrup and polypeptone via freeze drying. The flavor mixture was used at about 12% of the solids level. The retentions following drying were reported to be greater than 92% for all the test compounds.

If freeze drying is employed, the product must be pulverized and screened if uniformity of particle size is important.

iii. Other dehydration methods Moisture equilibration has been employed as a dehydration method for the encapsulation of flavors (Palmer 1975). The flavor oil, gum arabic and corn syrup solids are emulsified to obtain a smooth and extremely viscous mass. This emulsion is subjected to dehydration by mixing it with a large quantity of dry gum arabic. The product is then screened to recover flavor capsules. This particular product contained 16% oil following encapsulation. This illustrates an innovative approach to encapsulation via dehydration employing a simple technique and mild processing conditions.

4. Extrusion

Extrusion processes for encapsulation are gaining in popularity and have been found to be very useful for the encapsulation of fruit flavors in general and citrus flavors in particular. The basic process involves incorporation of the flavor into a molten carrier matrix followed by extrusion and solidification of the mixture. The carrier material almost invariably consists of a very high proportion of mono and disaccharides or other polyhydric alcohols. The high sugar content is essential to minimize the tendency of the product to crystallize and result in a more porous structure. The individual processes are unique and instead of a generalized process description, individual processes will be presented.

Fruit flavored sugar candy is one of the earliest uses of entrapping flavors in a sugar matrix. However, it was not considered an encapsulation process. Schultz *et al.* (1956) translated this old concept into a very versatile encapsulation process. This process consisted of emulsifying orange peel oil in a melt of glucose or sucrose, and corn syrup, solidifying the mass and then pulverizing the solid to obtain the "locked in" orange flavor. This product needed to be stored under vacuum to minimize oxidative degradation of the surface oil.

A modified process (Swisher 1957) was developed whereby the surface oil was eliminated by washing the final product in isopropanol. Removal of isopropanol was accomplished via vacuum drying. Further modification of the same process resulted in the extrusion process (Swisher 1962) of encapsulating flavors. An emulsion containing corn syrup solids, mono and diglycerides and citrus peel oil was hot (115°C) extruded at 32 psi pressure through $\frac{1}{64}$-in. holes into a cold (20°C) isopropanol bath. A contact time of 48 h with isopropanol followed by vacuum drying of the filaments resulted in an encapsulated citrus flavor.

Since sugars are hydroscopic and the major portion of the carrier happens to be sugars, hygroscopicity can pose a practical problem during storage. Coating the particles or filaments with an anticaking agent such as tricalcium phosphate can minimize this considerably. An alternative approach to this problem is through changing the formulation. If corn syrup of lower dextrose equivalent (20 or less) is used in the formulation, the tendency to cake is substantially minimized (Crocker and Pritchett 1978).

Gergely (1980) described an encapsulation process that does not use a solvent in the final stages. Orange oil is dispersed into a melt (145°C) of a mixture of mannitol, sorbitol and glucose under nitrogen atmosphere in a turbine mixer (4000–6000 rpm) and the resulting dispersion is sprayed through a heated nozzle onto a cold (−10°C) metal surface. Solidification is accomplished nearly immediately by contact with the cold metal surface. A nitrogen atmosphere excludes any entrapped oxygen and extends shelf life of the product.

It may appear that the extrusion process is more applicable to flavors having low volatility. Even though there is a risk of losing the low boiling flavors while adding it to a hot melt (>100°C) of sugars, process modifications yield favorable results. Encapsulation of apple aroma in a mixture of sucrose, glucose and maltodextrin has been accomplished without substantial losses of the flavor even though the melt was maintained at 145°C during flavor addition. This was achieved by using glycerol solvent for the flavor.

High-pressure high-temperature extrusion processes can considerably minimize the loss of very volatile flavor components. Encapsulation of acetaldehyde (boiling point 21°C) in sucrose is shown to be possible by this process. This particular process was intended to yield capsules with low volatiles content. In this case, a single component (acetaldehyde) was being encapsulated so it may be added to a food product to supplement the major flavor system. Acetaldehyde is a difficult flavor com-

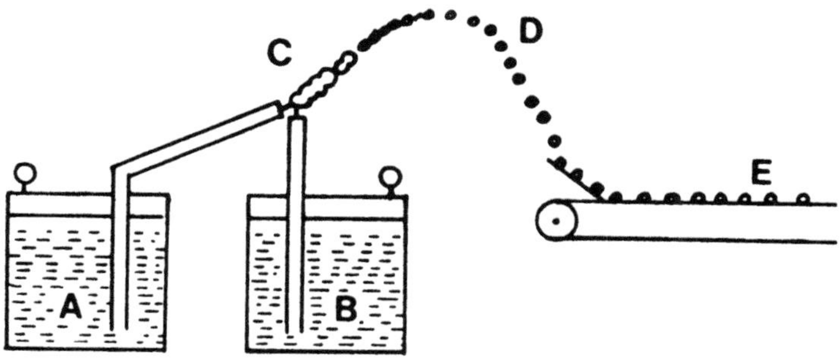

Fig. 11.3. Two fluid encapsulation process.
Arens and Sweeny 1969

ponent to encapsulate since it is extremely volatile, flamable, water soluble and toxic at high concentrations.

An interesting and versatile extrusion process has been described by Arens and Sweeny (1969). A biliquid column is extruded through a nozzle to yield a string of capsules or individual capsules. A schematic arrangement of the system is shown in Figure 11.3. The core material is pumped through the inner orifice and the wall material through the outer orifice. The particle size produced by this method follows a normal distribution about the mean. The mean particle size is very dependent on the orifice size, viscosity, surface tension and speed of rotation if the nozzle is rotated. Vibrating the encapsulation nozzle can yield particles with a very narrow size and weight distribution. It can also minimize particle size variation due to viscosity and surface tension differences over a wide operating range. Vibration frequencies in the range of 200–400 Hz have been found to be beneficial in atomization during this extrusion process (Dietert 1979).

References

ARENS, R. P. and SWEENY, N. P. 1969. Biliquid column of shell material around fill liquid. U.S. Patent 3,423,489. *In* Microcapsules and Microencapsulation Techniques, M. H. Gutcho, ed., 1976, p. 149. Noyes Data Corp., Park Ridge, NJ.

BAKAN, J. A. 1973. Microencapsulation of foods and related products. Food Technol. *27*, 36.

BALASSA, L. L. 1970. Flavor encapsulation. U.S. Patent 3,495,988. *In* Microcapsules and Microencapsulation Techniques, M. H. Gutcho, ed., 1976, p. 220. Noyes Data Corp., Park Ridge, NJ.

BALASSA, L. L. 1973. Polyglycol used to desolventize shell composition, U.S. Patent 3,780,195. *In* Microcapsules and Microencapsulation Techniques. M. H. Gutcho, ed., 1976, p. 232. Noyes Data Corp., Park Ridge, NJ.

BANGS, W. E. 1980. Spray drying of model food systems: Retention of selected lipid

oxidation and nonenzymatic browning compounds. M. S. thesis, University of Minnesota, St. Paul, MN.
CROCKER, D. C. and PRITCHETT, D. E. 1978. Improved encapsulated citrus oils—Modified matrix increases moisture resistance. Food Technol. *32*, 36.
DIETERT, R. D. 1979. Application of vibration to the encapsulation process. Amer. Lab. (June), 19.
GERGELY, G. 1980. Solid dispersion of solid or liquid materials in sugars. Austrian patent 359,978. Chem. Abstr. *94*, 13813g.
GIBBS, J. A., CALDWELL, C. R., LEACH, A. A., PRALT, A. N. and SULTON, D. A. 1968. Microscopic capsules, Ger. Patent 1,279,648. Chem. Abstr. *70*, 21022u.
GIEL, T. H. 1977. Particulate aroma-producing materials. U.S. Patent Appl. 640,189. Chem. Abstr. *88*, 16869r.
GIEL, T. H. 1981. Particulate lactose–starch hydrolysate based flavoring materials. U.S. Patent 4,271,202. Chem. Abstr. *95*, 78775n.
GYOGYSZER, C. and TERMEKEK, V. 1978. Aroma and flavoring compositions containing cyclodextrin. Belg. Patent 868,576. Chem. Abstr. *90*, 70806z.
HIROHARU, M., SAEKI, K. and SHIMADA, T. 1972. Forming microcapsules containing hydrophobic oil droplets. Ger. Offen. 2,138,842. Chem. Abstr. *76*, 145304f.
KING, W., TRUBIANO, P. and PERRY, P. 1976. Modified starch encapsulating agents offer superior emulsification. Food Prod. Devel. *10*, 54.
KITSON, J. A. and SUGISAWA, H. S. 1974. Sugar captured flavor granules. Canad. Inst. Food Sci. Technol. J. *7*, A15.
LEAHY, M. M., ANANDARAMAN, S., BANGS, W. E. and REINECCIUS, G. A. 1985. A comparison of encapsulating agents for the drying of artificial flavors. *In* Progress in Flavour Res. 1984, J. Adda, ed. Elsevier, Amsterdam.
MAROTTA, N. G., BOETTGER, R. M., NAPPEN, B. H. and SZYMANSKI, C. D. 1969. Encapsulating water insoluble substances. U.S. Patent 3,455,838. Chem. Abstr. *71*, 126285f.
MERRIT, C. G. 1981. Encapsulation of materials, U.S. Patent Appl. 693,563. Chem. Abstr. *95*, 95818w.
MOXEY, P. 1964. The encapsulation of flavors. Manuf. Chem. *35*, 68.
NATIONAL CASH REGISTER CO. 1967. Microencapsules containing an emulsion of oil in a hydrophilic liquid. U.S. Patent Appl. Chem. Abstr. *67*, 109438h.
NOZNICK, P. P. and TATTER, C. W. 1971. Spray-dried whippable food compositions. U.S. Patent 3,628,968. Chem. Abstr. *76*, 84674c.
NOZNICK, P. P. and TATTER, C. W. 1973. Spray-dried essential oil composition. U.S. Patent 3,764,346. Chem. Abstr. *80*, 26087k.
ONO, F. and SAIGUSA, T. 1980. Solubilization of fats and oils by solid components of soy sauce: XI. Encapsulation and solubilizations of volatile substances into a protein–carbohydrate matrix. Nippon Shokuhin Kogyo Gahkaishi *27*, 205. Chem. Abstr. *93*, 68843c.
PALMER, E. 1975. Encapsulation of volatile flavoring agents with dry nonfriable films. U.S. Pattent 3,903,295. *In* Microcapsules and Microencapsulation Techniques. M. H. Gutcho, ed., 1976, p. 226. Noyes Data Corp., Park Ridge, NJ.
PITCHON, E., SCHULMAN, M. and CHALL, S. B. 1980. Solid Aroma Composition. Ger. Offen. 2,840,511. Chem. Abstr. *92*, 196654h.
REINECCIUS, G. A. and BANGS, W. E. 1985. Spray drying of food flavors III: Optimum infeed concentrations for the retention of artificial flavors. Perf. Flavorist *9*, 27.
REINECCIUS, G. A., ANANDARAMAN, S. and BANGS, W. E. 1982. Spray drying of food flavors. Part 1. Theory of flavor retention. Perfumer and Flavorist *7*, 1.

SALDARINI, A. V. and DOERIG, R. 1981. Fixing a volatile flavoring agent in starch hydrolysate. U.S. Patent 4,285,983. Chem. Abstr. *95*, 185760a.
SAEKI, K., KATO, H. and MATSUKAWA, H. 1976. Encapsulation of perfumes. Japanese Patent 7,613,387. Chem. Abstr. *84*, 184796c.
SANERARI, N. and SHOZO, A. 1975a. Microencapsulation process. Japan Kokai 75,97,584. Chem. Abstr. *84*, 6128d.
SANENARI, N. and SHOZO, A. 1975b. Microencapsulation process. Japan Kokai 75,97,586. Chem. Abstr. *84*, 6127c.
SCHULTZ, I. H., DIMICK, R. P. and MACKOWER, B. 1956. Incorporation of natural fruit flavors into fruit juice powders. Food Technol. *10*, 57.
SHAFFER, J. R. and BURKE, C. E. 1974. Encapsulated flavor preparations for shortenings. Brit. Patent 1,376,870. Chem. Abstr. *82*, 138045t.
SWISHER, H. E. 1957. Solid essential oil containing compositions. U.S. Patent 2,809,895. Chem. Abstr. *52*, 1508c.
SWISHER, H. E. 1962. Solid essential oil flavoring composition. U.S. Patent 3,041,180. Chem. Abstr. *57*, 8980g.
SZEJTLI, J. 1982a. Cyclodextrins and Their Enclusion Complexes. Akademiai Kaido, Budapest.
SZEJTLI, J. 1982b. Cyclodextrins in foods, cosmetics and toiletries. *In* Proc. 1st Int. Symp. on Cyclodextrins, 1981, Budapest, J. Szejtli, ed. Reidel Publishing, Boston, MA.
TAKAYOSHI, B. 1978. Drying of single drop of O/W emulsion containing volatile flavoring material. Kagaku Kogaku Rombunshu *4*, 515. Chem. Abstr. *90*, 12685y.
THEODORE, M., ALLEN, R. J. and FOGLE, M. V. 1974. Capsule with water-soluble wall of modified gelatin. Canadian Patent 947,652. Chem. Abstr. *81*, 150729f.
VENCZEL, G. 1982. Production and application of β-cyclodextrin complexes in the food industry. *In* Proc. 1st Int. Symp. Cyclodextrins, 1981, Budapest, J. Szejtli, ed. Reidel Publishing, Boston, MA.
WYLER, R. and SOLMS, J. 1981. Inclusion complexes of potato starch with flavor compounds. *In* Flavour 81. P. Schreier, ed., p. 693. de Gruyter, Berlin.

12

Application of Flavorings in Food Processing[1]

The term *flavor* is widely used throughout this book and has several meanings depending on its context. Here, flavor is regarded as the summation of sensations induced by chemical compounds present in what we eat and drink and in equilibrium at the time of consumption (Hall 1968; Hall and Merwin 1981). Some of these flavor components arise from normal bisynthetic processes of animal and plant metabolism and hence are present in the basic nutritional constituents of our normal diet. Other components exist only as precursors and develop characteristic flavoring effects during subsequent cooking or processing due to chemical reactions induced by the effects of heat or fermentation. Some may be intentionally incorporated as flavorings at an appropriate stage of product processing or used as condiments when the product is served. Whatever the source, the observed odor and flavor impact of the end product is the total effect of the individual flavoring components present which in turn is determined by their relative proportion and their flavor rating.

A. Flavors in Foods

The flavor of foods may be classified as (a) *natural flavor*: preexisting in the diet, particularly in fruits, vegetables and spices; (b) *process flavor*: arising in end products as a result of conventional processes including heat and fermentation; (c) *compounded flavor*: intentionally added flavorings formulated to produce a desired sensory effect using selected flavorants of natural and/or synthetic origin; (d) *taste modifiers*: additives which affect the basic taste sensations (e.g., salt, natural or synthetic sweeteners, food acids and bittering agents) and (e) *abnormal flavors and*

[1] First published in Heath, H. 1982–1983. Perfumer & Flavorist, 7 (Dec./Jan.), 27–38.

taints: off-odors and off-flavors arising in products as a result of degradation, adventitious contamination or package/product interaction.

The flavor of what we eat and drink is not, in most instances, a static attribute but one which is in dynamic equilibrium, capable of change depending on many factors. In raw materials of both animal and plant origin, it changes during growth and maturation and further during postmortem or postharvest handling and preparation for market. Flavors arising from cooking depend on the time/temperature ratio employed in the preparation of the food; the flavor of the freshly prepared product may undergo further modification during subsequent storage. Microbial growth in products may produce significant flavor changes; some of these are desirable (e.g., dairy products), most are detrimental to product acceptability (e.g., souring of meats).

Flavor changes within the end product may be due to (a) chemical transformation induced by pH, Maillard and related reactions, hydrolysis, oxidative rancidity, enzymatic or microbiological activity, etc.; (b) volatile losses which upset the relative concentration of aromatic constituents; (c) the removal of flavor components by selective adsorption onto solid surfaces within the product and (d) differential partition between aqueous and lipid phases which significantly affects flavor perception.

Because there are so many variables, the formulation and production of foods and other comestibles is far from a precise science and depends to a large extent on subjective trial and error assessments to achieve a product profile having either maximum consumer acceptance or minimum consumer rejection (Heath 1978). It is well recognized that flavor plays a significant part in product acceptance by inducing hedonic responses and hence consumer satisfaction or dissatisfaction. Poor flavor is a major cause of product rejection (Muskowitz and Chandler 1978; Blanchfield 1978).

B. Achieving Flavor Balance

The control of flavor in an end product is one of profile alteration, which includes (a) the selection and balancing of existing or potential flavor factors working within the constraints of nutritional necessity, the nature and sources of the basic raw materials and supplementary ingredients and the total concept of the end product; (b) the adjustment of the flavor profile resulting from the method of processing employed to suit particular palates or consumer anticipations; (c) corrections to overcome any preexisting or developed flavor defects and (d) the imposition of an entirely new flavor in products which are bland or relatively flavorless.

Each of these actions calls for individual judgment on the part of the product development team and involves the knowledge of available raw materials, minimum and optimum processing conditions, legal constraints and probable consumer response to the final product. This is the true basis of the culinary art and applies equally to domestic and industrial-scale food preparation.

1. Consumer Acceptance

The real judge is, of course, the consumer, but in accepting this one has to appreciate that the majority of the consumers neither knows nor really cares about the many problems encountered within the food industry from product conception to ultimate consumption. Not least of these problems are in the realm of flavorings, for example, the form in which the flavoring is to be incorporated, the flavor contributions of its constituent parts, the interaction of flavoring effects within the product, variability of sensory perception, the problems of compatibility, the complexity of flavoring systems necessary to achieve optimum flavor balance under far from ideal processing conditions, payload protection, and satisfactory release characteristics which liberate locked-in flavors. These and many other problems all affect the ultimate flavor profile and determine the success or failure of the end product.

2. Flavor Defects

It is the correction of flavor defects which poses the greatest problems for the industrial food processor who must always aim at the maintenance of product quality which gives consistent consumer satisfaction. The following are available methods: (a) regulation of the processing conditions as flavor quality is often reduced by excessive heat or agitation; (b) incorporation of flavorings and/or other permitted food additives to achieve either flavor modification, intensification or suppression and (c) use of garnishes, sauces and condiments to suit individual palates at the time of eating; action more suited to domestic food preparation than commercially produced food products which are judged on their own attributes.

3. Flavor Intensification

Flavor attributes may be intensified by concentration or addition of a concentrate of the same flavor (e.g., tomato paste added to canned tomatoes); the incorporation of ingredients of a similar or supporting flavor profile to enhance and extend an existing flavor attribute (e.g., use of almond essence in cherry pie filling); the addition of compounded flavorings as topping notes to replace those lost during processing (e.g., imitation coffee aroma added to instant coffee); or the judicious use of flavor potentiators (e.g., MSG, ribonucleotides or maltol).

4. Flavor Suppression

Flavor attributes may be suppressed by removal of the unwanted character by further processing or maturation; incorporation of adsorbent materials (e.g., starch); neutralization or conversion of the flavor impact by adjustment of the pH or of the salt/sweet/acid balance; dilution with nonflavorful ingredients; masking or disguising the undesirable flavor attributes by using stronger flavorings (i.e., spices and seasonings in meat products). The most appropriate technique to use may have to be established by trial and error and even then may only be partially successful.

C. Criteria for Application of Flavorings

Flavorings may be added to food and other consumable products for various reasons, but mainly (a) to impart flavor to an otherwise bland product—the flavoring may be in imitation of an existing natural flavor or may be created to give some desirable flavor experience; (b) to impose a different flavor character from that arising from basic ingredients; (c) to boost weak intrinsic flavors or replace flavor notes lost during processing; (d) to modify or complement an existing flavor profile; (e) to disguise or cover undesirable flavor attributes; (f) to overcome seasonal variability in natural flavoring materials or constituents; (g) to impart a flavor where the use of a natural flavoring material is technologically impracticable; (h) to make available at an economical price the flavor of natural materials which are either of limited availability or are unacceptably expensive and (i) to make available flavor types where the natural product poses toxic or other hazards.

One of the major functions of intentionally added flavorings is to extend the range and flexibility of food products and processing technology, but their specific application is determined by factors which are not exclusively technical in character. These include the following.

1. Acceptability to the Consumer

The flavor of food, drinks, candy, chocolate confectionery and snack foods is open to wide hedonic interpretation. Preferences display a wide spectrum of response depending on such factors as ethnic origins, education and upbringing, age, sex, environment and even one's personal mood at the time. The strength and quality of flavors in end products are often regionalized. This poses a big problem to the manufacturer aiming at national distribution and even greater problems if international markets are being considered. Indeed, this desire to please everybody often results in the development of less than top quality products aimed at minimum rejection over a wide market area rather than products having maximum acceptance in a smaller regional market.

2. Legal Acceptability

This is of increasing concern to the food manufacturer and in most developed countries, the use of flavorings is controlled by legislation aimed at safeguarding the consumer from real or supposed health hazards arising from the ingestion of materials intentionally added to the natural diet. A secondary aim is the prevention of fraud and deception as to the true nature of the products which the consumer must, of necessity, take on trust. It is essential that any food product, beverage, etc., should comply with the legislation of the country in which it is offered for sale and this does not necessarily equate with that of the country in which it is manufactured. Ignorance of the law is no defense.

3. Nature of Product as Sold and as Consumed

Today the range of consumable products is enormous resulting from a rapidly advancing processing technology based on computerization and automation, imaginative product conception to meet consumer needs for speed and convenience without loss of nutritional values, versatile packaging and efficient and safe distribution and marketing. The form of the product will obviously determine the form in which flavorings may be incorporated—dry goods call for powdered flavorings and wet goods enable one to use flavorings in liquid form. Many products require further preparation and cooking by the consumer. Here, one generally has little or no control and the chances of product failure are great, hence preparation instructions must be simple and precise. Even so, some allowance should be made for indifferent domestic handling and it may be desirable to set flavoring levels a little on the high side as excessive cooking, which is the usual problem, can seriously reduce available flavor content to the detriment of the product when consumed.

4. Processing Conditions

These are of major importance, but from the point of view of deciding what is the most appropriate form of flavoring as well as the flavor profile of the end product, the product development team must be able to produce a facsimile of the end product under laboratory or pilot plant conditions closely similar to those encountered in full-scale production. This is the only sure way of establishing the technological and aesthetic acceptability of the end product. Evaluation of flavorings in alternative test media may be adequate as a first-stage screening but such assessments may be quite misleading in relation to the final product.

D. Available Flavorings

Each of these areas of constraint is complex and makes very special demands on flavor manufacturers in the industry's service to food processors. Individual flavoring compositions must reflect consumer tastes and indeed prejudices, as well as cater to unpredictable and often short-lived demands, particularly in the fields of snacks, confectionery, ice cream and soft drinks. Flavorings must be compatible with other prime constituents in the end product, be resistant to often adverse processing conditions and be stable before, during and after incorporation into the finished product.

Very few flavorings are suitable for all applications as processing conditions vary widely in physical parameters, particularly temperature. In addition the physical character of the end products often pose unique problems of flavoring incorporation and subsequent flavor stability. Few generalizations are possible in such a diverse field so that a study should first consider the general processing constraints common to the food industry and then examine products by groups in which the processing parameters are broadly similar. By this means one can highlight the problems and indicate the methods currently used to achieve satisfactory flavoring application. An in-depth study of each product group is outside the scope of this present work and reference should be made to the Additional Reading at the end of this chapter.

E. Processing Parameters

Unit operations encountered in the processing of foods, beverages, baked goods, sugar and chocolate confectionery, meat and other consumable products are relatively few and details can readily be obtained from the many excellent texts on process engineering (Brennan *et al.* 1982; Heldman and Singh 1981; Toldeo 1980). These are mainly associated with raw material preparation, mixing and blending, thermal processing involving various methods of cooking, the retorting of cans or pouches, refrigeration and deep freezing, dehydration, gamma irradiation and packaging. Within these processes the following conditions are the most important determinants likely to affect the incorporation of flavoring materials and the flavor profile of the end product.

1. Temperature and Time

By their aromatic nature, most flavoring materials are to some extent thermolabile or heat sensitive. At elevated temperatures, particularly in the presence of water, they may be lost to the system through evaporation or steam distillation. Less stable compounds may change due to chemical interactions with other constituents. The degree of change is usually a function of both temperature and time. The effects of ultra-high temperatures for very short time intervals may result in significantly less flavor loss or degradation than occurs with much lower temperatures employed over a longer period, although this will depend on other factors internal to the product (e.g., pH, the presence of proteins). A knowledge of the temperatures to which the total flavoring system will be exposed and the dwell-times involved are most important in deciding the nature of the flavoring to be used particularly with respect to any solvents present in systems subjected to temperatures in excess of their boiling points. It should be remembered during laboratory or small-scale product development that large bulks often retain their heat disproportionately to small quantities and this too may have a deleterious effect on the flavor level and profile.

The effects of heat processing are most obvious in products containing both sugars and amino acids as these are susceptible to nonenzymatic browning, which has been discussed elsewhere. Also, those containing high levels of lipids become prone to oxidative rancidity at even moderately elevated temperatures and these changes generally result in quite unacceptable off-flavor notes.

2. Open or Closed System

The incorporation of flavoring materials in an open system (i.e., blending in an open vat) is likely to result in greater volatile losses than in a closed system (i.e., in-line processing or retorting in sealed containers). Where open handling cannot be avoided, every precaution should be taken to minimize exposure by using covered containers and avoiding exposure to direct heat. The bigger the unit bulks and the higher the temperatures employed, the greater the chances of flavor loss or degradation.

3. The Mixing Sequence

The guiding principle is to expose any added flavoring to the minimum of treatment. Obviously, in some products addition into the primary mix cannot be avoided, but wherever possible, flavorings should be added at as late a stage as consistent with uniformity of dispersion in the end product.

4. pH

Most fruits contain natural organic acids (e.g., citric, malic, tartaric acids) which contribute significantly to the flavor profile. The use of acidulants is necessary when one is using an imitation fruit flavoring the taste of which is intrinsically neutral; otherwise the correct flavor impression is not achieved in the end product. Some flavorings and certain spices (e.g., turmeric) contain ingredients which are sensitive to changes in pH and it is essential that this particular condition be carefully reproduced during the product development stage and subsequent shelf-life testing to ensure that undesirable effects do not occur.

5. Pressure

Both positive and negative pressure changes are likely to endanger added flavorings by altering the relative concentration of aromatic components in the headspace vapors. Vacuum filling may result in the more volatile components being preferentially lost from the system resulting in an unbalanced profile in the end product.

6. Contact with Air

This is of particular concern in products which are aerated (e.g., ice cream, marshmallow). High-speed mixing operations can result in considerable volatile losses, but, more importantly, any occluded air produces conditions conducive to oxidation of any unsaturated lipids present. The pneumatic conveyance of powdered flavoring ingredients and food mixes containing flavorings (e.g., soup mixes, instant puddings and dessert powders, sauce mixes) may also result in significant volatile losses unless encapsulated flavorings or seasonings are used. A similar consideration applies to flavorful products predispensed into vending machine containers as the exposed surface area is comparatively large.

F. Specific Flavoring Applications

It is necessary now to review some of the practical implications of the above generalizations as they affect specific product groups. The food and related processing industries cover an enormous field of processing technology particularly when one appreciates that there are many subgroups calling for individual consideration. The details of flavor application, particularly the plant involved, must be left for specialist study.

1. Meat Products

This group of products embraces all types of meat, meat by-products, poultry and fish. Although the greater proportion of these commodities is sold for domestic preparation it is that which is industrially processed which most concerns us here. The resulting products include fresh, semi-dried, dried, fermented, deep-frozen and canned meats which may be eaten directly either cold, after reheating or after some further domestic preparation or cooking. The opportunities for using added flavorings are almost limitless but the diversity of product types, many of which are now traditional, imposes considerable constraints on the nature of any added seasoning or flavoring materials. In addition, in many developed countries the meat industry is controlled by legislation which is either separate from or supplements food legislation so that special regulations may govern the nature and quality of any additives.

In savory foods, the prime taste adjunct is salt and its level is significant to the shelf life of many products and in the total flavor profile as it affects product palatability. Additional flavoring effects may be achieved by using seasonings. This is an inclusive term applied to any ingredient which by itself or in combination adds flavor to a meat product. Most seasonings are blends of natural herbs and spices, or products processed directly from them, often admixed with other flavoring ingredients such as MSG, the ribonucleotides and hydrolyzed vegetable protein all of which may enhance or impart characteristic meaty notes to the product.

Different animal tissues vary considerably in moisture:protein ratios;

fat: lean ratios, the amount of pigment present and in their water and/or fat "binding" properties. This spectrum of variables means that the careful selection of meat ingredients by type and quality is basic to the manufacture of most meat-based products and particularly to the production of sausages and other comminuted meat products.

The basic technology of meat processing is complex and often specific to a country, a manufacturer, or, indeed, to an individual product line. Whatever the process, the following factors must be taken into account when selecting an appropriate seasoning: (a) the nature of the raw materials used, particularly the lean: fat: water ratio; (b) the nature of any pretreatment, particularly the use of curing agents; (c) the stage and method of incorporating the seasoning; (d) the degree and nature of any comminution stage; (e) postmixing treatment, particularly that involving heat (i.e., cooking, smoking, drying or retorting); (f) the temperature and times involved at any stage, particularly in an open system; (g) the nature of any added preservatives, particularly sulfur dioxide; (h) methods of packaging, particularly if this involves exposing the product directly to a vacuum and (i) postpackaging handling and storage, particularly refrigeration or deep freezing.

This, you will appreciate, implies a precise knowledge of the end product and its method of manufacture and on this one cannot generalize.

The principal aim in incorporating seasonings into meat products is to impose added flavoring notes which will enhance the natural meat flavors developed during cooking; maybe modifying them to suit individual tastes but not swamping them. Traditionally, seasonings have been prepared from herbs and spices ground to varying degrees of fineness. For most purposes such products should pass through United States standard sieves of No. 20–No. 60 mesh. There is a demand for much finer products or what are called "micro milled" spices for use in certain sausage emulsions. Numerous articles have appeared in the technical literature and their several advantages and disadvantages are listed in Table 6.10.

Increasingly, the industry has adopted seasonings based on spice extractives—the dispersed or "soluble" spices which give a standardized flavoring effect. Such commercially compounded seasonings are generally in the form of dry powders comprising not only the appropriate herbs and spices but also other permitted additives which may include flavor enhancers, hydrolyzed vegetable protein, yeast extracts, salt, phosphates and colorants. They are frequently supplied preblended in unit packs to facilitate addition to a single processing bulk at the chopping stage. To overcome volatile losses from such products several spice houses now offer seasonings based on spice extractive encapsulated by spray drying. These have a considerably longer shelf life and in many uses are preferred.

With an increase in automated processing with computer-controlled dosing of ingredients, the use of seasonings in the form of liquid emulsions is rapidly gaining in popularity. These have all the advantages of the dry processed products but can be accurately metered in-line and almost instantly and uniformly dispersed into the prepared meat emulsion. They are more concentrated than powdered seasonings and many are designed for use at 2 g/kg of meat mix. Precise usage level should be established with the supplier. The presence or absence of added preservatives will dictate whether or not the product is supplied in a single unit or multidose pack. The processing of poultry and fish calls for a very specialized technology which lies outside the scope of this text (Mountney 1976; Martin *et al.* 1982), but the manufacture of comminuted products such as "chicken burgers" or "fish cakes" involves considerations similar to those just described (Long *et al.* 1982).

2. Baked Goods and Bakery Products

This important branch of the food industry embraces such widely different products as bread and rolls, sweet yeast dough products, biscuits, cookies and crackers, pies and pastries, cakes and even breakfast cereals. All of these are based on flour, mixed as necessary with sugar, eggs, milk, shortenings, leavening agents or yeast, antistaling agents and flavorings. The industry covers an enormous range of specialty products many of which involve purpose-designed plant and processing techniques (Matz 1972; Matz and Matz 1978; Tressler and Sultan 1975; Sultan 1985).

The following constraints will determine the choice of any flavoring: (a) the effects of temperature and time; (b) flavor balance, particularly acidity and sweetness; (c) stability on storage, particularly related to the moisture content of the end product and (d) the nature of any leavening agents used, particularly if this is active yeast.

Within this product group, flavorings may be incorporated in one of four ways: (a) mixing into the dough or batter prior to baking; (b) spraying onto the surface of the product as it emerges from the oven; (c) dusting on to the surface after cooking and oiling and (d) introduction into the cooked product as a cream filling, glaze or coating.

Of these, baked-in flavors pose the greatest problems both of volatile losses and flavor change. Baking involves very high temperatures and relatively long dwell-times and these conditions place a great constraint on the use of any flavorings in the dough or batter. Since flavorings are volatile, usually a considerable proportion of an added flavoring is lost during the baking process. The extent of these losses is usually determined experimentally and an additional quantity added to give the desired flavor level in the end product. The dosage level of any added flavoring should be adjusted carefully as overflavored baked goods are much less attractive than underflavored. The color balance in the end product is also important and the use of combined flavorings/colorants is not recommended as this may restrict dose adjustment. The number of really satisfactory flavors is strictly limited. In practice, it has been found that the use of fat or oil-based flavors reduces losses. With liquid flavorings it is advisable to use a nonvolatile solvent such as propylene glycol (where this is permitted) and to premix the flavoring with any fat present in the formulation. Heat resistant powder flavors made by multistage encapsulation are commercially available and are very effective under these harsh conditions. The secondary capsule, being water insoluble, ensures that such products resist breakdown during the preparation of the wet dough and melt to expose the primary capsule only during the later stages of the baking cycle. At this stage the water content of the product is too low to result in a full release of flavor which only occurs when there is an excess of water, such as in the mouth during chewing. Although these encapsulated products may appear initially expensive they have a reduced use level which makes them a viable flavoring source.

Flavor emulsions are particularly suitable for use in bakery mixes although care must be taken in their incorporation into such products as foam-type cakes (i.e., sponge and angel cakes) as the essential oils present may cause the foam to fall back if beaten with eggs or egg whites. Flavoring emulsions should be added by gentle stirring or folding just before the flour is added. Fillings for use in baked goods may be either dairy cream, butter cream, fondant, custard, glazes or jellies (Heath 1978).

3. Snack Foods

Snack foods in this context will be taken to include both potato chips and extruded products based on various farinaceous materials, as biscuits, chocolate-based snacks and instant snack meals are better considered in other product groups. In the case of chips, flavoring is limited to surface dressing or dusting and many specialty flavors such as salt and vinegar, barbeque and smoky bacon are widely popular. Chips made from reconstituted potato starch offer a great opportunity for including the flavor in the dough prior to cooking (Blake and Renwick 1978).

Extruded products pose additional problems and flavoring is accomplished in two ways: (a) incorporation of the flavoring into the dough or premix prior to extrusion, which will be influenced by whether it is a hot or cold extrusion process and (b) external application by dusting or coating (Matz 1984).

Cold extruded products such as pasta are best flavored by using emulsions added to the system with the formula water, although preblending of a spray-dried flavor with the flour portion is also quite acceptable if somewhat more tedious.

Hot extrusion is a very efficient method of manufacturing a wide range of foods based on corn, oat or wheat flour or vegetable proteins or mixtures of both, and the technique is described by Smith (1978, 1985). The processing parameters are very severe as in most extruders the mix is exposed to high pressure (500–900 psi) and very high temperatures (120°–176°C or even higher) followed by a rapid steam flash-off as the product expands on leaving the extruder nozzle. It is necessary for the product then to be dried to about 8% residual moisture. Such conditions cause evaporation of a significant percentage of flavor volatiles and impose severe constraints on the type of flavoring that can effectively be used (Lyon 1980; van Polen 1980). Acceptability is generally one of trial and error under the exact processing conditions envisaged.

In the case of thermal reaction flavors (e.g., meatlike flavor) based on the Maillard and related reactions between sugars and amino acids, the precursors may be added to the premix and allowed to react during the extrusion, although variations in processing conditions may well result in unacceptable nuances in the achieved flavor profile. The preparation of meat analogs based on soy flour poses somewhat different problems as here it is essential that no residual soy flavor should remain in the end product (Ritter 1978).

The application of surface dustings is very widely used but the method is not without its problems. Numerous ingenious application techniques have been described in the literature (Mathewson and Swientek 1985). Savory flavors, particularly those containing hydrolyzed vegetable protein, yeast extract, cheese and onion powders, are usually hygroscopic and under factory conditions readily absorb atmospheric moisture and tend to cake and become sticky. This leads to uneven distribution and a blocking of the application machinery. From a consumer's point of view, surface-applied flavors are only satisfactory where the base product is acceptably flavorful and far less so when the product is flavorless, as in the case of most extruded snacks.

The use of residual surface oil or oil sprays poses limitations, not only on the nature of the dusting but on the shelf life of the product as rancidity readily sets in. Even in well-sealed packs the shelf life of such products is strictly limited. The use of electrostatic applicators, which avoid the use of oil, is now proving to be a much more acceptable method than tumbling or air blowing.

4. Sugar and Chocolate Confectionery

The processing constraints which determine the most suitable flavorings for use in sugar confectionery are so diverse that each warrants separate consideration (Weekes 1978). A detailed examination of each product group is outside the scope of this text and reference should be made to one of the standard works covering this industry (Alikonis 1979; Minifie 1980). The main sectors involved are high-boiled confectionery (i.e., hard candy); low-boiled confectionery (i.e., chewy caramels); starch-deposited confectionery (i.e., pastilles); chewing gum and chocolate.

The list of available flavorings is comprehensive and most flavor manufacturers readily supply usage data applicable to the various end-product groups concerned. Some flavorings on their own are quite flat and lifeless even if they do have an impressive aroma. Fruit flavors generally need the support of citric, tartaric, malic or lactic acids in amounts varying between 0.25 and 5% depending on the nature of the end product.

In confectionery products where added flavor is paramount, choice is critical and one must try for the most natural even though the use of entirely natural flavorings is not practicable. Excellent imitations of most natural flavors are now available with a wide choice of nuances to give any desired profile. Using flavor in unusual combinations is rarely successful as consumers are very conservative.

In modern confectionery manufacture, continuous and automated processing is replacing the long-established batch methods and the new parameters impose special constraints on any added flavoring ingredients. In most plants, the flavoring is injected into the cooked sugar mass as it moves through the system, hence mixing time is strictly limited and, for uniform flavor effect, ease of dispersion is critical. Flavorings must withstand 154°C for a relatively short period but much longer dwell-times at 140°C are not uncommon. Such conditions can induce undesirable profile changes. The need to use vacuum for deaeration also leads to flavor loss and it is usual to add up to 25% more flavoring than would be necessary in batch processing in order to achieve the same sensory impact in the end product.

Chocolate manufacture is very much something for the specialist but even here added flavorings have a part to play (a) to modify the flavor of the basic chocolate mass (e.g., the use of vanilla extract, vanillin or ethyl vanillin) to give a rounded smoothness to the profile; (b) to impose an overriding but compatible flavor (e.g., orange, rum, peppermint), and (c) to flavor fondant-based or other centers.

Chewing gum is notorious from a flavor application point of view. The base itself, usually with a significant content of gum chicle, has the ability to absorb very large quantities of flavor with no noticeable effect on the flavor sensibility in the mouth. Dosage rates of up to ten times those used for candy are not unusual. The actual physical state of the gum can be significantly altered and its chewing properties destroyed by certain flavoring ingredients and solvents used in the manufacture of flavorings. Flavors for chewing gum are always very concentrated, relatively cheap and are formulated to be free from deplasticizing ingredients.

5. Pickles and Sauces

This group of products is characterized by its high content of acetic acid (vinegar) and the use of herbs and spices to achieve the desired flavor profile having a good aromatic impact. Variety of end products is obvious, the main quality attribute being the correct balance between piquancy, spiciness, acidity, saltiness, sweetness and fruity character to provide the appropriate flavor adjunct to the dishes with which they are to be eaten. The group embraces thick sauces (e.g., fruit sauces, prepared mustard, salad creams and dressings, mayonnaise, tomato ketchup); thin sauces (e.g., Worcestershire); thick pickles (e.g., mustard, relish, chutney) and clear pickles (e.g., single vegetables, sweet mixed pickles).

The processing technology involved is specific to each product subgroup hence one cannot generalize on flavor usage. However, in broad terms, it is the judicious use of herbs and spices, coupled with the correct degree of acidity which largely determines the acceptability of the end product. In developing a suitable flavoring system two factors are determinants: (a) the extent to which the flavor should complement or override the predominant flavor of the base and (b) the necessity for the flavoring to be completely soluble to give a clear liquid phase.

Processed spices, in the form of oleoresins and/or essential oils, are widely used for this purpose in the form of solubilized flavorings containing an acceptable solvent or as an emulsion. These can readily be solubilized by admixture with a polysorbate (e.g., Tween 80) but the use of these ingredients is restricted by legislation in many countries. Flavorings based on acetic acid are widely available. Other methods of dispersion generally give a pickling liquor which is more or less cloudy, calling for filtration before use. The level of flavoring must be determined experimentally.

6. Soups

This form of nutriment is probably the most widely used worldwide, for almost anything that can be eaten can be used in making a soup. Apart from those prepared domestically from fresh ingredients, manufactured soups occur in three forms: (i) canned—either single strength ready to take or concentrate, (ii) dry mix for reconstitution as required with or without the need for cooking and (iii) frozen. Each of these calls for a different approach to flavoring so as to achieve consistency in the end product.

In the manufacture of canned soups, concentrates and quick frozen soups, the end product is liquid and hence it is advantageous to use liquid flavorings and seasonings although dispersed spices, blended with other dry ingredients, are quite satisfactory. In automated processing, a total liquid flavoring system is advantageous as this allows accurate metering and a very ready dispersion throughout the bulk.

Dehydrated soups require the use of dry powder flavorings and seasonings, but the question of seasoning blends and the flavor profile of the end product are matters of opinion depending on the raw materials, consumer expectations and price. Dispersed or plated spices are adequately stable when the product is packed in laminated sachets but, wherever possible, flavoring products in which the aromatic fraction has been encapsulated should be selected to ensure maximum shelf life. It should, however, be remembered that such dry products do not have very much intrinsic smell and may be misjudged in consequence. Most flavor manufacturers offer a wide range of spray-dried products and advise on their optimum application. Encapsulated flavorings are of particular value in soup powder mixes formulated for automated vending in predosed containers as unencapsulated aromatics are not only liable to be lost from the product but may also contaminate other products within the machine.

7. Ice Cream and Frozen Goods

The choice of flavorings for ice cream, sherbet and water ices is wide and their selection depends not only on the quality of the end product but more on regional and age preferences and popular demand. The flavoring of products intended to be distributed under chilled or deep-freeze conditions calls for careful attention to application techniques and the use of flavorings which have been purpose designed for these types of products. The following are the determinants: (a) accuracy of dosage to achieve flavor consistency; (b) economics, as flavorings contribute significantly to product quality; (c) ready dispersion in the product mix;

(d) the percentage of lipids present in the product, as partition can significantly affect flavor perception; (e) hygiene, as flavors are often added after pasteurization they must be microbiologically acceptable, and (f) stability under pasteurizing conditions.

It must be remembered that these products are eaten cold and that this has a marked impact on flavor perception due to "freeze-out" and resulting fading of flavor impact (Grabb *et al.* 1979). There is an enormous range of flavorings available for use in ice cream, including naturals such as cocoa powder, fruit pastes, natural extracts and essences, particularly vanilla, and, of course, imitation flavorings. The profiles of these are legion and synthetic chemicals such as vanillin and ethyl vanillin together with ingredients such as fruit pieces, nuts or crystallized fruits are almost universally used.

When checking the usage level of any flavoring material it is essential to establish the degree of overrun which, for ice cream, is generally 70–80%, but in water ices may be as low as 25%. This obviously will have a considerable impact on the flavor level in the end product (Arbuckle 1985).

Ice cream contains milk solids so that highly acidic flavors will tend to precipitate casein and so ruin the texture of the end product. Frozen mousse is in effect an overstabilized ice cream containing considerably more gelling agent than would normally be required in a standard ice cream mix. This extra gelling power means that casein precipitation is not a major problem hence mousse formulations can contain much higher levels of acids than are possible in ice cream in order to achieve a truer, full fruity, flavor profile.

8. Soft Drinks

This product group includes carbonated beverages, both clear and cloudy; noncarbonated or "still" products such as squash and cordials usually concentrated and requiring dilution by the consumer; specialties such as ginger beer, root beer; and the various cola products and "crystal" beverages which are in powder form requiring reconstitution with water (Downer 1973; Ratcliffe 1978).

Flavorings for such a diverse product group must (a) impart the characteristic profile implied by the name; (b) be technologically compatible; (c) be stable to heat, light, acids and preservatives, particularly sulfur dioxide; (d) impart the correct physical appearance to the end product; (e) be free from spoilage organisms and (f) comply with existing legislation.

Their application is, of course, dictated by the nature of the beverage as ready for drinking. Many flavoring materials are either insoluble or only very sparingly soluble in water, so that special techniques have to be employed to ensure a uniformly flavored and stable product. Two widely used methods are the following: (a) They can be dissolved in a permitted solvent so that when added to the concentrated bottling syrup and ultimately diluted with water they remain in solution. (b) They may be emulsified. Such flavorings are generally used to produce a cloud in a drink which would otherwise be clear.

Fruit and berry juices are widely used as flavor bases for soft drinks and most of these are concentrated by the removal of water under vacuum to give a commercial product which is between four and six times stronger than the original juice. The flavor value of these concentrates depends not only on the degree of concentration but on the precise processing conditions used in their manufacture. From an application point of view, fruit products may be offered in sealed containers which have been pasteurized, in which case the whole contents must be used once the container has been opened, or they may be in multidose containers the contents of which contain a permitted preservative, usually sodium benzoate. All fruit-based flavorings are best stored under refrigeration or in a cold store.

Finished soft drinks have a specific gravity of about 1.05, whereas the flavorings usually contain essential oils having a gravity of less than 1.00. To achieve a stable suspension on dilution it is necessary to "weight" the flavoring components. Formerly this was done by the use of a much heavier brominated vegetable oil but the use of this and other weighting agents (e.g., ester gum) is now restricted by legislation which varies considerably between countries. Flavoring manufacturers have attempted to overcome the problems of "ringing" in various ways and can best advise on the optimum usage of their flavoring emulsions.

Crystal beverages offer few problems in flavor application. They fall into two broad categories: (i) high quality products made from spray-dried fruit and (ii) cheaper, lower quality products, based on imitation flavorings. A typical formulation contains sugar (62%), dextrose (24%), citric acid (7.5%), sodium citrate (1%), ascorbic acid (0.2%) together with a dry flavoring and an appropriate colorant (Heath 1978). The use of encapsulated flavors is recommended as these give a long shelf life. Where a cloudy product is desired, a spray-dried vegetable oil may be incorporated into the above formulation at about 4%.

G. Conclusion

Each product group within the food and related industries poses its own specific flavoring problems to a point where generalizations are of limited value. Within each segment of the industry there are technologists whose responsibility is to ensure that the end products are of a consistent quality and correctly flavored to satisfy customer expectations, technological needs and legal requirements. Within the flavor industry there are also technical service experts available to give advise on the choice of flavorings and their optimum application. The value of their advice depends on a knowledge of the manufacturer's precise flavoring problems. The better the communication between the development technologist, the creative flavorist and the flavorings applications technologist, the better the chances of success for the end product.

References

ALIKONIS, J. J. 1979. Candy Technology. AVI Publishing, Westport, CT.

ARBUCKLE, W. S. 1985. Ice Cream, 4th edition. AVI Publishing, Westport, CT.

BLAKE, A. and RENWICK, J. 1978. Technology of flavouring snack products. Food Trade Review *48*(8), 459–462.

BLANCHFIELD, J. R. 1978. The importance of flavours in product development—introduction of symposium. Food Trade Review *48*(8), 459–473.

BRENNAN, J. G., BUTTERS, J. R., COWELL, N. D. and LILLY, A. E. V. 1982. Food Engineering operations, 2nd edition. Appl. Sci. Publ, Barking, Essex.

DOWNER, A. W. E. 1973. The application of flavours in the soft drinks industry. Flavour Ind. *4*(11), 488–490.

GRABB, W., BECK, E. and BERNEGGER, T. 1977. Flavourings for fat-containing foods. Int. Flavours Food Additives *8*(Mar./Apr.), 63–66.

HALL, R. L. 1968. Flavor and flavorings, seeking a consensus of definition. Food Technol. *22*, 1496.

HALL, R. L. and MERWIN, E. J. 1981. The role of flavors in food processing. Food Technol. *35*(6), 46–52.

HEATH, H. B. 1978. Flavor Technology, Profiles, Products and Applications. AVI Publishing, Westport, CT.

HEATH, H. B. 1982/1983. Application of food flavorings. Perf. Flavorist *7*(Dec./Jan.), 27–38.

HELDMAN, D. R. W. and SINGH, R. P. 1981. Food Process Engineering, 2nd edition. AVI Publishing, Westport, CT.

LONG, L., KOMARICK, S. and TRESSLER, D. K. 1982. Food Products Formulary: Vol. 1. Meats, Poultry, Fish and Shellfish, 2nd edition. AVI Publishing, Westport, CT.

LYON, L. 1980. Popularity of extrusion processing places added demands on flavorings. Food Product Dev. *14*(1), 58–61.

MARTIN, R. E., FLICK, G. J., HERBARD, C. E. and WARD, D. R. 1982. Chemistry and Biochemistry of Marine Food Products. AVI Publishing, Westport, CT.

MATHEWSON, S. and SWIENTEK, R. J. 1985. Redesign of potato chip lines reduces seasoning costs 30–40%. Food Processing *46*(11), 110–111.

MATZ, S. A. 1972. Bakery Technology, 2nd edition. AVI Publishing, Westport, CT.

MATZ, S. A. and MATZ, T. D. 1978. Cookie and Cracker Technology, 2nd edition. AVI Publishing, Westport, CT.

MINIFIE, B. W. 1980. Chocolate, Cocoa and Confectionery: Science and Technology, 2nd edition. AVI Publishing, Westport, CT.

MOUNTNEY, G. J. 1976. Poultry Products Technology, 2nd edition. AVI Publishing, Westport, CT.
MUSKOWITZ, H. R. and CHANDLER, J. W. 1978. Consumer perceptions, attitudes and trade-offs regarding flavor and other product characteristics. Food Technol. *32*(11), 34–57.
VAN POLEN, F. 1980. Snacks-flavorings before extrusion. Food FIPP *2*(1), 29.
RATCLIFFE, R. B. 1978. The role of flavour in drinks development. Food Trade Rev. *48*(8), 470–473.
RITTER, W. J. 1978. Flavoring systems for meat analogs and extenders. Food Product Dev. *12*(9), 60, 62–63.
SMITH, A. C. 1985. Extrusion—Product and process development. Food FIPP *7*(9), 17–22.
SMITH, O. B. 1978. Extrusion-Cooked Foods. *In* Encyclopedia of Food Science, M. S. Peterson and A. H. Johnson, eds. pp. 245–250. AVI Publishing, Westport, CT.
SULTAN, W. J. 1985. Practical Baking, 4th edition. AVI Publishing, Westport, CT.
TOLEDO, R. T. 1980. Fundamentals of Food Process Engineering. AVI Publishing, Westport, CT.
TRESSLER, D. K. and SULTAN, W. J. 1975. Food Products Formulary: Vol. 2. Cereals, Baked Goods, Dairy and Egg Products. AVI Publishing, Westport, CT.
WEEKES, J. 1978. Selection of flavours for new confectionery products. Food Trade Rev. *48*(8), 462–465.

Additional Reading

Process Engineering

CHARM, S. E. 1978. Fundamentals of Food Engineering, 3rd edition. AVI Publishing, Westport, CT.
FARRELL, A. W. 1976. Food Engineering Systems: Vol. 1. Operations. AVI Publishing, Westport, CT.
McKENNA, B. M., ed. 1984. Engineering in Food, Vols. 1 and 2. Elsevier, London and New York.

Meat Products

PEARSON, A. M. and TAUBER, F. W. 1984. Processed Meats, 2nd edition. AVI Publishing, Westport, CT.
PRICE, J. F. and SCHWEIGERT, B. J. 1971. The Science of Meat and Meat Products. W. H. Freeman, San Francisco, CA.

Bakery and Baked Goods

MANLEY, D. J. R. 1983. Technology of Biscuits, Crackers and Cookies. Wiley, Chichester, England.

Snack Products

GUTCHO, M. 1973. Prepared Snack Foods. Noyes Data Corp., Park Ridge, NJ.

Pickles and Sauces

BINSTED, R., DEVEY, J. D. and DAKIN, J. C. 1971. Pickle and Sauce Making. 3rd edition. Food Trade Press, London.

Soups

BINSTED, R. and DEVEY, J. D. 1970. Soup Manufacture, 3rd edition. Food Trade Press, London.

Soft Drinks

WOODROOF, J. G. and PHILLIPS, G. F. 1981. Beverages: Carbonated and Non-carbonated, rev. edition. AVI Publishing, Westport, CT.

General

RANKEN, M. D., ed. 1984. Food Industries Manual. Leonard Hill, Glasgow and London.

13

Flavors and the Law

A. Need for Legislation

Within our regulated social society consumers have a right to expect that the food available to them should not only be organoleptically attractive but also nutritious, wholesome and present neither short- nor long-term health hazards. In the distant past, most of man's diet was of natural origin, much of it being either from the wild, or cultivated or reared on the homestead. As communities developed and the pattern of living became progressively urbanized and interdependent, consumers ceased to be self-supporting and were forced to rely on foods produced and marketed by others. Initially such products were made by traditional methods and any flavorings were confined to the well-known herbs and spices. But with advancing technology and in response to consumer needs, foods were increasingly manufactured and marketed not merely locally but regionally, nationally and even internationally. There was probably small need for legislation in the control of locally produced food ingredients and products as the customer could quite effectively evaluate and select at the time of purchase, but in the modern food chain, legislation covering raw materials and end products has become essential in the interests of the consumer who can no longer readily assess quality and must accept products on trust. In all communities, the need for an adequate and safe food supply is undisputed but with the increasing complexity of food processing and distribution the means of achieving this may differ widely. The "Food Law Handbook" (Schultz 1981) provides a unique source of information on the complexities of the food legislation in the United States.

The acceptability and effective nutritional value of food is determined to a considerable extent by its sensory appeal, but primarily by its odor and flavor. This calls for the use of a large number and variety of flavoring ingredients which must be regarded as essential to the food formulation. It is this wide range and complexity of flavoring components which poses legislative problems both nationally and internationally.

Legislation covering flavoring materials is in a specialized area of the general regulations governing foodstuffs, the main objects of which are (a) the protection of the consumer's health, in particular the assurance that flavoring ingredients are safe to use; (b) the prevention of fraud by misrepresentation or deception and (c) the unification of standards necessary for well-regulated trade. In a booklet entitled "Basic Features of Modern Flavour Regulation," the International Organization of the Flavour Industry (IOFI 1976) stresses that legislation covering flavorings should be (a) technically feasible—enforcement must be practical, which implies the need for analytical techniques which are sensitive, accurate and specific to measure the critical attributes; (b) administratively feasible—regulations should be easily understood and invite maximum compliance rather than complex and difficult either to understand or administer and (c) equitable—all regulations, whether they be restrictive or permissive, should respect the principle of "equality before the law" irrespective of where they are applied.

B. Nature of Flavoring Ingredients

The IOFI defines flavorings as concentrated preparations, with or without solvents or carriers, used exclusively to impart flavor. They are not intended to be consumed as such. Internationally there is some measure of agreement regarding definitions for "food additives" and "flavorings" given in the *Codex Alimentarius* and other international publications but further classification of these food ingredients is open to wide interpretation. It is generally accepted that flavorings are indeed intentional food additives but should be considered as a quite separate group requiring specific legislation. The IOFI lists the following classes of substances within this category:

a. Natural aromatic plant or animal materials—used for their flavoring properties either as such or processed for human consumption (e.g., herbs, spices, aromatic vegetables).

b. Natural flavors—concentrated preparations obtained exclusively by physical means from aromatic raw materials (e.g., fruit juices).

c. Natural flavoring substances—components obtained from natural materials exclusively by physical means (e.g., distillation, expression, extraction).

d. Nature-identical flavoring substances—organic chemicals obtained by synthesis or isolated by chemical means from aromatic raw materials. These are chemically identical to substances present in natural products either as such or processed for human consumption (e.g., vanillin from wood lignin equates with vanillin in vanilla beans).

e. Artificial flavoring substances—synthesized chemicals which have as yet not been identified in natural materials intended for human consumption, either processed or not (e.g., ethyl vanillin, allyl hexanoate).

In the United States only two subcategories of flavorings are recognized and defined as follows:

a. Natural flavoring—a substance obtained from vegetable and sometimes animal sources exclusively through the appropriate physical processes. Those biological processes which occur spontaneously and roasting are assimilated to physical processes.
b. Artificial flavoring—a substance which has flavoring properties and which has been obtained by chemical means. This category includes (i) substances which exist in natural products and (ii) substances not present or as yet undiscovered in natural products.

Based on the preceding, the following definitions are legally accepted in the United States.

1. *Natural flavorings.* The essential oils, oleoresin, essence, extractive, protein hydrolysate, distillate, or any product of roasting, heating or enzymolysis which contains the flavoring principles derived from a spice, fruit, or its juice, vegetable or its juice, edible yeast, herb, bark, bud, root, leaf or similar plant material, meat, fish, poultry, eggs, dairy products or the fermentation product of any of these.
2. *Artificial flavorings.* Any flavoring which is not derived from the sources which yield natural flavorings.
3. *Spice.* Whole, broken or ground aromatic vegetable substances traditionally recognized as foods (e.g., onions, garlic, celery); no portion of the volatile oil or flavoring principles may be removed.
4. *GRAS.* Means "generally recognized as safe."

Most national legislation governing flavoring materials includes a preamble giving the appropriate definitions applicable to the ensuing regulations. This is an essential part of the legislation and requires careful study to ensure compliance.

C. Systems of Flavor Legislation

The principal industrial countries have developed regulatory systems which differ widely in their scope and application. It is well recognized that in the interests of international trading there should be a harmonization of such legislation as it applies to foods and food ingredients and additives, including flavoring materials, and this is very slowly taking place. Inevitably history and precedent play a major part, as flavorings have been used for very many years in most countries of the world and their use reflects national needs and attitudes which are generally slow to change.

Central to all food additive legislation are two criteria: the "need to use" and "safety in use," and it is the interpretation of these which dictates the method by which the use of flavoring materials is regulated. Conventional food additives (e.g., emulsifiers, stabilizers, preservatives) generally have well-defined technological functions and their use is governed by "good manufacturing practice" to achieve the desired end product. Flavorings, on the other hand, are judged on an entirely different basis as the "need to use" is determined not technologically but by considerations of palatability and hedonic acceptance of the food product. In most countries permitted conventional food additives are strictly limited both in their nature and usage conditions. In total the number probably does not exceed 200 whereas the number of individual flavoring components available to the flavoring manufacturing industry is in excess of 3000. A large proportion of these organic chemicals occur in nature and are consumed in natural foodstuffs without question. It is only when they are intentionally added as flavorings that controls are applied. The major difference between conventional food additives and flavoring substances is that the latter are used at very low levels of concentration and are self-limiting, as excessive dosing almost always results in nonacceptable products (Stofberg and Stoffelsma 1980).

Any legislation implies a system of enforcement so that the method adopted may be dictated by the technical and economic resources available to ensure compliance without either restricting trade or exposing consumers to unwarranted health hazards. Basically, flavorings may be controlled in one of three ways:

a. *Negative list.* This restrictive system defines all flavoring substances which are prohibited for inclusion in foodstuffs and other comestibles or may be used only to a very limited and defined extent. All other nonlisted flavoring materials may be used without limitation. This system is generally regarded as being oversimple and does not offer adequate protection to consumers. It poses no problems for the use of natural and nature-identical materials but is unsuitable for controlling artificial flavoring substances. Despite its limitations, this system is widely used. In the United Kingdom, the Food Additives and Contaminants Committee (FACC 1976) proposed a list of 16 flavoring substances which it recommended be prohibited in foods but this has not been accepted into law and the list acts only as a guide to manufacturers.

b. *Positive list.* This defines all the individual substances which are permitted for use in food flavorings. Any substance which is not listed is automatically prohibited. Such a system poses real problems in practical application. Natural flavoring materials are included even though their chemical composition is not fully known. Any components which are newly identified are not acceptable unless the listings are regularly updated in the light of current research findings. There is also the difficulty of enforcement as each peak in a very complex chromatogram would have to be positively identified to the point where it could stand up in a court of law. When one notes that there are some 300 components in natural strawberry aroma and in excess of 600 in coffee the immensity of this task is obvious and the costs unacceptably high. In spite of the sensitivity and precision of modern instrumental analysis, the finite analysis of a complex flavoring composition is still almost impossible to achieve with any degree of certainty and precise reproducible results would be essential in any legal enforcement action.

The inclusion of any chemical in a positive list implies that it is entirely safe for use under the conditions stated. Such positive lists are necessarily limited, involve considerable and very expensive toxicological testing and indeed seriously limit research and technological development unless there is a system for the regular admission of new substances. Toxicological testing would have to be carried out by the researching company who would probably be unable to recover these on any commercial basis. Too restricted a list of permitted substances would result in a serious limitation of available flavoring products to the detriment of the consumer.

c. *Mixed system.* This offers the best of the former options being based on a positive list for artificial flavoring substances and restrictive lists for natural and nature-identical flavoring substances. The system gives adequate consumer protection and is enforceable using efficient analytical control without excessive costs, and the method of approval does not restrict research into new flavoring substances. This is certainly the system best suited to modern flavoring legislation.

The advantages and disadvantages of these various systems are further discussed by Vodoz (1975, 1978); Burke (1976); Lenane (1976); IOFI (1977); Knowles (1978); and Guglielmina (1982). The various systems in use worldwide are listed in Table 13.1.

D. Flavor Legislation in the United States

The United States has always been the leader in legislation aimed at the protection of consumers and the first federal legislation governing

Table 13.1. Types of Flavoring Legislation in Use Worldwide in 1979[a,b]

No Specific Flavor Regulations— General Food Laws Apply	Negative List of Prohibited Substances	Positive List of Permitted Substances	Both Positive and Negative Listings	Flavorings Requiring Prior Government Authorization
Austria	Australia	Japan	Argentina	Bulgaria
Belgium	Brazil	Romania[c]	East Germany	Czechoslovakia
Denmark	Canada	Soviet Union[c]	West Germany	Finland[d]
Eire (Ireland)	Chile	Switzerland	Italy	Hungary
France	India	United States[e]	Netherlands	Peru
Greece	New Zealand		Spain	Poland
Jamaica	Singapore		Yugoslavia	
Luxembourg				
Norway				
Portugal				
South Africa				
Sweden				
United Kingdom[d]				

[a] From Heath (1981).
[b] Legislation is frequently changed and reference should always be made to the appropriate authority for the up-to-date position.
[c] Very short positive list coupled with prior government authorization.
[d] Positive list in preparation but not yet accepted into law.
[e] GRAS list with very short negative list and product group specifications.

food and drugs was the Food and Drug Law (1906). This relatively simple piece of legislation was replaced by the Federal Drug and Cosmetic Act (1938) which has remained the basic law although updated regulations were incorporated by the Food Additives Amendment (1958). These amendments have a threefold purpose: (a) the protection of the public health by requiring proof of safety in the use of any food additive, (b) regulation of the safe use of substances deemed technologically necessary to safeguard and improve the food supply, (c) the forbidding for use in food of any substance, in any amount whatever, which is found to induce cancer when ingested by man or animal (the Delaney clause). This latter clause has been the subject of much debate over the past several years but is still the accepted principle. These amendments influenced the legislation governing flavoring substances and led to the introduction of the concept of "generally recognized as safe" (GRAS).

Flavoring materials which are and are not permitted for use in foods in the United States are listed in the USA Code of Federal Regulations, Title 21, Chapter 1, as amended in the Federal Register 42 (1977). In addition, flavoring substances included in the Flavoring Extract Manufacturer's Association (FEMA) GRAS lists 4–13 may also be used (Hall and Oser 1965, 1970; Oser and Hall 1972; Oser and Ford 1973a, 1973b, 1974, 1975, 1977, 1978, 1979; Oser *et al.* 1984, 1985).

1. Legally Permitted Flavoring Substances

The listings of flavorings permitted in the United States appear in the following subsections of Title 21:

Part 172. Flavoring agents and related substances.

172.510. Natural flavoring substances and natural substances used in conjunction with flavors—131 items, listed under common name, scientific name and limitations in use.

172.515. Synthetic substances and adjuncts—comprising 633 GRAS items included in the FEMA GRAS lists 1–3.

Other sections cover such items as cocoa for manufacturing, flavor enhancers and hop extract used in the manufacture of beer.

Part 182. Substances generally recognized as safe.

182.10. Spices and other natural seasonings and flavorings.

182.20. Essential oils, oleoresins (solvent free) and natural extractives (including distillates).

182.30. Natural substances used in conjunction with spices and other natural seasonings and flavorings.

182.40. Natural extractives (solvent free) used in conjunction with spices, seasonings and flavorings.

182.50. Certain other spices, seasonings, essential oils, oleoresins and natural extracts.

182.60. Synthetic flavorings substances and adjuncts.

The remaining sections are devoted to substances which migrate from packaging and pesticide chemicals.

In the GRAS listings of synthetic flavorants no tolerance levels or restrictions in use are included beyond the requirement that they be used in the minimum quantity necessary to produce the intended effect and otherwise in accordance with all the principles of good manufacturing practice. However, limitations are imposed on certain natural flavoring substances when used in alcoholic beverages.

2. Prohibited Flavoring Substances

Nothing which does not appear in one of the permitted lists may be used in foods in the United States. Substances which are specifically prohibited by the legislation include calamus oil, coumarin and safrole.

3. Evaluation of Materials for Inclusion in Permitted Lists

In 1971 (Federal Register dated June 25) the FDA published the criteria to be used in the assessment of a chemical for inclusion in the GRAS list and brought into line the acceptance of flavoring materials and other food additives in spite of the many dissimilarities in use levels. The term GRAS with respect to flavorings has acquired a special connotation. Such materials are regarded as having a low priority for exhaustive toxicological evaluation and quite safe to use in foods at levels within the normal technological needs and sensory requirements of the end product.

Both the FDA and FEMA admit that there is a degree of overlapping and duplication in their activities in the assessment of toxicological safety, but the FDA has stated that it is the sole judge of what substances are to be classified as GRAS and this has posed problems of the basis for such decisions (Stofberg 1980). Work is currently in hand in both organizations to produce a single listing of all flavoring materials having an established safety in use. Lenane (1976) has described the cross-checking procedure used. This is complicated and the progress to date has been slow. The work of the FEMA expert panel is ongoing and regular update articles appear in *Food Technology* and specific legislation is published in the *Federal Register* and usually noted in the same journal.

The Select Committee on GRAS Substances of the Federation of American Societies for Experimental Biology, after 10 years of evaluating GRAS substances, has issued a report entitled "Insights on Food Safety Evaluation" (1983) in which there are specific recommendations for improvements in the legislation governing flavoring substances. These include phasing out the GRAS list, modifying the Delaney clause and improving human and animal test protocols. Undoubtedly these suggestions will lead to considerable discussions and may ultimately result in changes in the present legislation. The future of artificial flavoring ingredients is discussed by Stofberg (1980).

E. Flavor Legislation in Europe

Each of the European countries has its own unique national legislation on food additives although there is presently little specific legislation governing the use of flavorings in food beyond the requirement that foods containing flavorings should neither endanger public health nor be used in such a manner as to mislead the consumer. The influence of intranational directives emanating from the EEC Council of Ministers is having an increasing influence on national attitudes towards the use of flavorings in food in the interests of trading relationships. Such directives have no effect until they have been incorporated into the national legislation of each member country. In consequence there are still many differences which remain unresolved. In some countries, the use of artificial flavorings, even those compounded from nature-identical chemicals, in designated food products is either prohibited or requires that the food be declared as artificial. Only Germany, Italy and the Netherlands have a mixed system of flavor legislation. In Germany there is a restrictive list of five natural and nature-identical flavoring substances (calamus, quinine, coumarin, quassia wood and thujone); a negative list of 20 natural and nature-identical flavoring substances and a positive list of 18 artificial flavoring substances. The Netherlands legislation is similar but contains a positive list of about 200 artificial flavorants all with maximum use levels. In Italy the positive list contains only 17 items which may be used in only a few specified foods. Denmark originally introduced a fully positive list system but found that it was unenforceable and suspended its operation.

1. The European Economic Community (EEC)

Presently there is no formal EEC proposal on flavoring legislation although a discussion document was published in 1976. There is general agreement among the member countries that artificial substances should be strictly controlled but there is a difference of opinion about the need for or practicality of a positive listing of flavoring materials. The situation is made more complex by the activity of the Council of Europe (CE) which is a United Nations body and not a part of the EEC (although all member states also belong to the Council of Europe). This body has approached the problem somewhat differently and in a much more restrictive way. The Working Party on Flavoring Materials has recognized only two categories of flavoring substances: (a) natural flavoring substances—obtained from vegetable and sometimes animal sources exclusively through the appropriate physical processes and (b) artificial flavoring substances—obtained by a chemical process including substances which exist in natural products (nature-identical flavoring substances) and substances not present or as yet undiscovered in natural products (artificial flavoring substances).

These two main groups are further divided into materials which are (i) admissible, (ii) temporarily admissible and (iii) not fully evaluated.

Based on this classification the Council of Europe has drawn up positive lists embracing some 1700 items each of which is serially numbered though it should be noted that the numbers do not coincide with those published by FEMA. These lists appear in what is called the "Blue Book"—Natural Flavouring Substances, Their Sources, and Added Artificial Flavouring Substances—published by Council of Europe in 1974.

A Commission of the EEC, working on the proposals of the CE, recognizes three categories of flavoring materials—natural, nature-identical and artificial. In February, 1982, the European Parliament accepted that there be a positive list of artificial flavoring substances and their source materials and a negative list of nature-identical flavoring substances. In addition there should be a negative list of source materials for the production of natural flavoring substances.

2. International Organization of the Flavour Industry (IOFI)

Although not a government or legislative body IOFI, which represents the flavor industry worldwide, does make a significant contribution to the data available for flavor legislation wherever this is required. Its Expert Committee on Flavouring Substances constantly reviews the lists of artificial flavoring materials approved by experts in the United States (FDA and FEMA), the Council of Europe (CE), the UK (FACC, BEMA, BIBRA) and other recognized bodies and publishes a positive list of these materials as a section in its Code of Practice for the Flavour Industry. It also publishes a restrictive list of natural and nature-identical flavoring substances the use of which in foods is limited to specified levels.

In the positive lists only substances for which use levels have been established are included. The IOFI is particularly active in establishing consumption figures based on the ratio between the quantity consumed as an ingredient of traditional foods and the quantity consumed as an intentionally added flavoring (Stofberg 1981, 1983).

3. The Codex Alimentarius Commission

The Codex Alimentarius Commission (CAC) was established in 1962 under the aegis of the Food and Agriculture Organization of the United Nations (FAO) and the World Health Organization (WHO). The aims of the organization include protecting the health of consumers, ensuring fair practices in the food trade and coordinating the preparation of internationally acceptable food standards. The Joint Expert Committee on Food Additives (JECFA), an advisory body, have grouped food additives and flavoring materials into three toxicological categories: (A) A1 flavoring substances which have been fully cleared by JECFA and A2 flavoring substances with incomplete evaluation but which have been accepted for use on a provisional basis; (B) a working list of flavoring substances on which evaluation is pending but in which technological interest has been confirmed; and (C) C1 substances which are considered by JECFA as unsafe for use in food and C2 substances restricted in use for reasons of safety to health.

Categories A1 and A2 are not strictly positive listings as the Codex Commission does not regard flavorings as unacceptable unless they appear in list C1 or C2.

4. International Flavor Legislation

The flavor legislation as it applies to the major trading countries worldwide has been reviewed by Heath (1981). Obviously, regulations are constantly being amended and new ones introduced and it is essential to make certain that one's information is regularly updated and to confirm the current legislative position and regulations to ensure compliance with the law.

F. Labeling of Flavors

The whole purpose of labeling is to inform the user or consumer of the contents of any given package, hence legislation is directed towards ensuring that the information is correct, clear, precise and given in a manner which neither directly nor indirectly can lead to deception as to the true nature of the contents. Although the basic premise is simple, national legislation differs considerably in detail both for the labeling of flavoring materials and for foods containing flavorings (Heath 1981).

In the case of bulk flavoring preparations, as opposed to flavoring packed for direct sale to the domestic user, it is always necessary for the label or container itself to be clearly marked in bold letters with (a) the correct designation of the contents, including any distinguishing product code numbers or quality classification; (b) the net weight or volume of the contents (in the case of drums it is usual to stencil the gross, tare and net weights as an aid to checking); (c) a distinguishing batch number by which the contents can be directly related to a specific manufacturing formulation and (d) the name and address of the manufacturer (or supplier, if different). Other details may be required (e.g., country of origin, fire hazard data). Information given should wherever possible be the same as that given in any accompanying documentation. Labels should be affixed with an adhesive which is not water soluble. Many practical aspects of labeling flavorings and foods are discussed by Bauer 1984.

U.S. Labeling Regulations

The labeling regulations of the United States are published in the Code of Federal Regulations, Title 21, sec 1.12(g)(2). These require that mixtures of flavoring materials shipped to food processors shall be fully declared on the label. However, in place of the full listing the following declaration may be made as appropriate: *All flavor ingredients contained in this product are approved for use in a regulation of the food and drug administration*. Any nonflavor or nonapproved flavorants must be listed individually by name.

Flavoring ingredients listed in FEMA GRAS lists Nos. 3–9 are approved under this section until such time as the FDA has completed its review procedures. Items included in the FEMA GRAS lists Nos. 10–14 are not automatically approved unless they are the subject of a petition or are covered by a SLR (Scientific Literature Review), in which case they need not be declared.

References

BAUER, K. T. 1984. Legal considerations in the development and application of flavors. *In* The Development and Application of Natural and Artificial Flavor Systems, pp. 37–46, E. J. Mervin, ed. Allured Publishing, Wheaton, IL.

BURKE, C. 1976. Flavourings legislation in the EEC. Int. Flavours Food Additives *5*(Nov./Dec.), 265–267.

FACC. 1976. Report on the Review of Flavourings in Food. H. M. Stationery Office, London.

GUGLIELMINA, F-C. 1982. Some aspects of flavourings legislation. Perf. Flavorist *7*(Oct./Nov.), 61–68.

HALL, R. L. 1960. Recent progress in the considerations of flavoring ingredients under the Food Additives Amendment. Food Technol. *14*(10), 488.

HALL, R. L. 1975. GRAS—Concept and application. Food Technol. *29*(1), 48, 50, 52–53.

HALL, R. L. and OSER, B. L. 1961. Recent progress in the consideration of flavoring ingredients under the Food Additives Amendment. 2. GRAS substances. Food Technol. *15*(12), 20.

HALL, R. L. and OSER, B. L. 1965. Recent progress in the consideration of flavoring ingredients under the Food Additives Amendment. 3. GRAS substances. Food Technol. *19*(2), 151.

HALL, R. L. and OSER, B. L. 1970. Recent progress in the consideration of flavoring ingredients under the Food Additives Amendment. 4. GRAS substances. Food Technol. *24*(5), 25.

HEATH, H. B. 1981. Source Book of Flavors. AVI Publishing, Westport, CT.

IOFI. 1976. Basic Features of Modern Flavour Regulation. Int. Organization of the Flavour Industry, Geneva, Switzerland.

IOFI. 1977. Basic features of modern flavour regulation. Perf. Flavorist *1* (Dec./Jan.), 7–14.

KNOWLES, M. E. 1978. Flavouring legislation. Int. Flavours Food Additives *7*(Jan./Feb.), 33–36.

LENANE, G. 1976. USA legislation on flavourings. Int. Flavours Food Additives *7*(Nov./Dec.), 267–270.

OSER, B. L. and HALL, R. L. 1972. Recent progress in the consideration of flavoring ingredients under the Food Additives Amendment: 5. GRAS substances. Food Technol. *26*(5), 35.

OSER, B. L. and FORD, R. A. 1973a. Recent progress in the consideration of flavoring ingredients under the Food Additives Amendment: 6. GRAS substances. Food Technol. *27*(1), 64.

OSER, B. L. and FORD, R. A. 1973b. Recent progress in the consideration of flavoring ingredients under the Food Additives Amendment: 7. GRAS substances. Food Technol. *27*(11), 56.

OSER, B. L. and FORD, R. A. 1974. Recent progress in the consideration of flavoring ingredients under the Food Additives Amendment: 8. GRAS substances. Food Technol. *28*(9), 76.

OSER, B. L. and FORD, R. A. 1975. Recent progress in the consideration of flavoring ingredients under the Food Additives Amendment: 9. GRAS substances. Food Technol. *29*(8), 70.

OSER, B. L. and FORD, R. A. 1977. Recent progress in the consideration of flavoring ingredients under the Food Additives Amendment: 10. GRAS substances. Food Technol. *31*(1), 65.

OSER, B. L. and FORD, R. A. 1978. Recent progress in the consideration of flavoring ingredients under the Food Additives Amendment: 11. GRAS substances. Food Technol. *32*(2), 60.

OSER, B. L. and FORD, R. A. 1979. Recent progress in the consideration of flavoring ingredients under the Food Additives Amendment: 12. GRAS substances. Food Technol. *33*(7), 65.

OSER, B. L., FORD, R. A. and BERNARD, B. K. 1984. Recent progress in the consideration of flavoring ingredients under the Food Additives Amendment: 13. GRAS substances. Food Technol. *38*(10), 65.

OSER, B. L., WEIL, C. S., WOODS, L. A. and BERNARD, B. K. 1985. Recent progress in the consideration of flavoring ingredients under the Food Additives Amendment: 14. GRAS substances. Food Technol. *39*(11), 108, 110, 112.

SCHULTZ, H. W. 1981. Food Law Handbook. AVI Publishing, Westport, CT.

STOFBERG, J. 1980. The future of artificial flavoring ingredients. Perf. Flavorist *5*(Aug./Sep.), 17–22.

STOFBERG, J. 1981. Setting priorities for safety evaluation of flavoring materials. Perf. Flavorist *6*(4), 69–72.

STOFBERG, J. 1983. Safety evaluation and regulation of flavoring substances. Perf. Flavorist *8*(Aug./Sep.), 53–62.

STOFBERG, J. and STOFFELSMA, J. 1980. Consumption of flavoring materials as food ingredients and food additives. Perf. Flavorist *5*(7), 19–35.

VERNIN, G. 1980. The regulation of food flavours in the European Community. Rivista Italiana, EPPO *62*(5), 249–252. (In French.)

VODOZ, C. A. 1975. Flavour legislation—New developments and trends. Flavours *4*, (Jul./Aug.), 219–220, 236.

VODOZ, C. A. 1978. Flavour and taste in food legislation. Int. Flavours Food Additives *7*(Mar./Apr.), 77–79, 85.

14

Quality Control

Quality control is a very important function in the flavor industry. The flavor house must provide a good quality consistent flavoring material for the consumer. It must also be certain that the flavor meets legal restrictions on ingredients and their qualities. The Quality Control (QC) department meets these needs by performing both analytical tests and sensory evaluation of incoming ingredients and finished flavors.

A. Analytical Tests

1. Overview of Physicochemical Tests

Heath (1981) has provided a detailed outline of the analytical tests used in a QC lab. This outline follows below.

a. Natural Plant Materials

Sample may represent whole unground, broken or previously ground material and should first be judged on overall appearance and visual quality. In most cases, whole samples must first be reduced to a uniform, moderately fine powder before subjecting them to any of the following tests.

i. General tests

Ash. Total, acid insoluble, water soluble.

Crude fibre.

Extractive matter. Alcohol soluble (in alcohol of nominated strength), other nominated solvents, nonvolatile ether extractive. The test may be followed by an examination of the resulting extract after removal of the solvent.

Extraneous matter and filth.

Moisture content. Oven drying, infrared drying, Karl Fischer method, xylene codistillation method.

Volatile oil content. This test may be followed by an examination of the recovered essential oil by GLC and/or other test methods.

ii. Tests of limited application

Copper-reducing substances.

Starch content. Followed by microscopic examination of the starch grains.

Sieve analysis of particle size distribution (ground material).

Microscopic examination (for insect parts, etc.).

Undesirable plant parts or other foreign matter.

Limit tests as specified.

iii. Additional specific tests

With specified limits in many countries.

Allspice. Quercetannic acid.

Capsicum (cayenne, red pepper). Capsaicin content, Scoville pungency, color content.

Clove. Quercetannic acid, clove stems and other foreign matter.

Curry powder. Sodium chloride.

Ginger. Gingerine content (or other measure of pungency), starch, lime (CaO), cold-water extractive, water-soluble ash.

Marjoram. Stems and other foreign matter.

Mustard. Allyl isothiocyanate, sodium chloride, starch and other farinaceous matter.

Paprika. Color index, iodine value of fixed oil.

Pepper, black and white. Piperine content (Kjeldahl nitrogen or instrumental method), pepper-starch content.

Saffron. Yellow styles and other foreign matter.

Tumeric. Curcumin content (or color index), starch.

Vanilla beans. Vanillin content, lead number on alcoholic extract.

b. Essential Oils

Essential oils are liquids some of which may crystallize and solidify on cooling. Such samples should be gently warmed until completely liquid before carrying out any tests. Samples should first be evaluated for appearance (i.e., color, clarity), preferably being compared with a reserve sample, followed by an evaluation of the odor and the flavor in a diluted sugar syrup.

i. General tests

Specific gravity (normally at 25°C/25°C or corrected for any other temperature) or weight per milliliter at 20°C.

Refractive index (RI) at 20°C.
Optical rotation (OR) at 20°C.
Solubility in alcohol of a specified dilution.

ii. Tests of limited application

Boiling range (°C).
Melting point (°C).
Congealing point (°C).
Flash point. Open or closed cup (°F or °C).

iii. Instrumental tests

Gas–liquid chromatography (to establish identity/origin, detect adulteration, determine compositional changes and for production quality control).
Infrared spectroscopy.
Ultraviolet spectrophotometry.

iv. Specific tests for constituents

Acetals.
Acid value.
Alcohols. Aceto-formylation method, total alcohols, tertiary terpene alcohols, citronellol by formylation.
Aldehydes and ketones. Bisulfite method, neutral sulfite method, phenylhydrazine method, hydroxylamine method.
Chlorinated compounds.
Esters. General, high-boiling solvent method.
Heavy metals.
Phenols. Free, total.

v. Tests specific for citrus oils

UV absorbance.
Evaporation residue.
Peroxide value.
Foreign oils.

vi. Specific tests and procedures The special tests applicable to all the major essential oils are described in the Standards and Specifications of the Essential Oil Association of USA. Where standards are set for essential oils, it is usual for the test method to be described. Variations in the test method may lead to different test results so that the method prescribed should be adhered to in detail.

c. Oleoresins

In physical form, oleoresins range from mobile to viscous fluids, from semisolids to waxy or sticky pastes. Samples should be well mixed to ensure uniformity before testing. Initially, the sample should be evaluated for appearance (often characteristic), odor and flavor after dilution in a suitable medium.

i. General tests

Volatile oil content, followed by an examination of the recovered essential oil.
Solvent residues.
Solubility in a specified solvent.

ii. Specific tests

Capsicum (cayenne, red pepper) oleoreins. Capsaicin content, Scoville pungency index, color index.
Ginger oleoresin. Gingerine content (or other pungency index).
Paprika oleoresin. Color index, pungency test.
Pepper (black and white) oleoresin. Piperine content.
Turmeric oleoresin. Curcumin content (or color index).

d. Plated or Dispersed Spices

Manufactured flavorings which comprise oleoresins and/or essential oils plated or spread on an edible carrier (e.g., salt, dextrose, flour). If stored in clear glass, the sample may show surface color fading. Samples should be well mixed before carrying out tests.

i. General tests

Volatile oil content.
Extractives soluble in a nominated solvent.
Carrier base constituents. Specific tests as appropriate.
Sieve classification or test for particle size distribution.

ii. Tests of limited application

Microbiological examination. Total plate count, freedom from coliforms, salmonella, etc.

iii. Specific tests
See preceding Oleoresins.

e. Synthetic Chemicals

Samples may be liquids or crystalline solids. The nature of the testing is often determined by the intended end use. Routine physicochemical tests are necessary for control of purity but odor and flavor evaluations are of more importance when the chemical is intended for use in a flavoring or fragrance composition.

i. General tests—liquids

Specific gravity (normally at 25°C/25°C or corrected for any other temperatures) or weight per milliliter at 20°C.
Refractive index (RI) at 20°C.
Optical rotation (OR) at 20°C.
Solubility in alcohol of stated dilution or in other specified solvent.
Boiling point/range (°C) at specified pressure (mm Hg).
Freezing/congealing point (°C).
Flash point (°F or °C). Open or closed cup method.

ii. General tests—solids

Melting point (°C).
Congealing point (°C).
Solubility in specified solvent.
Freedom from insoluble matter.

iii. Specific tests for chemical identity and purity—instrumental methods

Gas–liquid chromatography.
Infrared spectroscopy.

f. Flavorings

Flavorings may be either liquid, emulsions or dry powders. Samples should first be evaluated for appearance (i.e., color, clarity, opalescence) then for odor and flavor after dilution in an appropriate medium. They should preferably be compared directly with a standard sample which has been maintained under optimum conditions and changed at regular intervals.

i. General tests—liquid flavorings

Specific gravity (normally at 25°C/25°C or corrected for other tempertures) or weight per milliliter at 20°C.

Alcohol or other specified solvent content.

Refractive index at 20°C.

Color intensity or shade (as appropriate).

Flash point (°F or °C). Open or closed cup method (if required).

ii. General tests—emulsions

Specific gravity (25°C/25°C) or weight per milliliter (20°C).

Volatile oil content. Steam distillation method.

Microscopic examination of particle size.

Microbiological examination. Total plate count, yeasts, molds.

Ringing test (for beverage emulsions).

Specific tests for stabilizers, antioxidants, etc.

iii. General tests—encapsulated dry flavorings

Particle size analysis.

Moisture content. Xylene codistillation method.

Volatile oil content. Steam distillation method or organic solvent method.

Microbiological examination. Total plate count, coliforms, salmonella, molds and yeasts.

g. Vanilla Extract

Natural vanilla extract is subjected to a wide range of tests to ensure its authenticity. These include:

Specific gravity (25°C/25°C) or weight per milliliter at 20°C.

Alcohol content.

Glycerin content.

Vanillin content.

Neutral lead number (Wichmann).

Total solids.

Ash. Total, water soluble, water insoluble (followed by test for alkalinity of ash).

Acidity (pH).

Coloring matter soluble in amyl alcohol.

Thin-layer chromatographic examination.

Paper chromatographic examination.

h. Fruit-Based Products

Fruit juices and fruit-based flavorings should be kept chilled unless the soluble solids content exceeds 65° Brix; or they should be suitably preserved. Samples should be made uniform and should be subjected to sensory evaluation before applying any of the physicochemical tests.

i. General tests

Alcohol content.
Total solids.
Ash.
Fruit and sugar content.
Sucrose content.
Pectins.
Microbiological examination. Total plate count, molds, yeasts.

ii. Special tests

Detection of β-ionone in fruit extracts.
Detection of strawberry aldehyde.

The following discussion will highlight some of the more commonly performed tests.

2. Specific Gravity

One of the most commonly performed analyses on ingredients and finished flavors is measuring specific gravity (or density). Specific gravity (SG) is a physical property of a raw material (liquid) or finished flavor which is dependent on composition. Specific gravity may be determined using a pycnometer or hydrometer. However these methods often are not rapid enough for a busy QC lab. Therefore, a rapid instrumental method is used to increase productivity. One such instrument is the Paar DMA45 Digital Density meter. To determine density, the sample tube is electromagnetically excited to vibrate at its natural frequency. The tube is then filled with sample and the tube is again excited to its new resonance frequency. The difference in frequency between the tube loaded and empty is related to the mass difference. Therefore, mass and subsequently density can be calculated by the instrument. Some advantages to this technique are that neither volume nor weight of sample need be measured. Also, the determination is independent of sample viscosity. The instrument is sensitive to three decimal places which is adequate accuracy for QC purposes.

3. Refractive Index

The second analysis which is performed on most ingredients and finished flavors is refractive index (RI). Refractive index is typically done using an Abbe refractometer. The Abbe refractometer is accurate to four decimal places, covers the RI range typically encountered in flavoring materials and is rapid. The RI of a flavor is a function of all the components and their proportions. Therefore, this determination is very sensitive to mixture composition. RI and SG will detect most compounding errors made in the formulation area.

4. Optical Rotation

Optical rotation is commonly used as a QC for essential oils and oleoresins. The significant flavor constituents in these oils very often are optically active. Therefore, optical rotation may be used to quantify specific flavor components in the oleoresin or essential oil.

5. Alcohol Content

Alcohol (ethanol) content is determined on all flavors which use alcohol as a significant part of the flavoring. That is, where alcohol may be one of the flavor solvents. Flavor companies typically use large volumes of ethanol in manufacturing. They must pay tax on the alcohol (to the Bureau of Alcohol Tobacco and Firearms—BATF) unless the flavor company can document that the alcohol was used in a product which was not potable, as is. Since most flavorings are not potable at production concentrations, the flavor house can get a drawback (tax rebate) on the alcohol tax. This amounts to a substantial dollar value in most flavor houses so alcohol usage is monitored very closely.

Alcohol is typically determined by gas chromatography. The official method (AOAC) involves directly injecting a sample of the flavor into the GC. A Chromosorb 102 column and flame ionization detector are typically used for the determination.

6. Residual Solvent

Residual solvents must be determined on any flavoring material which is obtained by solvent extraction. Oleoresins and absolutes compromise the largest group of flavorings in this category. Table 14.1 presents the legal limits on residual solvents in oleoresins. Note that residual solvents are permitted only in the low ppm levels.

***Table 14.1. Permitted Levels of Residual Solvents in Oleoresins*[a]**

Methylene chloride residues. Methylene chloride may be present in food under the following conditions: In spice oleoresins as a residue from the extraction of spice, at a level not to exceed 30 ppm (0.003%). Provided that, if residues of other chlorinated solvents are also present, the total of all residues of such solvents shall not exceed 30 ppm.

Ethylene dichloride residues. A tolerance of 30 ppm is established for ethylene dichloride in spice oleoresins when present therein as a residue from the extraction of spice. Provided, however, that if residues of other chlorinated solvents are also present the total of all residues of such solvents shall not exceed 30 ppm.

Trichloroethylene residues. Tolerances are established for residues of trichloroethylene resulting from its use as a solvent in the manufacture of foods as follows: Spice oleoresins . . . 30 ppm (0.003%). Provided that if residues of other chlorinated solvents are also present, the total of all residues of such solvents in spice oleoresins shall not exceed 30 ppm (0.003%).

Acetone residues. A tolerance of 30 ppm is established for acetone in spice oleoresins when present therein as a residue from the extraction of spice.

Isopropyl alcohol residues. Tolerances are established for residues of isopropyl alcohol resulting from its use as a solvent in the manufacture of spice oleoresins at 50 ppm (0.005%).

Methyl alcohol residues. A tolerance of 50 ppm is established for methyl alcohol in spice oleoresins when present therein as a residue from the extraction of spice.

Hexane residues. A tolerance of 25 ppm is established for hexane in spice oleoresins when present therein as a residue from the extraction of spice.

[a] Code of Federal Regulations (1984).

Gas chromatography is the method of choice for residual solvents (EOA Method No. 1-1D-3-1). The oleoresin is often subjected to an initial distillation step to isolate the volatiles (including residual solvents). This distillate is then analyzed for residual solvent.

7. Particle Size of Emulsions

As has been pointed out earlier in this text, emulsions are typically very unstable. Since a primary determinant of stability is particle size, the QC department routinely determines particle size of emulsions.

Particle size is generally determined using a microscope. A 0.1–1.0% dilution of the emulsion in water (or glycerin) is made and then two drops are placed on a glass slide. The emulsion slide is covered with a cover slip, then mineral oil and viewed under the oil immersion lens of a light microscope. The microscope is fitted with a stage micrometer so that particle size can be estimated. A quality control decision at this point becomes rather subjective. Since the particles are not fixed but will move across the field and only one layer of particles are in focus, it is difficult to obtain an accurate estimate of particle size. In addition, statistical validity requires that several hundred particles must be sized in order to obtain an accurate estimate of particle size distribution. Since time limitations and eye fatigue do not permit the sizing of such a large number of particles, often a trained technician visually examines several microscope fields and either accepts or rejects the emulsion. While this would appear to be a rather poor method, practical constraints often require that this approach is taken.

Alternatively, there are instrumental methods to determine particle size distributions of emulsions. The two most commonly used techniques involve either light scattering or electrical resistance measurements. Light scattering is a function of the composition of the emulsion (both continuous and particulate phases), number of particles in the light path and the size of the particles. An average particle size is quite readily obtained using this approach.

The Coulter Counter utilizes changes in electrical resistance as the emulsion (diluted in electrolyte) is drawn through an orifice. One electrode is placed on one side of an orifice while the second electrode is placed on the opposite side of the orifice. Since the only path for electrical conductivity is through this orifice, the passage of an emulsion particle through the orifice changes electrical current flow. The degree of change in current flow is proportional to the volume of the particle. Therefore, this technique both counts and sizes each particle drawn through the orifice. Particle size distributions can be obtained in only a few minutes.

8. Volatile Oil

The volatile oil content of various botanicals is a measure of flavor strength. It is typically the volatile components which carry the flavor of a spice or herb (exceptions being those materials that have a "bite"). Therefore, the QC department will run volatile oil content on many botanicals used as flavoring materials.

Volatile oil is also run on some spray dried flavorings (Kernik *et al.* 1985; Anandaraman and Reineccius, 1986). Spray-dried citrus oils, mint oils, spice oils, etc., are commonly analyzed for volatile oil to determine the efficiency of the encapsulation procedure.

Volatile oil is generally measured using a distillation method. The Clevenger trap (Fig. 14.1) is designed to volumetrically measure volatile oil content of water insoluble flavorings. Sufficient material is placed in the sample flask to yield 10–15 ml of volatile oil. Water and antifoam are

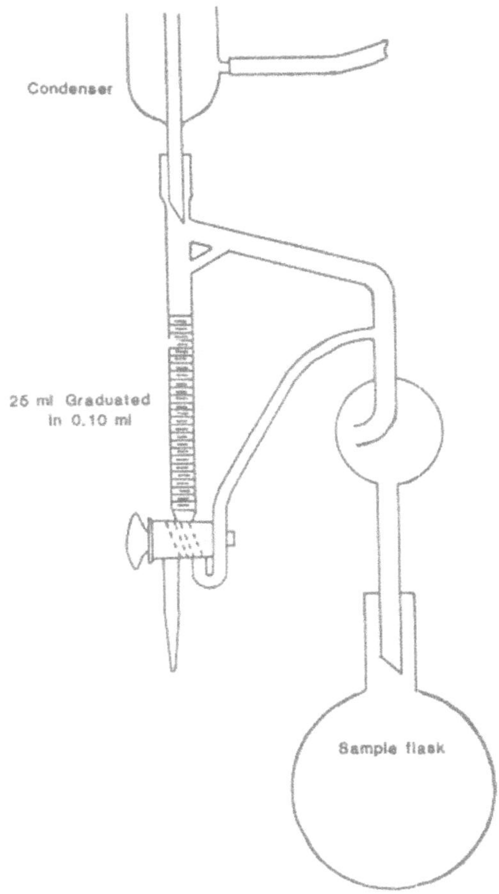

Fig. 14.1. Clevenger trap for the analysis of volatile oils lighter than water.

added to the sample flask and then it is heated under reflux for 2–4 h. The volume of oil present in the sample can be read directly from the side arm.

An alternative method for measuring volatile oil is the Fosslet method. This method involves placing the sample (a botanical or spray-dried powder) into a sample holder which has a sliding hammer inside it. Perchloroethylene is added and then the sample holder is placed in the "vibrating reactor." The vibrating reactor is similar to a paint shaker and subjects the sample to vigorous shaking plus the smashing action of this hammer. Extraction is completed in 1½–2 min. The sample holder is then opened, a dessicant added and the perchloroethylene/essential oil mixture is filtered into an electronic device for the measurement of specific gravity of the filtrate. Since perchloroethylene has a very high specific gravity, specific gravity will be lowered depending upon the concentration of essential oil in solution. The instrument may be calibrated to read out the percentage of extractable oil. This is a very rapid method (5–10 min per test) which is gaining use in the industry.

9. Surface Oil

Surface oil refers to the flavoring oil which is on the surface of an encapsulated flavoring material. This surface oil is of particular interest since it is exposed to light, oxygen and package environment and will readily become oxidized and produce an off-flavor in the product.

Surface oil is typically determined only on encapsulated essential oil products. This is because some of the essential oils (e.g., citrus oils) are extremely labile to oxidative deterioration. This analysis is not performed on a routine basis but only occasionally to monitor the spray drying process.

Surface oil is most commonly determined gravimetrically. This involves placing the encapsulated product in an Erlenmeyer flask and adding organic solvent. The flask would be shaken for several minutes, the powder filtered from the solvent and then the solvent removed by evaporation. The residue, or surface oil, would be determined by weighing the residue. Residue may also be quantified by gas chromatography (Anandaraman and Reineccius 1986). In this procedure, the organic solvent would contain an internal standard. Evaporation of the filtrate would be stopped at a concentration suitable for gas chromatographic determination.

10. Moisture Content

Moisture content is of concern in dry flavorings (e.g., dry whole spices and encapsulated flavors). Since most flavoring materials contain volatile components in addition to water, evaporative methods for moisture determination are not suitable. Toluene or xylene codistillation procedures are often the methods of choice (Anandaraman and Reineccius 1986).

11. Gas Chromatography

Gas chromatography is not routinely used in the analysis of finished flavors. Refractive index, specific gravity and sensory analyses are much faster than GC and will detect most problems encountered in flavor production. If a problem were found by one of the screening methods, then GC would probably be used to determine the source of the problem.

Gas chromatography is commonly run on incoming flavor ingredients. The purpose may be to determine flavor quality, purity or detect adulteration in the ingredient. Many flavoring materials are prone to deterioration during storage. Citrus oils are exceptionally susceptible to oxidative degradations. Even pure chemicals may undergo undesirable reactions when stored. Benzaldehyde is readily oxidized to benzoic acid and limonene both oxidizes and polymerizes. Therefore a GC profile may be used as one method to determine quality of the ingredient.

To assist in determining both flavor and authenticity, the Analytical Methods Committee of the Analytical Division of the Royal Institute of Chemistry has been establishing gas chromatographic profiles of commercially important essential oils. These profiles have been obtained on both Carbowax 20M and methyl silicone phases (Analytical Methods Committee 1980, 1981, 1984a, 1984b). This data is useful in providing gas chromatographic profiles of authentic oils as standards for the QC laboratory.

Purity may also be monitored by GC methods. The Essential Oil Association publishes a compilation of chromatographic data on flavor ingredients. This provides the purchaser with an idea of purity and amount of active components that should be expected in each flavor ingredient. Examples of data available on methyl ionone are presented in Figs. 14.2 and 14.3. Most of the GC data is from packed column GC. This is being replaced gradually by data on fused silica capillary columns. It may be worthwhile to comment on the purity of the methyl ionone used as an example. Chemicals used as flavoring materials most often are less "pure" than reagent grade chemicals used in the laboratory. However, these materials are produced using methods which yield food grade quality. "Food grade" is a statement of safety rather than purity.

Some forms of adulteration may also be monitored via GC methods. Adulteration is different from purity since adulteration is the *intentional* addition of cheaper ingredients to make greater profit. Adulteration has been a substantial problem for many years in the essential oil area. Some essential oils will sell for several hundred dollars per pound. The dilution of a very expensive oil with a cheaper oil is unfortunately too common. This cutting of expensive oils with cheaper oils often can be detected by running a GC profile of the oil. One approach is to search for components in the expensive oil which are not commercially available and are unique to this oil. An example is β-selinene in oil of celery (Fig. 14.4). Good quality oil of celery should contain 7–7.5% β-selinene (Straus and Wolstromer 1974). Oils containing less than 7% are suspected of being adulterated. A second example would be the adulteration of oil of pep-

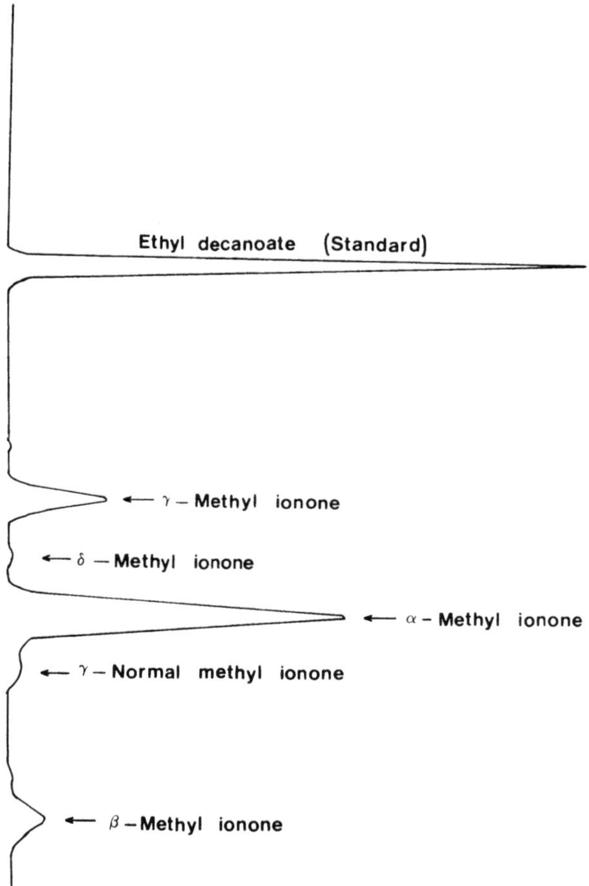

Fig. 14.2. Gas chromatogram of commercial methyl ionone.
EOA Book of Standards and Specifications 1975

permint with oil of cornmint. Oil of peppermint contains sabinene hydrate at a level of 1%. Cornmint does not contain this component, nor is it commercially available.

Unfortunately, GC is not always adequate to detect adulteration. The most adulterated flavoring material in the world is vanilla. Pure vanilla is a very expensive product. The active ingredient of vanilla, vanillin, is readily available at minimal cost. Natural variations in vanilla components accentuated by different processing techniques, makes adulteration exceedingly difficult to detect. Over the years numerous methods have been developed in QC labs to detect adulteration of vanilla. As new procedures have been developed, adulterators have kept pace.

One of the more recent methods for the detection of adulteration of

Sample: Methyl Ionone α and γ Type EOA

Column: Material: Copper Length: 10 ft O.D.: $\frac{1}{4}$ in. I.D.: $\frac{3}{16}$ in.
Stationary phase: Silicone D.C. 710, 10%
Solid support: Chromosorb W, 60–80 mesh, silicone treated

Carrier gas: Helium

Conditions: Column temperature—isothermal: 160°C
Inlet temperature: 225°C
Inlet pressure: 35 PSIG Column flow rate: 135 ml/min
Sample size: 1.0

Detector: Type: Thermal conductivity—Hot Wire
Temperature: 275°C

Recorder: Span: 1 mV Chart speed: 24 in./h

Chromatogram

Principal Components[a]	Relative Retention Time	Attenuation
1. Ethyl Decanoate	1.00	4
2. γ-Methyl Ionone	1.89	4
3. δ-Methyl Ionone	2.07	4
4. α Methyl Ionone	2.26	4
5. γ Normal Methyl Ionone	2.41	4
6. β Methyl Ionone	3.02	4

Retention time of last component: 23 min (β-methyl ionone)
Typical curves: The conditions given here refer specifically to the curves attached.

[a] Peak 2 is also called α-isomethyl ionone; peak 3, also called β-isomethyl ionone; peak 4, also called α normal methyl ionone; and peak 6, also called β normal methyl ionone.

Method of Calculation: Internal Standard (See EOA Instrumental Determination 1D-3, Gas Chromatography, Section 11-D) using ethyl decanoate in 30% mixture with sample.
A factor of 1.30 was obtained using the procedure described herein.

This analysis may also be accomplished with columns containing: Silicone DC 550

The preceding conditions are provided for guidance only.
Note: The analyst is cautioned to purify the knowns used in the determination just prior to use, by distillation and/or preparative GC followed by GC analysis for percentage of purity.

Fig. 14.3. Example of data from Essential Oil Association.
EOA Book of Standards and Specifications 1975

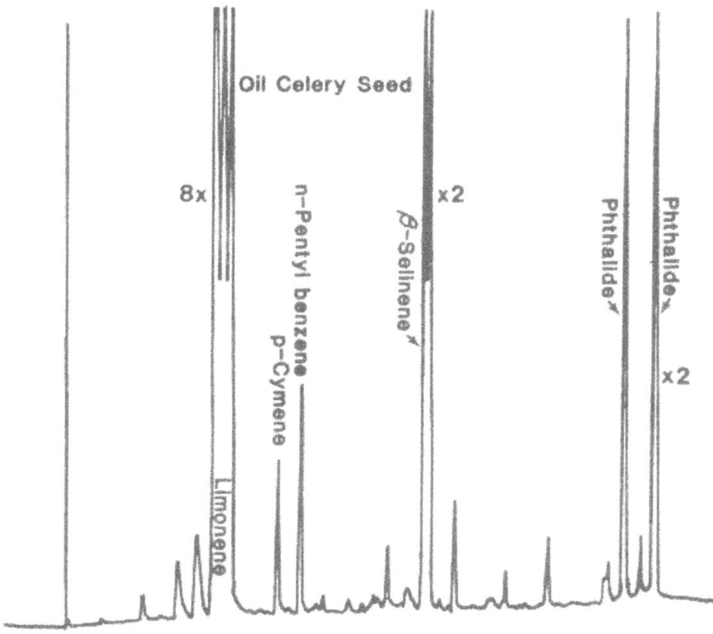

Fig. 14.4. Gas chromatogram of oil of celery.
Straus and Wolstromer 1974

vanilla is via stable isotope ratios. The carbon-13 (^{13}C) ratio in vanillin from vanilla ($^{13}C/^{12}C$ − 20.9) is significantly higher than in vanillin which has been synthesized from lignin ($^{13}C/^{12}C$ − 27.8) or gaiacol ($^{13}C/^{12}C$ − 31.5) (Bricout and Koziet 1978). Unfortunately, adulterators have now synthesized a ^{13}C enriched vanillin which is added in trace amounts to boost the ^{13}C level up to that normally found in natural vanillin. The ^{13}C enrichment is typically done in the aldehyde group of the vanillin molecule. Therefore, the determined analyst can still detect adulteration by splitting the aldehyde group from the vanillin and determining the ^{13}C level in the remainder of the vanillin molecule. This technique, however, is quite tedious and each step adds error which limits adulteration detection limits.

The determination of ^{14}C levels in food flavorings has also been used as a technique for monitoring adulteration. All living plant tissues contain some ^{14}C due to ^{14}C in the CO_2 of the atmosphere (nuclear testing). The only way for a flavor compound to show no ^{14}C is if it has been synthesized from petroleum sources. Carbon-14 has a half-life of about 5730 years. Unfortunately this technique can therefore only be used to determine whether a pure chemical has been synthesized from petroleum based chemicals or from recent plant metabolism. This technique will not detect an artificial chemical made from plant starting materials. A further limitation of this technique is that it is suitable only when large pure samples are available (approximately 1 g). This technique is useful for the analysis of pure ingredients, not finished flavors or flavors in food products.

Bricout and Kaziet (1978) demonstrated how ^{14}C analysis could be used to determine the source of citral. They reported 20.1 dpm/g C for citral from lemon grass and 0.25 dpm/g C for citral synthesized from fossil fuels. As would be expected, there is not a significant difference in ^{14}C determinations for synthetic citral synthesized from pinene and citral from lemon or lemon grass.

Quality control labs have another type of adulteration to monitor. That is the addition of synthetic chemicals to natural flavorings. Most natural flavorings (other than essential oils) are quite weak in flavor strength. It is common to need to use up to 1–2% of a natural fruit flavor in the finished product in order to produce an acceptable flavor strength. Some companies will "fortify" natural flavorings with a significant amount of artificial character impact compounds and illegally sell the product as a natural flavoring. The QC lab is therefore called in on occasion to analyze a competing flavor in order to determine if the flavor is natural as labeled. The QC lab falls back on the analytical techniques described in Chapter 1 to obtain a GC profile of the suspect flavor. The identification of a compound such as allyl caproate (character impact compound in artificial pineapple), ethyl vanillin or ethyl maltol would indicate without question that the flavor contains added synthetic compounds. The analyst would also search for typical flavor solvents such as propylene glycol, triacetin and triethyl citrate as an indication of fortification via the addition of an artificial flavor. These compounds are not found in nature. If no compounds were found that do not occur naturally in food products, then the analyst must look at the natural compounds present in the flavoring. Is the profile "in balance" with expectations? For example, a natural raspberry flavoring has a typical flavor profile. An excessive amount of ionones or irones (character impact compounds) may suggest adulteration. However a flavor imbalance may come from the use of WONF components. If the compounds present in greater proportions than expected are not commercially available from a natural source, then QC can strongly suggest adulteration but this is still difficult to prove without reasonable doubt. An interesting case study on the detection of adulteration in coconut flavoring is presented by Eberhardt *et al.* (1981).

A new approach for detecting adulteration is via the ability to determine individual optical isomers in a flavor. Considerable progress has been made towards separating D and L isomers of optically active flavor compounds (Tressl and Engel 1984, 1985; Chang 1986; Mosandl and Heusinger 1984; Konig 1984). Nature typically will produce only one optical isomer while most chemical syntheses produce the racemic mixture. Therefore the QC lab could analyze the optically active flavor constituents for the proper stereoisomer.

For example, plants typically produce D-limonene. The presence of L-limonene would indicate that the molecule has been chemically synthesized and is artificial.

Adulteration is unfortunately more common than it should be. Quality control labs spend considerable effort to ensure that they get the flavor ingredients for which they paid. They also tend to "police" their own industry.

12. Microbiological Analysis

The majority of flavorings produced in a flavor house will not support the growth of microorganisms. As mentioned earlier, they are toxic to most microorganisms. Therefore, the QC lab does not typically run a large number of microbiological assays. They will run analyses on the spray-dried products, emulsions and some of the natural flavorings. The microbiology lab generally is set up to run standard plate count, yeasts and molds, and lactobacilli. Salmonella and other pathogens are typically sent out for analysis.

B. Sensory Analysis

Sensory analysis is performed on all raw materials before they are used to make a flavor and on all finished flavors before they go out to the customer. Sensory is the final word on acceptance. The following description of sensory analysis as a QC function is reasonably typical of the industry as a whole. There will be some differences in operations from company to company, however, the basic approach is similar.

Finished flavor samples come to the quality control lab directly from the compounder in production. Along with each sample is a form with the name and number of the flavor, the amount produced, date of production, compounders name and customers name. This information allows QC to retrieve information on that particular flavor.

The standard for comparison on finished flavors is a sample of the most recent lot of the flavor which was sent to that customer. All of these flavor samples are stored in full bottles in a refrigerator. Sample retention is based on how frequently each customer purchases each flavor. Typically the last three samples sent to a customer are retained. When a new lot is produced, the oldest sample is discarded.

The samples are set up in a side-by-side comparison in order to evaluate color, odor and flavor. There are basically three different tasting mediums which are used for flavor evaluation. They are (1) water, (2) sugar syrup and water solution and (3) sugar syrup, citric acid and water solution. Water is used for tasting flavors such as meat, onion, garlic, butter, coffee and most spicing compounds. The sugar and water solution is made by mixing 250 ml of 68°Brix sugar syrup with 750 ml tap water. This gives approximately a 17% sugar solution. It is used to taste flavors such as mint, spearmint, peppermint, wintergreen, vanilla and banana flavors. The sugar/acid/water solution is made by mixing 250 ml of 68°Brix sugar syrup, 0.3 ml of a 50% solution of citric acid and 750 ml water. Most of the fruit flavors are tasted in this sugar–acid solution. The reason for using these different mediums for evaluation instead of plain water for all flavors is to put the flavor into a medium that may be similar to the finished product it is meant for. The sugar syrup and citric acid act as flavor enhancers.

Levels used for evaluation depend on the strength of each flavor. However, there are certain general guidelines which can be followed. Natural flavors are generally set up at 0.5–1% concentration. This level is obtained by adding 0.5–1 ml of flavor to 100 ml of tasting medium. Natural flavors refer to fruit juice-based flavors, most cordial flavors, savory flavors and WONFs. Most artificial flavors are tasted at 10–50 ppm. To get to this level, a 1% solution is first made by adding 0.1 ml of the flavor to 10 ml of alcohol. Then 0.1–0.5 ml of this solution is added to 100 ml of the tasting medium. Natural and WONF spray dries will usually be set up at about 1% in the proper tasting medium. Artificial spray-dried flavors will usually be tasted in the range of 50–100 ppm. The 1% solution can be made with water instead of alcohol since spray dries are water soluble. Slightly higher levels are required for spray dries than for liquid flavors. This is since spray dries are generally 80% inert carrier and only 20% active flavor. It is very important that the same level be used for both the standard and the new sample. A difference of 0.1 ml here could be interpreted as a difference in flavor strength during evaluation.

There are certain flavors which are evaluated by smell only. These are flavors used for tobacco which might contain an item not approved for food use but allowed in tobacco products. A few examples of such items are coumarin, safrole, and oil of cade.

Once the samples are set up at the proper levels in the proper medium, they are ready to be evaluated. Most flavor companies use one or two individuals to make the sensory decision in QC, although some labs will panel the flavors going to certain customers. While paneling would be more desirable on a routine basis, the large volume of samples that come through the QC lab make this nearly impossible. It is common for over 50 samples a day to be evaluated in a larger flavor house.

Sensory evaluation is concerned with differences in color, odor and flavor. When evaluating the flavor, flavor character and strength must both be considered. Certain color and flavor changes occur normally as the flavor ages. Fruit juice concentrates and fruit juice based flavors are often affected by nonenzymatic browning. Thus, a newly produced natural strawberry flavor may be light red in color and taste fresh, while an older lot of that same flavor may be turning brownish red and have a more pruney flavor. Citrus flavors, become lighter in color with age due to oxidation. They may also have a very terpeney flavor. Flavors containing vanillin may turn yellow or pink with age. This is due to Schiff's base formation (chemical reaction between the vanillin and amines in the flavor).

If the stock sample is substantially older than the new sample, any one or combination of changes may have occurred in the older sample. The discrepancy between the retained sample and new sample then is not necessarily cause for rejection of the new sample. If, however, the new sample shows such signs of color or flavor changes, this may be cause for concern. Assuming that the retention sample has undergone little deterioration, the criteria for rejection of a flavor is a difference in color and/or flavor between the newly produced flavor and the older retention sample. Flavor character and strength are most important. Slight color differences are often tolerated, depending on what the finished product will be. If a flavor seems to have a problem, the first step is to pull the second most recent retention sample and compare it to the new flavor and the first retention sample used to make the original comparison. It is always possible that this first retention sample was off for some reason. If this does not seem to be the case, then the flavor must be rejected.

There are at least four reasons why the sample may have been bad. The first is that the compounder might have improperly sampled the batch or even sent the wrong sample. Therefore, a second sample is taken from the production batch for evaluation. The second cause of a flavor problem may be the use of old stock materials. If old juice concentrates or citrus oils have been used, the finished flavors would exhibit flavor and color changes characteristic of browned or oxidized stock material. Since the compounders keep very extensive data on each flavor they make, it is easy to determine the age of the stock materials used through lot numbers. When old materials have been used, a color difference is usually obvious. If the proper flavor components have been used and the flavor meets all other specifications, it may be worked off. The flavor is compounded again using fresh, new stock materials and small amounts of the old flavor can then be added to this new flavor without negative effects. In this way, it does not have to be dumped. Work-off is generally done at no higher than 5% of the old flavor added to the new flavor.

A third alternative would be that the compounder used an incorrect ratio of aromatics to solvents. This would result in a difference in flavor strength. For example, if too much alcohol solvent were added, the flavor would be weaker than the stock sample. An excess of alcohol would make the specific gravity lower than the specified range. This problem can be solved by calculating the amount of flavor materials needed to bring the finished flavor back to the proper ratio of aromatics to solvent. The result is a larger batch of properly proportioned material. Once again, it is not usually necessary to dump the material.

The fourth alternative is one in which a compounder has used the wrong flavor components. In this case, it is almost always necessary to reject the material, throw it away and begin again.

Once a sample has been approved, it is recorded by date, lot number and customer on a card (or computer file) for that flavor. It is also recorded in a book with the same information. The sample is retained and becomes the newest stock sample of that flavor. The form is sent to the shipping department and the flavor may now be sent to the customer.

Some items are used in flavors primarily to impart color. Two examples of these are caramel coloring and special grape extract. Quality control on these items is a color check done in a 50-ml Nestler tube. The tube is filled with 50 ml of water to which 0.1 ml of the carmel color is added. This sample is compared to a retention sample for color, intensity and clarity. Special grape extract is set up the same way, except that three drops of 50% citric acid are added to the Nestler tube. Special grape extract is made from grape skins and gets its color from anthocyanins. Color from anthocyanins is pH dependent, therefore, a small amount of citric acid must be added to get the characteristic reddish purple color. In a basic solution, the special grape extract would appear blue. It is usually used in fruit flavors which are acidic or which will go into a finished product that is acidic.

Special grape extract and caramel color are not considered flavors. They are raw materials. Another function of a quality control lab is to evaluate and approve raw materials from a sensory standpoint. Aromatic chemicals and essential oils would either be manufactured in-house at a raw materials plant or purchased from an outside supplier. In either case, they would go through quality control for approval before becoming stock material. The method of evaluation used for raw materials is very similar to that used for finished flavors. The major difference is in the tasting medium.

All raw materials are set up for tasting in plain water. It is important to evaluate the true character of the raw material without interference from sweetness or pH. Most essential oils and aromatic chemicals are set up at 10 ppm. Certain oils such as lovage, onion and garlic and most oleoresins are tasted at lower levels (1–5 ppm). Raw materials are evaluated by looking for differences in color, flavor strength and character. Since most of these are clear or turbid, color becomes less important.

One of the few situations requiring a QC panel evaluation is the determination of Scoville heat units. While there are analytical methods available for the determination of capsaicins (Johnson *et al.* 1979), the sensory approach (ASTA 1985: van Gemert *et al.* 1981) is generally used. The Scoville heat unit is a measurement of the pungency or biting of products such as oleoresin red pepper. This test involves making a initial dilution of the flavoring material in alcohol (0.20 g flavoring plus 50 ml 95% alcohol). This solubilized material (0.15 g) is diluted with 140 ml of 5% sucrose in water. If 5 ml of this solution produces a sensation of pungency in at least 3 of 5 tasters, a minimum value of 240,000 Scoville units is given to the flavoring material. The solution is then diluted further to determine the dilution at which 3 of 5 tasters can detect pungency. Scoville units are proportioned according to dilution. That is, a one-to-one dilution gives the flavoring 480,000 Scoville units while a one-to-two dilution yields 720,000 Scoville units.

The QC lab combines both analytical and sensory evaluations to assure that good quality ingredients are used in manufacturing and that finished flavors meet the specifications selected by the customer. The QC lab is a very significant part of a typical flavor house since the flavor house will maintain an inventory of somewhere between 2000 and 4000 raw materials and at least the same number of finished flavors (Dorland and Rogers 1977). The QC lab also gets involved in problem solving. It is the QC department that typically detects the problem and is assigned the task of problem solving.

Other perspectives of the QC department in a flavor house may be gained from Heath (1978, 1981), Bruhn (1981) and Dorland and Rogers (1977).

References

ANALYTICAL METHODS COMMITTEE. 1980. Application of gas–liquid chromatography to the analysis of essential oils: Part VII. Fingerprinting on essential oils by temperature-programmed gas–liquid chromatography using a Carbowax 20M stationary phase. Analyst *105*, 262.

ANALYTICAL METHODS COMMITTEE. 1981. Application of gas–liquid chromatography to the analysis of essential oils: Part VIII. Fingerprinting on essential oils by temperature-programmed gas–liquid chromatography using methyl silicone stationary phases. Analyst *106*, 448.

ANALYTICAL METHODS COMMITTEE. 1984a. Application of gas–liquid chromatography to the analysis of essential oils: Part X. Fingerprinting of specified essential oils by standardized procedures. Analyst *109*, 1339.

ANALYTICAL METHODS COMMITTEE. 1984b. Application of gas–liquid chromatography to the analysis of essential oils: Part XI. Monographs for seven essential oils. Analyst *109*, 1343.

ANANDARAMAN, S. and REINECCIUS, G. A. 1986. Analysis of encapsulated orange peel oil. Perf. Flavorist. (In press.)

ASTA. 1985. Official Analytical Methods, 3rd edition. American Spice Trade Association, New York.

BRICOUT, J. and KOZIET, J. 1978. Characterization of synthetic substances in food flavors by isotopic analysis. *In* Flavor of Foods and Beverages, G. Charalambous and G. E. Inglett, eds., p. 199. Academic Press, New York.

BRUHN, W. 1981. Problems of the industrial quality control of flavours. *In* Flavour '81. P. Schreier, ed., p. 635. de Gruyter, Berlin.

CHANG, Y. I. 1986. Resolution, Quantitation, and Reactions of Chiral Flavor Compounds. Ph.D. Thesis. Department of Food Science and Nutrition, University of Minnesota, St. Paul, MN.

CODE OF FEDERAL REGULATIONS. 1984. Solvents, lubricants, release agents and related substances. Subpart C. Title 21. Food and Drugs §173.210–§173.290, p. 113.

DORLAND, W. E. and ROGERS, J. A., JR. 1977. The Fragrance/Flavor Industry. W. E. Dorland Co., Mendham, NJ.

EBERHARDT, R., WOIDICH, H. and PFANNHAUSER, W. 1981. Analysis of natural and artificial coconut flavouring in beverages. *In* Flavour '81, P. Schreier, ed., p. 377. de Gruyter, Berlin.

EOA. 1975. EOA Book of Specifications and Standards. Essential Oil Association of USA, Inc., 900 17th Street N.W., Washington, DC 20006.

EOA. 1978. EOA Book of Specifications and Standards. Supplement 1. Essential Oil Association of USA, Inc., 900 17th Street N.W., Washington, DC 20006.

GEMERT, L. J. VAN, NIJSSEN, L. M. and MAARSE, H. 1981. Improvements of the Scoville method for the pungency determination of black pepper. *In* Flavour '81. P. Schreier, ed., p. 211. de Gruyter, Berlin.

HEATH, H. B. 1978. Flavoring Technology: Profiles, Products and Applications. AVI Publishing, Westport, CT.

HEATH, H. B. 1981. Source Book of Flavors. AVI Publishing, Westport, CT.

JOHNSON, E. L., MAJORS, R. E., WERUM, L. and REICHE, P. 1979. The determination of naturally occurring capsaicins by HPLC. *In* Liquid Chromatographic Analysis of Food and Beverages, G. Charlambous, ed., p. 17. Academic Press, New York.

KERNIK, K., REINECCIUS, G. A. and SCIRE, J. P. 1985. An evaluation of methods for the analysis of encapsulated flavors. *In* Progress in Flavour Research 1984. J. Adda, ed. Elsevier, Amsterdam.

KONIG, W. A. 1984. New developments in enantioner separation by capillary gas chromatography. *In* Analysis of Volatiles: New Methods and Their Applications, P. Schreier, ed. de Gruyter, Berlin.

MOSANDL, A. and HEUSINGER, G. 1984. Stereoisomers of fruit flavour substances—some aspects of synthesis and analysis. *In* Analysis of Volatiles: New Methods and Their Applications, P. Schreier, ed. de Gruyter, Berlin.

STRAUS, D. A. and WOLSTROMER, R. J. 1974. The examination of various essential oils. Proc. VI Int. Congress of Essential Oils.

TRESSL, R. and ENGEL, K. H. 1984. Formation and analysis of optically active aroma compounds. *In* Analysis of Volatiles: New Methods and Their Applications. P. Schreier, ed. de Gruyter, Berlin.

TRESSL, R. and ENGEL, K. H. 1985. Biogenesis of chiral aroma components and their analytical characterization at trace level. *In* Progress in Flavour Research 1984, J. Adda, ed. Elsevier, Amsterdam.

Index

D

G

O

P

Q

R

S

T

www.ingramcontent.com/pod-product-compliance
Ingram Content Group UK Ltd.
Pitfield, Milton Keynes, MK11 3LW, UK
UKHW020131290926
15471UKWH00003B/405
9789401197618